JAPANESE CONSTRUCTION

An American Perspective

JAPANESE CONSTRUCTION
An American Perspective

Sidney M. Levy

VNR VAN NOSTRAND REINHOLD
New York

Softcover reprint of the hardcover 1st edition 1990

Library of Congress Catalog Card Number 89-14745

ISBN 978-1-4684-6667-6 ISBN 978-1-4684-6665-2 (eBook)
DOI 10.1007/978-1-4684-6665-2

Van Nostrand Reinhold
115 Fifth Avenue
New York, New York 10003

Van Nostrand Reinhold International Company Limited
11 New Fetter Lane
London EC4P 4EE, England

Van Nostrand Reinhold
480 La Trobe Street
Melbourne, Victoria 3000, Australia

Nelson Canada
1120 Birchmount Road
Scarborough, Ontario M1K 5G4, Canada

16 15 14 13 12 11 10 9 8 7 6 5 4 3 2 1

Library of Congress Cataloging-in-Publication Data
Levy, Sidney M.
Japanese construction: an American perspective/written by Sidney M. Levy; reviewed by Thomas M. Craig, Jr.
p. cm.
Includes index.

1. Construction industry—Japan I. Title.
HD9715.J22L48 1989
338.4'7624'0952—dc20 89-14745
CIP

Dedicated to the Japanese people who responded to an American outstretched hand of friendship with warmth and sincerity.

Contents

Preface

The 1970s and 1980s have been marked by turbulent times for certain portions of America's industrial base, as their dominance of many domestic and foreign markets has eroded. During such times of stress it is tempting to create scapegoats in order to rationalize shortcomings.

Much is heard about the Japanese in this regard. How they have contributed to the deterioration of specific segments of American industry, how jobs in the U.S. are being lost to foreign competition, and how the resulting trade deficit will be the downfall of us all. Much of this rhetoric has been directed against the Japanese automobile manufacturers and the Japanese electronic industry, which has been accused of "dumping" product into the United States.

It was not until Japan unveiled its plan to build the multi-billion dollar Kansai Airport project that Japanese restrictive bidding practices in their domestic construction market became headline news. Construction then became a popular subject for "Japan Bashing" and attention was focused on the activities of Japanese contractors around the world, and, more particularly, on their involvement in the U.S. construction market. Well, the Japanese construction companies *are* in the United States and have been for some time. They have been awarded many contracts for federal and municipal construction projects and they have negotiated a significant number of construction contracts in the private sector. These world-class contractors have built quite a few projects for the U.S. subsidiaries of their long standing clients from Japan, and in addition, these Japanese builders have participated in a number of joint venture projects in this country, bringing with them engineering and construction expertise and money. Like it or not, the Japanese builders appear to be here to stay. We can either ignore them, hoping they will go away (which they won't), or we can legislate against them (a politically attractive but otherwise irrational act), or we can try to learn something from them (a more sensible approach). And learn we must.

We have entered an era of global interdependance, in which major powers of the world can no longer act independently on crucial matters without considering the consequences of their actions. West Germany sneezes and Washington says "Gesundheit." The rapid deforestation of South America's Amazon region coupled with the wide spread use of fossil fuels in Europe may accelerate the global warming trend, and put that beach house in Malibu or Southhampton under water within the next 50 years.

The United States, shortly after World War II was the dominant economic force on earth and was responsible for more than 50% of the world's gross national product. Today the U.S. must share the spotlight with a partnership in which Japan and West Germany play key roles, and we must all learn how to work together. America seems to react more kindly to its European partners, and eyebrows are raised only when Mercede-Benz and BMW introduce their new models each year, and prices continue to inch toward the stratosphere.

Our attention is more closely focused on the Japanese who have visibly influenced the lives of the American consumer with products made by Sony, Casio, Panasonic, Minolta, Nikon, Kawasaki and Sharp—just to name a few. The shock waves created by the Japanese automobile manufacturers have shaken the foundations of more than one basic American industry, and the consequences have been the subject of intense re-evaluation and soul-searching within these industries. The Japanese style of operation, vis a vis the automobile industry, could very well be looked upon as a case study for all businessmen to study. When Toyota introduce the Toyopet in 1956, that car was an uninspiring sedan with a little 48 horsepower, 4 cylinder engine capable of 60 miles per hour, if there wasn't too much headwind. This boxy, underpowered automobile contrasted sharply with the big fin chrome-bedecked behemoths being produced in Detroit, and since this little Toyopet offered little to the American public, its sales were dismal.

In the years that followed, the Big Three Automakers kept right on producing the kinds of cars they insisted were being demanded by U.S. buyers, and Toyota began to research the American market as only the Japanese seem to do. Now, just slightly more than 30 years later, Toyota has become a trendsetter and in 1987 was the Number One foreign import. And now Detroit has begun to work closely with the Japanese. General Motors and Toyota are now in partnership in California, producing "Japanese" and American cars side-by-side.

The U.S. auto makers have been joined by other American manufacturers in seeking out a new spirit in manufacturing that seems to pervade Japanese industry, and firms such as National Steel Corporation and Xerox are pushing the limits of high quality production. For survival purposes, if for no other reason, several basic American industries have reacted to the onslaught of foreign competition by shaking themselves out of their malaise, and becoming stronger for it.

This just might be the proper time for the American construction industry to look over its shoulder and see who is gaining on it.

When Admiral Matthew Perry sailed into Tokyo Bay in the mid 1800s, the Japanese were fascinated with examples of American technology that he brought with him in the form of gifts. A working model of a steam locomotive and a telegraph set presented to the Japanese dignitaries created the awareness that these Americans were "people to learn from."

Americans haven't lost the qualities and values that made this country great, they have just misplaced them somewhere, and there has not been a pressing need to look for them—until now. We should not feel awkward or embarrassed about learning from our friends from the East because they have had no hesitation about learning from us.

The Japanese construction industry has a saying—Wakon Yosai—Japanese Spirit—Western Technology, and they have become formidable competitors by embracing this theme. As the globalization of the construction industry accelerates, the chances of our contact with Japanese contractors increases, whether it be as a potential competitor or as a potential partner. The time is ripe for us to know as much as we can about these master builders from the other side of the globe.

1

The Japanese Presence in the United States Today

Driving across town in Manhattan these days, the traffic congestion caused by midtown construction activity is by no means new or rare. What is unusual, however, is to see construction signs on the fences around these projects written in both English and Japanese. This same phenomenon is being repeated in other major cities around the country as well as in a whole host of smaller towns such as Georgetown, Kentucky; Bloomington, Indiana; Marysville, Ohio; and Smyrna, Tennessee.

The Japanese presence in this country cannot be denied. First came the automobiles and consumer goods such as calculators, cameras, television sets, audio equipment, digital watches, and the like. Following this influx of product, the Japanese started to build offices and distribution warehouses and then began to build the factories to assemble or manufacture their products in this country.

A flurry of construction activity built a Toyota factory in Georgetown, Kentucky, a Nissan automobile plant in Smyrna, Tennessee, a manufacturing facility in Bloomington, Indiana to produce Mitsubishi cars, and a Honda assembly plant in Marysville, Ohio.

The Japanese auto industry has not been the only source of construction activity generated by Japanese manufacturers. Kawasaki Heavy Industries, Ltd., has converted a disused Otis Elevator factory in Yonkers, New York into a sophisticated subway manufacturing complex. Japan's Pilot Pen Corporation built a factory in nearby Connecticut, and down South a plant opened in Chattanooga, Tennessee to house Komatsu's heavy construction equipment assembly and manufacturing plant.

Along with familiar names such as Seiko, Sony, Toyota, Minolta, and Hitachi, there are many new names being added to the Japanese roster that might not be so familiar to most Americans: Toda, Shimizu, Aoki, Kumagai Gumi, Fujita, Daiwa, and Mitsui Fudosan. These are the names of some of the Japanese construction giants, along with their development companies and architectural and engineering design firms that have established subsidiary companies in the United States within the past fifteen to twenty years. Following closely on the heels of these construction companies, the Japanese financial institutions and banks have also begun to make their move into the American marketplace.

WHY ARE THEY HERE AND WHAT ARE THEY DOING?

That these large Japanese builders, developers, designers, and engineers are here is a fact; the manner and extent to which they plan to participate in America's $400 billion construction market has not yet been fully played out.

Several Japanese contractors have come to this country at the request of long standing clients planning to build a new U.S. distribution facility or office building or manufacturing plant. Following a time honored tradition of long, cordial relationships between builder and client, some of which span 100 years or more, the Japanese businessmen wanted their Japanese contractors to look out for their interests no matter where they build throughout the world. According to figures published by International Construction Week in 1985, fully $1.3 billion of the total $1.8 billion in construction contracts issued to Japanese contractors in the U.S. that year came from Japanese clients.

The globalization of the world's construction market is another reason why these Japanese contractors are here. Either in anticipation of or because of the collapse of the massive construction market in the Mideast, the high volume Japanese contractors had to search out other markets in order to feed their voracious billion dollar appetites. What better area to focus on then the stable American market?

These Japanese contractors can either be viewed as potential partners or potential adversaries or a little of each. In either case it behooves American builders to know more about them.

There was a joke circulating around Washington, D.C. shortly after President Bush's inauguration: Just a few months after Mr. Bush had assumed office, he was struck by a rare malady and spent the next three years in a coma. When he awakened, he immediately summoned his vice-president, Dan Quayle, to his bedside. "Tell me Dan, what's been happening to the country over the past three years?" he asked. "Well, Mr. President," Dan replied with a grin, "inflation is down to 1.5%, interest rates are about 4%, the trade deficit has not only disappeared, but you'll be happy to know that this year we expect a $200 billion trade surplus" "By golly, Dan, that's fantastic," replied the newly awakened President, "but is everything that rosy?" "Well, Mr. President," the Vice President answered grimly, "there is one bad piece of news. Gasoline is now up to 280 yen per gallon."

Japan is a land of contrasts, the new alongside the very old, the big shadowing the very small. A nation smaller than the state of California, where there is one general contractor for every third of a square mile, 75% of whom are capitalized for less than $37,000, but where there are a half-dozen companies, Kajima, Takenaka, Shimizu, Taisei, Ohbayashi, and Kumagai Gumi, known as the Big Six, who have yearly sales totalling $35–40 billion. Japan is a nation rocked by speculative land fever from time to time, where the total market value of real estate in Tokyo today is estimated to be $8 trillion, more than the worth of *all* land in the continental United States.

Japan is a land of savers where each day their financial institutions scoop up 1.4 billion investment dollars that can be directed by government agencies in order to shore up or bolster any sagging part of the economy. This is in sharp contrast to the corporate raiders in the U.S., who have been responsible for creating corporate *debt* totalling more that $1.78 trillion by mid-1988 as the frenzied buyouts, acquisitions, and stock repurchases of take-over mania spread through the country.

Because Japan has been prohibited by their Constitution from creating nothing more than a national defense force, the government has spent very little on defense and been able to focus its full attention and resources on economic growth while other great powers have been committing 5–9% of their gross national product to military uses. Along with the strain imposed by defense expenditures upon a government's budget and the taxpayer's pocketbook, the drain of human resources has been just as telling, but perhaps more subtle. Engineering's best and brightest in the United States have been lured to the aerospace and related industries with promises of huge and exotic research programs and cutting edge technology. The construction industry and its related engineering disciplines have had to settle for second best, while in Japan, devoid of the magnet created by defense research, many bright young engineers have been drawn to the construction industry, where the major contractors maintain sophisticated research institutes, staffed by 200–300 research engineers, funded by annual budgets in the $35–40 million range, and where exciting new discoveries are being made. Although the United States retains a commanding lead in pure research, the Japanese have developed an uncanny ability to develop the commercial potential of American discoveries.

And as they continue to discover America, will these Japanese contractors develop the potential of their well organized construction and design firms?

RELUCTANCE TO GO IT ALONE

A senior executive from one of Japan's billion dollar construction companies told the writer that the U.S. construction market contains many elements that run contrary to the way business is conducted in Japan.

The adversarial relationship prevalent in most labor-management negotiations, as well as between some general contractors and their subcontractors, and often between contractor and architect-engineer and contractor and client, is not fully understood by Japanese builders who maintain harmonious relationships with all parties to the construction process in Japan.

The complexity of the construction contract in this country, with emphasis on legal interpretations, and the entire legalistic approach to the construction process, is truly "foreign" to them. In Japan so much more reliance is placed upon good faith and long standing, cordial relationships.

Although quite willing to enter into joint ventures with American builders or to assume the role of construction manager, the spokesman for this particular corporation said that they were not quite ready to jump into the fiercely competitive private sector market in the classical role of a general contractor.

DIFFERING APPROACHES BY OTHER JAPANESE COMPANIES

Ohbayashi Corporation, with headquarters in Tokyo and branch offices in both New York and California, is operating contrary to this "wait and see" philosophy. Ohbayashi has taken a more aggressive stance through their U.S. subsidiary, Ohbayashi America. The parent company has had world-wide sales in the billions of dollars for many years and has had extensive experience in tunnel technology and public works projects. Ohbayashi America opted to put this experience to use and submit bids on several U.S. public works jobs, and their experience has paid off: they garnered at least ten government funded projects, the last being a $67 million, 249-foot-high by 2580-foot-long dam known as Elk Creek built for the U.S. Army Corps of Engineers in Trail, Oregon. At this Elk Creek project, an engineer borrowed from the home office in Japan designed a conveyorized concrete plant to aid in the placing of more than 1.1 million cubic yards of roller compacted concrete used in the project. Ohbayashi crews, using this new system, were able to place a record breaking 12,392 cubic yards of concrete in two-ten hour shifts. Other Japanese construction companies have approached the U.S. construction scene in a different way. Giant Kumagai Gumi (pronounced Koom-a-guy Goo Me) has been interested in obtaining equity positions in construction contracts in this country. Over the past several years Kumagai Gumi has committed about $1.2 billion in funding to several undertakings in Manhattan, including those developed by William Zeckendorf, Jr., the son of the famous Webb and Knapp real estate magnate. These large investments in Zeckendorf projects have made Kumagai Gumi the largest Japanese real estate investor in the City of New York.

THE OTHER JAPANESE REAL ESTATE INVESTORS

The entry into the U.S. real estate market by the Japanese has moved slowly but inexorably forward. The total Japanese real estate investment in this country in 1986 was estimated to be $4.5 billion, while just two years earlier it was valued at about $630 million.

This rush to buy real estate has been due partially to the fluctuations in international currencies, and more particularly to the strength of the yen against the dollar. In 1981 it took 220 yen to buy one dollar, in 1982 it took 249, and in 1983 and 1984 the yen was pegged at about 237 to the dollar. It remained at this level through 1985 and then started its rise to 136 toward the end of 1987 as the dollar continued to fall resulting in the yen increasing to 120 early in 1988 before settling down to 125–130. Discounting for inflation and real estate appreciation this fluctuation in the value of the dollar meant that an office building for sale to the Japanese in 1985 for $100 million could be bought for the equivalent of slightly more than $50 million in 1988, when the purchasing power of the yen is taken into account.

The U.S. real estate market also represents an investment to the Japanese that has both political and economic stability. Add to this a fair amount of liquidity, and the reasons for the Japanese investment philosophy begin to crystallize.

With land values in downtown Tokyo in the neighborhood of $29,000 per square foot, costs of $200–300 per square foot in some American cities represent bargain prices to these Asian investors. While real estate investments in Japan have historically yielded rates of return from 1% to 4% annually, the average U.S. real estate investment has returned 5% to as high as 9% per annum in some cases. Quite possibly these are some of the reasons that have triggered many of the recent Japanese real estate transactions, most of which had been directed toward high quality and high price urban properties.

Shuwa Corporation, a privately held Japanese development company purchased the American Broadcasting Corporation headquarters building in New York City for $175 million, $365 dollars per square foot, which was the highest square foot price paid for any office building in the country at the time. This record was surpassed in 1987 when Kato

Real Estate bought Tower 49 in Manhattan for $301 million, raising the record to $500 per square foot. The Tiffany Building on Fifth Avenue in New York, housing the world famous jewelry store of the same name, was sold during this same period to Dai-Ichi Real Estate Company for $94 million, the highest square foot price paid to that date for any retail store space anywhere in the country.

Real estate purchases by Japanese investors are being recorded throughout the U.S., although California and Hawaii remain, along with New York, the strongest area for sales. In New England, One Washington Square, a downtown office building in Boston, was sold to a Japanese investor for $36 million in 1987 and nearby, 265 Franklin Street changed hands for $100 million. Outside of Boston, in nearby Burlington, Massachusetts, the Burlington Business Center was sold to a Japanese corporation in late 1987 for $37 million.

A CHANGING PHILOSOPHY?

In mid-1988, Japanese investors seem to have adopted a more conservative approach to real estate investment in this country. Their record breaking purchase prices for premium urban properties appear to have been toned down. In California, Japanese investors failed to come up with the highest price during a bidding contest for Fox Plaza, one of the most prestigious buildings in the Century City section of Los Angeles. In previous years, the Japanese would probably have submitted the winning bid and along the way establish a record high for comparables in that area. In New York City, Japanese investors dropped out of a bidding battle for the J.C. Penney building on the Avenue of the Americas because they felt the price was approaching an excessively high level. They also chose not to submit a bid on the Celanese building just a few blocks away when it was put up for sale.

The Penney building seems to have a aversion to Japanese ownership. Tishman Speyer of New York and Chiyoda Realty of Japan were going to join together and submit a bid on the building. It would have been a natural alliance—Chiyoda was anxious to buy prime urban real estate and Tishman Speyer had considerable experience in developing and managing New York real estate. For about one month, beginning in April 1988, non-Japanese-speaking Tishman Speyer executives communicated with non-English-speaking Chiyoda managers in Japan. Since there is a 13 hour time difference between the two countries, a phone call at the beginning of Chiyoda's work day would be received at the end of Tishman Speyer's work day. The cultural differences seemed to prove too much and at the last moment, the alliance split up, Chiyoda dropped out and Tishman Speyer teamed up with Dallas based Trammel Crow to buy the building for $350 million. Shuwa Corporation had been one of the most aggressive Japanese real estate investors in the United States, having previously purchased the ABC building in New York and the Arco Plaza complex in L.A., but they too were reluctant to pursue a high priced offering when three buildings were offered to sale in Chicago in early 1988.

The emphasis on real estate purchases by Japanese investors seems to have shifted away from high-quality, high-ticket urban properties and is now directed toward shopping centers, golf courses, and a diversified portfolio of commercial and industrial properties. Although there has been a decided shift in where the Japanese are spending their money, there does not appear to be a shift away from continued U.S. market investments. Some experts predict that the Japanese will invest $16.6 billion in real estate in 1988.

While the Japanese investors were paying above market prices for select properties and establishing benchmark square foot prices for office space, they were also creating lower returns on real estate investments. Since return on investment (ROI) is a function of cash flow as a percentage of purchase price, when cash flow does not increase but the purchase price is dramatically increased, ROI and total income from a property declines. According to some sources, total income from some of these overpriced purchases dropped by as much as 1½%. This phenomenon has had a domino effect within the real estate investment community as insurance companies, pension funds, and other institutional investors have had to adapt to lower anticipated returns on several real estate transactions.

The Japanese concede that they have matured as investors in the U.S. market and that the record highs created by their frenzied purchases of prime buildings in the past will be well modulated in the future.

INVESTMENTS IN THE RESIDENTIAL MARKET

Japanese interest in acquiring parcels of commercial real estate has recently been expanded to include residential properties. In early 1988, residential sales in Hawaii seemed to reach a fevered pitch with one Japanese buyer purportedly scooping up ten $1-million residences in one shopping spree. Tales abound of homeowners being offered prices so far in excess of market value that they couldn't resist selling, particularly after the Japanese buyers indicated their willingness to offer long term leaseback agreements at very low rental rates.

The interest in upscale single family homes and condominiums seems to mirror their opinion of the commercial market in this country—real estate values, no matter how high, seem to be viewed by the Japanese as being excellent long term investments. Early in 1988, two duplex penthouses on the exclusive upper East side of Manhattan sold for a total of $7 million, a pricey $1250 per square foot, and by one count it has been estimated that the Japanese have purchased about 20% of the apartments built during the latest wave of condominium construction in that city. Some Japanese buyers are coming to the U.S. in tour groups sponsored by some of the big Japanese banks such as Mitsui and Nomura. These visitors go methodically from New York to Dallas, with stops in Chicago, and may eventually end up on the West Coast in pursuit of $500,000 minimum investment properties for sale.

American developers, recognizing a virtually untapped market for high price residential sales, are moving in on the Japanese market rapidly. A newly formed glossy monthly magazine is now on the newsstands in Tokyo devoted to pictorial displays of posh U.S. condos. Fujisankei Communication Group televises a "high ticket" shopping program to viewers in Japan, and has begun to feature U.S. condominiums in the program which is being beamed to eleven major cities.

THE INVESTMENT HOUSES AND BANKS

Where there are substantial real estate transactions, there will be investment houses and banks. In early 1987, Shearson Lehman agreed to sell 13% of their operations to Nippon Life Insurance Company, Japan's largest insurance firm. The price for this interest in Shearson Lehman was set at about $530 million. Goldman Sachs, another investment firm, followed suit and sold a 12½% equity position in their company to Sumitomo Bank for about $500 million. The Industrial Bank of Japan entered the primary investment market in the U.S. by purchasing an interest in the Aubrey G. Lanston company for $234 million. U.S. subsidiaries of Japanese brokerage firms such as Nomura Securities, Daiwa Securities, Nikka Securities, and Yamaichi International had a combined capital investment in excess of $400 million in this country by the end of 1986, and it continues to grow. The Japanese banks have not sat by idly. With vaults bulging with dollars from the country's enormous trade surpluses, and the very high rate of consumer savings (Japanese save about 17% of their income, versus 6% for Americans), the big banks of Japan would naturally look to the safe haven provided by the United States in which to invest their funds.

Ten of Japan's largest banks have established branches in this country. Five of them, Dai-ichi Kangyo, Fuji, Sumitomo, Mitsubishi, and Sanwa have individual asset bases of $200 billion or more, and collectively represent a total asset structure of $1.1 trillion!

TOTAL JAPANESE INVESTMENTS IN THE U.S.

Japanese investments, in all forms, have been growing in this country from 1974 to 1985, when they took third place behind No. 1 ranked Canada, No. 2 ranked Great Britain, but well ahead of fourth placed West Germany. However, according to the U.S. Department of Commerce's Office of Trade and Investment Analysis, Japan jumped into the lead in 1986 when she accounted for 33.4% of all foreign investments in the U.S., dramatically ahead of Canada's 10.9% share, Great Britain's 16.9% portion and West Germany's 5.7%. We have briefly discussed Japanese involvement in real estate, investment financing, and banking activities. Now let us focus on some of the U.S. subsidiaries of their world-wide, highly regarded construction leviathans.

U.S. CONSTRUCTION SUBSIDIARIES

Shimizu Construction Company, Ltd.

On the twelfth floor of one of New York's newest, glistening office buildings, the Crystal Palace, Shimizu American maintains an office. A few blocks away, on Park Avenue, are located Shimizu Land Development (real estate development), Shimizu Equities (high-tech building information and venture capital investments), and Shimizu Development (NY) Corporation (property management and hotel operations). In nearby Teaneck, New Jersey, Shimizu maintains an active construction office, along with other offices in Portland, Oregon, Los Angeles, Dallas, Detroit, Atlanta, and in the state of Hawaii. Shimizu, the parent company, maintains offices in twenty-eight countries around the world in addition to their facilities in Japan.

According to Shimizu's 1985 Company Profile, their world-wide sales for that year were $6.19 billion, and according to McGraw-Hill's ENR Top Contractor's listing, Shimizu posted sales of $7.129 billion in 1986 and $9.37 billion in 1987. The company employs more than 13,000 people world-wide, including 214 consulting engineers, 2528 architects, 1474 civil engineers, and 46 engineers with doctoral degrees. Shimizu was founded in Tokyo in 1804 and started its first foreign venture in Pakistan in 1950. Since that time they have built throughout the free world. Shimizu has constructed housing projects ranging from 2300 units in Iraq to 100 city blocks of housing in Singapore with a total floor area of 19,798,000 square feet. They have built educational and recreational facilities in Indonesia, medical complexes in Africa, desalinization plants in the Arab world, and an entire city in Turkey with a population of 130,000.

In this country, Shimizu built Sharp Electronic's massive headquarters building in New Jersey and in Oregon constructed a 139,000 square foot assembly plant for Fujitsi America. They also built an assembly plant and testing facility for Epson, the computer company in Oregon, a project in excess of 1 million square feet.

Shimizu's considerable experience in clean room construction, along with their extensive ongoing research activities in ultra clean environments, may have been why IBM intially selected them to be the Construction Manager on a $100 million state-of-the-art facility in East Fishkill, New York. IBM declined to sign the construction phase of Shimizu's contract, and although many Shimizu engineers remained on the job site, Walsh Construction handled the construction of the four building complex.

Like most Japanese construction companies, Shimizu has a strong commitment to quality and research and development. In 1984 they donated $500,000 to Harvard University's Graduate School of Design for the establishment of a laboratory for construction technology and to be used to add courses to the school's construction technology curriculum. This donation will also be used to finance exchange studies between faculty from Harvard and employees from Shimizu.

In 1987, Shimizu became the first Japanese contractor to purchase an equity position in an American construction company when they bought a minority share in Dillingham Construction Corporation of Pleasanton, California. At the time Dillingham's annual volume was slightly in excess of $1 billion.

Kajima Corporation

Not too far away from Shimizu's New York office, Kajima Corporation has an equally impressive presence in Park Avenue Plaza, one of Manhattan's newest and most prestigious addresses.

Upon entering Kajima's office, it becomes apparent that this is a Japanese corporation where managers, assistants and staff work side by side in an open office environment. Across the corridor from the reception desk there is a separate and quietly sedate room set aside to receive visitors. After the important ritual of exchanging business cards has taken place, Mr. Takashi Takemoto, Chief Representative of the New York Representation Office talks freely about his company, a global network of architects, engineers, and construction people totalling slightly more than 13,000, with offices in China, Hong Kong, Thailand, Malaysia, Singapore, Indonesia, Egypt, the United Kingdom, Chile, and, of course, the United States.

According to Kajima's 1986 Annual Report, the company counts among its subsidiary organization's real estate companies, chemical grouting companies, video production facilities, hotel operations, leasing companies, book publishing and sales corporations. Kajima owns and operates a major research and development facility in Japan known as the Kajima Institute of Construction Technology. This facility, established in 1949, consists of several buildings with a total floor area in excess of 295,000 square feet and housing complete geotechnical labs, materials testing facilities, environmental and mechanical engineering laboratories, large structures testing equipment, and extensive computer operations. The entire complex would be the envy of any major U.S. engineering university, and there are not too many R & D projects that the Institute cannot handle. In the United States, Kajima's oldest office in located in Fort Lee, New Jersey, having been established there since 1964. On the West Coast, Kajima Development Corporation in Los Angeles opened in 1979 and in 1984, Kajima Engineering and Construction also moved into the L.A. area. The East-West Corporation, an affiliated development company, was established in 1973 and is also located in California.

Kajima has built many projects throughout the U.S. In New Jersey they worked for Fuji Electric, Toshiba America, and Shimano America; in Texas they built facilities for Masak and Bridgestone Tires; and in California, some of their clients have been Nissan Motors, Sony, Seiko, Mazda, Alpine, and Bridgestone.

Many of Kajima's contracts have been negotiated, but they have also entered the competitive bid market in this country, and rather successfully at that. They were low bidder among seventeen other contractors in the tough New Jersey market when they were awarded a contract to build a large plant for Berlux, a West German corporation. Just outside of Atlanta, Georgia, in Peachtree City, Kajima went head-to-head with local contractors and became the successful bidder on Phase 1 and Phase 2 of that city's high school building program.

In Minneapolis, Minnesota, Kajima worked with the Minneapolis Community Development Agency on a huge mixed-use waterfront development project known as Riverplace. Their expertise and involvement as architects' planners, and contractors played a significant role in the revitalization of downtown Minneapolis.

Kajima's Dallas, Texas office has been active in Southwestern construction activities even though the economy in that part of the country has suffered an oil-generated slump. They built a "super" clean room for Hitachi Computer Products new disk plant in Norman, Oklahoma, and remained in that area long enough to construct a 200,000 square foot addition onto Goodyear Rubber's Lawton, Oklahoma plant. Texas based Trammel Crow, one of the country's largest developers teamed up with Kajima to build several business parks in Atlanta, Georgia and nearby Cobb Country, both areas being very familiar to this Japanese constractor.

Kajima has enhanced its presence in the Southeast by opening an office in North Palm Beach, Florida, where they have pledged to continue their plans to "be a quality builder in one of America's most dynamic and growing markets." No newcomer to this area either, Kajima built two high rise condominium projects several years ago on nearby Singer Island. One building, Seawinds, had 182 condo units and the other, Martinique 2, contained 198 apartments and ocean villas.

One of Kajima's newly created divisions, the Process Mechanical Division has scored well on its first American venture, the Mazda Motor Manufacturing Corporation's 2.7-million-square-foot automobile plant in Flat Rock, Michigan. This Division works hand in hand with Kajima's Design and Construction Groups to coordinate building design with the process or manufacturing equipment that will be installed in the new facility. By identifying equipment requirements such as roof openings, miscellaneous steel and embedded items required, along with electrical and mechanical feeds and final connections, the Process Mechanical Division can purchase all of the required manufacturing equipment, install it, and provide start-up services. This puts Kajima in a position to provide an owner with a totally coordinated effort with both centralized direction and responsibility. This turnkey facility construction capability has caught the eye of more than one U.S. manufacturer, including General Motors, Ford, and Chrysler.

Toda Construction Company Ltd.

Toda established its first overseas subsidiary, Toda America, Inc., in 1972 and currently maintains an office on Fifth Avenue in New York along with offices in Irvine and Los Angeles, California.

Toda Development, Inc. is located in Los Angeles and T & F Construction, Inc., a joint venture with the George A. Fuller Company, was established in Florida in 1982. The United States is familiar territory to Toda because in 1919, Toda Gumi, the predecessor to Toda Construction, built the Japanese Exhibition Hall at the San Francisco World Exhibition, and Mr. Junnosuke Toda, the current president of the company, studied engineering and construction in the U.S. during the 1950s.

Like so many other Japanese contractors in the U.S., Toda has been requested to oversee many projects in this country for their long time Japanese clients. When Nippon Electric Corporation (NEC) decided to build an 89,000 square foot large screen TV assembly plant in McDonough, Georgia, they looked to Toda for construction management expertise. Since the construction of NEC's initial building in 1984, Toda has been called upon to design and construct a second phase expansion program, and they are currently working on Phase 3. Brother, the Japanese printer and sewing machine maker, has been a long time Toda client, and when it was decided to build facilities in the United States, Toda was called upon to design and construct Brother's $5 million plant in Tennessee in 1987. A partial list of some of Toda's other projects in the U.S. are listed below:

Expansion of the Japanese Embassy in Washington, D.C.	$22,000,000
Science Lab and Administration Building, Miami University	$8,000,000
Terminal Expansion—Miami International Airport	$25,000,000
Condo construction in California—Co-developer with Daiwa House of Japan	$10,000,000
Shimano Industries Office—Warehouse, Irvine, California	100,000 square feet
Biomedical Research Center—Showa Medical School, Florida	32,600 square feet

The Toda Group—Toda America, Toda Construction of California—has achieved a great deal since coming to the United States just sixteen years ago. Their project mix exhibits the same versatility as their parent company's in Japan. Toda is justly proud of its rating by the Ministry of Construction in Japan, a task performed yearly by the government to rate the performance of contractors doing business with them on public works projects. In 1985 the Ministry of Construction rated Toda's management and civil engineering capabilities as the second best in Japan.

Ohbayashi Corporation

Ohbayashi Corporation has had a highly visible presence in the United States via their overseas subsidiary, Ohbayashi America Corporation, with offices on Madison Avenue in New York and Wilshire Boulevard in Los Angeles, California. Ohbayashi has several subsidiaries in the U.S.: United Development Corporation in Mill Creek, Washington; James E. Roberts-Ohbayashi Corporation in Oakland, California; Citadel Corporation in Atlanta, Georgia; and two corporations in Hawaii.

Toyota has been a client of Ohbayashi for more than 50 years, so it was only natural for them to choose their long time Japanese contractor to come to Georgetown, Kentucky and assist them in building a new $800 million automobile factory. On the West Coast, Ohbayashi's successful bid for the Corp of Engineer's Elk Creek Dam in the state of Washington was probably a major contributing factor behind the passage of the Brooks-Murkowski Amendment banning Japanese contractors from future work on government funded projects. This was not Ohbayashi's only major work in the Pacific Northwest. Along with their record breaking performance at Elk Creek, they participated in a joint venture with Tokyu Corporation to create the 1100 acre Mill Creek housing development in the state of Washington. Housing for 4500 families was carved out of the surrounding fir and spruce forests while preserving 120 acres of natural preserve where deer, water fowl and salmon live in and around the crystal clear waters of the streams flowing through Mill Creek.

Further south in Los Angeles, Ohbayashi joint-ventured the construction of two one-mile tunnels for that city's subway stem expansion program utilizing their shield tunnel technology and extensive tunneling experience. On the East Coast, an Ohbayashi-Dillingham Corporation joint venture worked on a $19.54 million subway line expansion project for Washington, D.C.'s Metro system. On the New Jersey side of the Hudson River, an Ohbayashi-Sordoni joint venture resulted in the construction of a 16 story office building called Evertrust.

A wholly owned subsidiary, The Citadel Corporation, headquartered in Atlanta, Georgia, recently completed a $7.5 million elderly housing rental project just outside of the city.

In Japan, Ohbayashi Corporation maintains a major presence. Their 1987 sales volume world wide was reported to be $5.8 billion and came from their 16 offices within Japan as well as their 13 branches and subsidiaries in other parts of the world. Ohbayashi's total employment exceeds 10,000 and includes 483 architects, 3,181 construction engineers, and 154 researchers. From nuclear power plant construction, where they introduced pre-stressed concrete containment vessels, to their innovative opening and closing membrane domed roof technology, they remain on the cutting edge of construction technology. Like many of their major domestic competitors, Ohbayashi develops long range corporate plans and constantly monitors the progress of these goals. Their current five-year plan commenced in 1986, and after the first fiscal year, Mr. Yoshiro Ohbayashi, the Chairman of the Board, stated that "ten thousand people moving in a single direction is an almost irresistible force. That is why it is so vitally important to marshal our forces with a viable plan of action. We constantly apply our 'Three Rs' process to all facets of the execution of our plan—Review, Revise, and Renew."

Taisei Corporation

Taisei is one of Japan's Big Six construction companies along with Kajima, Shimizu, Ohbayashi, Kumagai Gumi, and Takenaka Komuten. In the United States, Taisei operates through its wholly owned subsidiary, North America Taisei Company, Ltd. with offices in New York and California.

North America Taisei has recently undertaken an ambitious project in Long Beach, California. The Long Beach Redevelopment project, begun in 1975, will ultimately encompass 421 acres for renewal within the downtown district when it has been completed by the year 2000. The total scheme will contain four million square feet of mixed-use space upon completion.

North America Taisei's involvement is with the Shoreline Square Development parcel of the redevelopment plan. They have joined up with a local developer to build, own and manage a number of facilities within the parcel, including a hotel and office building. This portion of the project will also include the creation of 25,000 square feet of retail space, a 1400 car parking area, and a swimming pool–health club complex, and will represent a total investment of more than $132 million.

Taisei is not a newcomer to America. Although North America Taisei was established in New York in 1983, the company came to this country in 1939 when it built the Japan Pavilion for the San Francisco International Exposition. The company was founded in 1873. By 1987 it reported sales slightly in excess of $8 billion. Building construction accounts to 65% of Taisei's yearly volume and civil engineering projects make up another 25% of their world sales. Taisei, like its other Bix Six associates maintains a sophisticated research and development facility in Yokohama known as the Taisei Construction Technology Research Center. The complex consists of six buildings with a total square footage of 178,000, situated on nine acres of land. With a total research staff of 354 working in materials labs, structural, acoustical, hydraulic, and environmental laboratories, Taisei engages in both theoretical and practical construction technology research.

North America Taisei, with highly developed technical and financial resources, is in a position to carve out a significant market share in the United States.

Other Contractors

Along with Shimizu, Kumagai Gumi, Ohbayashi, Taisei, and Toda, there are quite a number of other Japanese contractors doing business in America through their U.S. subsidiaries.

- *Aoki America Corporation.* Established December 1980 in Columbia Maryland.
  Business: General Contractor:
  Parent Company: Aoki Corporation. 1983 Sales $1.194 billion.
  Aoki Corporation in early 1988 announced a joint venture in Orlando Florida's Disney World. Along with Tishman Realty and Construction

and Met Life, they will be building a $375 million convention center and resort hotel.

- *Daiwa House Corporation of America.* Established August 1976 in California.
  Business: Residential contractor.
  Parent Company: Daiwa House Industry. 1983 sales $1.705 billion. Employs 5810.
- *Fujita Corporation U.S.A.* Established 1972 in California.
  Business: General contractor, developer.
  Parent Company: Fujita Corp. 1983 sales $2.588 Billion. Employs 5600.
- *Kumam.* Established April 1980 in California.
  Business: General contractor and developer.
  Parent Company: Kumagai Gumi. 1986 sales $6.7 billion. Employs 7748.
- *Mitsui Fudosan (USA) Inc.* Established May 1973 in California.
  Business: Industrial parks developer.
  Parent Company: Mitsui Real Estate Development. 1983 sales $1.47 billion. Employs 955.
- *Sumitomo.* Established December 1973 in California.
  Business: Contractor and Engineer.
  Parent Company: Sumitomo Construction Company. 1983 sales $1.294 billion. Employs 3315.
- *Takenaka & Associates Inc.* Established December 1960 in California.
  Business: Planners-Builders-Architects-Engineers.
  Parent Company: Takenaka Komuten Co., Ltd. 1983 sales $5 billion. Employs 8875.

THE JAPANESE APPROACH

The Japanese way of doing things in the construction industry might bear watching by U.S. contractors. Just as Toyota, Honda, and Nissan taught Detroit automakers a few things about building automobiles, so might some of the Japanese contractors teach American builders a thing or two.

Japanese builders are constantly searching for new ways to increase their rates of productivity and increase the quality of construction, and they are as cost conscious as any builder in this country. Because of the absence of craft unions as we know them here and the different relationship that exists between labor and management, many types of labor saving devices have been developed and put into use in Japan.

Robotics plays an important role in Japanese research and development projects and has resulted in a number of differing types of construction robots being developed and becoming operational. One contractor has developed a robot to spray fireproof structural steel framework, a messy and laborious job when performed by humans. Several builders have produced workable robots to steel trowel finish concrete slabs at a faster rate and with more accurate tolerances than most masons can operate. Another contractor has developed a robot that scales the exterior surface of skyscrapers looking for cracks in the skin and, upon finding them, accurately keeps track of their exact location. Technological advances such as these are an accepted part of the labor-management working arrangements in Japan, but can you visualize a team of these robots at work on their tasks in any major metropolitan high rise building in this country? Consider the concrete floor finishing robot.

The various trade unions claiming jurisdiction over the work would have negotiated contracts to protect the interests of their respective memberships. The masons' union would require that three standby concrete finishers accompany each robot, and the laborers union would require that they assist in moving the robot from floor to floor. A composite crew of electricians (the robot has internal wiring) and an operating engineer (the robot does have a stop-start button) would certainly be needed to fill out the crew. By the time the various International Brotherhoods finished with the poor robot, it would probably have made more sense to have it just fetch coffee for everyone at break time. On second thought, that wouldn't work because it certainly would displace a laborer. Well, Mr. Robot could always learn to sharpen pencils in the Field Office.

Along with their pursuit of technological advances geared toward productivity and enhanced quality, many Japanese contractors institute other procedures that might find limited acceptance within the American workforce.

Every morning, on most large construction job sites, the Project Superintendent assembles all of the workers, both his own and those of his subcontractors, and for about ten minutes leads them through a series of calisthenics. When the morning exercises are over, he spends another ten minutes or so advising everyone of the day's scheduled activities. At that time it might be pointed out that the tower crane will be hoisting materials on the east side of the building all morning and all workers are cautioned to avoid that area for safety reasons. The Japanese, among other traits, have become very safety conscious.

The super will review the work scheduled for various parts of the building, where specific trades will be working and what operations will be taking place. This reflects another Japanese trait, the need to be kept informed, and another purpose of the daily morning briefing is to let everyone know what is going on throughout the entire building. In Japan, a six day work week is standard and one construction company makes use of a practice that with certain adaptations, could be considered by American contractors. Each Saturday, work stops after the lunch break and the rest of the working day is spent by all trades cleaning up their debris. This makes for a clean building at the start of each week, and the safety conscious Japanese equate a clean building with a safe one. If U.S. contractors could somehow adopt a similar regimen, putting it in effect during the normal work week, the constant arguments between general and subcontractor about prompt cleanup might be avoided. The flow of charges and backcharges might be lessened or eliminated. More important, the unsafe conditions that exist on a number of construction sites might be reduced because of cleaner conditions and a few painful and costly injuries could be avoided.

Many of the common sense solutions to construction problems practiced by the Japanese would be difficult or downright impossible for them to implement while working in an union shop environment in the U.S., but just think what they might be able to accomplish if they applied themselves in an open shop or merit shop operation.

THE TRADING COMPANIES

Japanese trading companies, initially formed to sell Japanese products abroad, are now looking for new opportunities to make money, and the real estate market is a natural avenue for them to explore. Real estate development companies are not new to Japan; many trading companies formed them in the 1930s to create new investment vehicles for their owners. The allying of Japanese investment and banking houses with the development companies helped to create the large real estate holdings that exist today. Mitsubishi Estate and Mitsui Real Estate between them own almost 40 million square feet of property in Japan.

The methods employed in the U.S. today by Japanese real estate developers along with their construction and financing divisions might well point the way to their future plans.

Formulating a Plan

The major Japanese builder-developer subsidiaries in America seem to taking the same approach as their manufacturing brothers in Japan took before embarking on massive market penetrations in this country.

- Phase 1. Learn as much as you possibly can about the market in which you are planning to enter. Accumulate as much pertinent information as you possibly can.

- Phase 2. Constantly search for new opportunities in these markets.
- Phase 3. Formulate detailed plans and be able to implement them when the time and opportunity permits. Response must be rapid while remaining flexible to any changes in the market.

Japanese builders and developers, some of whom have been in this country since the 1960s, have certainly been implementing Phase 1 (the learning curve). They are proceeding rapidly to accumulate knowledge of our construction and development practices. These companies are on the alert for new opportunities (Phase 2) and with their bulging treasuries and favorable exchange rates, have already availed themselves of many sound investment offerings. Phase 3 is being approached with caution—so far.

To Look Ahead—Sometimes You Look Back

The philosopher Santayana said that those who ignore history are doomed to repeat it, and we might be able to see where some of these Japanese companies are going in this country by looking at where they came from. A look into the past, present, and future of one of Japan's largest trading companies may provide a few clues.

The House of Mitsui

The House of Mitsui was founded in 1637 by Hachirobei Takatoshi Mitsui, and from a humble beginning as a dry goods shop grew during the mid-1800s to be one of the largest merchant houses in Japan, serving as a worldwide sales agent for Japanese products. Mitsui's influence in the business world spilled over into the public sector when his company financed the Emperor's Meiji's armies as they battled with the Shoguns in the 19th Century. Mitsui's political influence continued to spread when it bankrolled one of the two leading political parties in Japan, the Seiyukai Party, prior to World War II, thereby forming a close alliance between big business and government.

Zaibatsu Companies

Several other large trading companies existed along with Mitsui, and they were all known as zaibatsu companies (literal translation, "financial cliques"), which had been formed to further the financial interests of the family owned trusts such as C. Itoh, Yasuda, Mitsubishi, and Sumitomo.

These zaibatsu companies were outlawed when General Douglas Macarthur, the Supreme Commander for the Allied Forces (SCAP), set up a provisional government in Japan at the close of the Second World War. Although banned by law, many say that these zaibatsu companies continue to exist.

When Mitsui Zaibatsu was broken up in 1945, hundreds of independent companies were formed and in 1950 a Monday Morning Club was formed which brought together the heads of 27 of the more prominent companies within the old Mitsui organization. This Monday Conference Club, now known as Getsuyo-kai, was the formation of the new Mitsui Group and is now limited to 24 of the Group's leading companies. A second get-together of group companies takes place on the second Thursday of each month and is called Nimoku-kai (Second Thursday Conference)

Mitsui Construction Company has membership in both Clubs, as does Mitsui Real Estate Development Corporation, but Mitsui Real Estate Sales is only invited to Getsuyo-kai. It is estimated that the six great trading companies existing in Japan today—Mitsui, Mitsubishi, Marubeni, C. Itoh, Sumitomo, and Nisshi-Iwai—control approximately 25% of the entire nation's natural resources. The strong alliance that still exists between the business community and the federal government in Japan further enhances the spirit of zaibatsu.

Of all the trading company subsidiary and affiliate companies operating in the U.S. today, Mitsui stands in the forefront.

The One to Watch

Mitsui is a diverse conglomerate consisting of 69 companies, excluding financial and insurance companies with a total employment of about 400,000 people. Its very structure demands that it take advantage of all possible growth opportunities. Just look at some of the companies which form the Mitsui Group:

Mitsui Construction Co., Ltd.	A construction and engineering company with seven offices throughout the world and 1986 revenues of $2.2 billion.
Mitsui Harbor and Urban Construction Co., Ltd.	Land reclamation, port construction. Annual revenue about $380 million.
Mitsui Home Co., Ltd.	Residential design and construction. Annual revenue approximately $680 million.
Mitsui Kanko Development Co., Ltd.	Commercial hotel management with yearly revenues of $197 million.

Mitsui Lumber Co., Ltd.	Building materials. Yearly sales $253 million.
Mitsui Matsushima Co., Ltd.	Sale/purchase of fossil fuels, building materials, machinery. Annual sales $606 million.
Mitsui Mining and Smelting Co., Ltd.	Zinc, lead, copper, nonferrous metals, industrial materials. Yearly revenues $2.13 billion.
Mitsui Real Estate Development Co.	Real estate development. Annual sales $2.02 billion.
Mitsui Real Estate Sales Co., Ltd.	Commercial-residential brokerage firm. $270 million revenue.
Mitsui Road Co., Ltd.	Road construction, paving, landscaping. $171 million in sales.
Sunland Company, Ltd.	Recreational development/management. $60 million annual revenue.
Mitsui and Company	Development, sales, purchases, distribution of a wide range of goods in Japan and worldwide. Annual trading transactions $135 billion. Total revenue $2.85 billion.

Then there is Mitsui Bank, a commercial and long term credit bank with an asset base of $140.8 billion. But that's not the end of this massive combine:

Onodo Cement	One of Japan's leading producers of cement and cement products. Annual revenue about $2.57 billion.
Mitsui Toatsu Chemical	Manufacturer of chemicals and electronic materials with yearly sales of $3.15 billion.
General Seikiyu Kabushiki Kaisha	Import, refining, marketing crude oil and LP gas. Annual revenues approximately $6.16 billion.
Toyo Engineering	Design/construction of chemical, petrochemical plants. Annual sales $1.33 billion.
Mitsui Mutual Life Insurance	Premiums received $5.74 billion.
Nippon Flour Mill	Flour milling, warehousing. Revenues $1.19 billion annually.

Mitsukoshi Ltd.	One of Japan's leading department store chains, with flagship store in Tokyo's Ginza district. Yearly sales $4.14 billion.
Toshiba Corporation	Needs no explanation to Americans. Contributes about $25.7 billion to Mitsui Group.
Mitsui O.S.K. Lines Ltd.	Shipping, freight forwarding. Annual volume $3.97 billion.

Last, but by no means, least is Toyota Motor Corporation with annual sales of about $46 billion.

In 1987, Mitsui Trading Company registered total sales of about $520 billion with a gross profit of just about $8 billion.

Mitsui Real Estate Development Company

Mitsui Real Estate Development activities initially centered around the leasing of buildings and the 13.7 million square feet of property which they own in Japan. A few years ago their chairman, Mr. Hajime Tsuboi, announced that he wanted his company's growth to be amplified by placing more emphasis on the development, construction, and sale of capital properties so that these short term gains could supplement their long term investments. This seemed to signal Mitui's efforts to accelerate and expand their search for new opportunities in real estate, and the United States is the country with the largest and most stable real estate market. In 1972, Mitsui Real Estate Development expanded its operations overseas and in the process entered the U.S. market. Mitsui looks to a continued involvement in the American market as a two-fold learning process. First, the Japanese have felt that the American real estate market is much more sophisticated than their own, and learning from this market would allow them to transport new ideas back to Japan. Second, this learning process would assist them in their efforts to negotiate future real estate deals in this country.

How do you learn about hotel construction and renovation procedures in the U.S.? Buy an older hotel, refurbish it and then watch how the new image is marketed and managed. Mitsui did just that when the bought an older hotel on the east side of Manhattan and set about rejuvenating it.

How do you learn first hand about New York building construction and finance packaging? Become a general partner in a medium size building being developed on Fifth Avenue and watch the deal as it is being pulled together before exercising a purchase option. How do you become more knowledgeable in financing construction projects in America? Provide the owners with a $48 million permanent mortgage on an office building in Chicago. Mitsui has pursued all of these courses of action as part of their long term learning process, while still becoming involved in other projects throughout the country:

Bonaventure Hotel	Los Angeles	Purchased as a going concern
AT&T Center	Los Angeles	Joint venture with Dai-ichi Insurance on this 1.1 million square foot building
Crocker Bank Center	Los Angeles	Purchased
LaneBryant Building	New York City	Purchased for renovation
Exxon Building	New York City	Purchased this 2.29 million square foot building for $610 million
Office Park	Phoenix, Arizona	Joint venture with Shimizu
Hotel	Tempe, Arizona	Joint venture, Shimizu and Westcor

Mitsui's West Coast activities have found their way East, in what may be an entirely new phase of Japanese real estate investment activity. Mitsui Real Estate Sales USA recently became a general partner in a 400,000 square foot shoreline office building in Stamford, Connecticut, known as Harbor Park. This twin-pod, eight story building with attached precast concrete garage was developed by Urbco, a subsidiary of Connecticut Mutual Life Insurance Company of Hartford. It was constructed by W.E. O'Neill from Chicago, Illinois.

This building was initially designed for lift slab construction. The foundations and steel columns were being erected in preparation for the multiple slab pours on grade prior to being jacked into place when another lift slab project, about 20 miles east in the city of Bridgeport, collapsed while under construction by San Antonio-based Textstar Construction Corporation. The horrible collapse of this structure, L'Ambiance Plaza, on April 23, 1987 killed 28 workers and caused Connecticut Governor William O'Neill to place a moratorium on all lift slab construction in the state. The owners of Stamford Harbor Park very wisely did not wait for the moratorium to be lifted, but quickly redesigned the structure from lift slab to cast-in-place structural concrete slab, and work commenced at a rapid pace. Stamford Harbor Park is being closely observed by Japan-watchers as a new, broad-band approach to Japanese investment activity—a venture into the speculative suburban market.

Mitsui held to their practice of bringing in a group of major Japanese banks as limited partners and syndicated some of its own investment in the project to other Japanese investors.

Given the Japanese propensity to observe and learn, and their ability to implement projects with considerable construction and financial resources at their command, Mitsui might be the one to watch on the U.S. construction scene.

PUTTING THINGS IN PERSPECTIVE

"Japan bashing" has become a popular sport in America. Nightly television news programs show astute politicians smashing Japanese-made appliances to pacify their constituent's concerns about local plant closings in the wake of increased sales of foreign made product. A ban on Japanese contractor participation in federally funded projects in this country, enacted early in 1988, was another case in point. Its primary purpose was to pressure the Japanese government into opening their domestic public works projects to U.S. construction company participation.

But Ohbayashi's recently completed federal project at Elk Creek should be seen in the proper perspective. It is not likely that precious U.S. dollars were being shipped back to Japan by the truckload. Possibly a few high level managers and engineers from Ohbayashi's home office in Japan were brought into the country to work on the project, but many other managers, supervisors, and workers were hired from the local work force. Local subcontractors were hired, and one nationally known concrete form supplier has been advertising the fact that they supplied Ohbayashi with forms for the job. Given the relatively low profit margins associated with projects of this kind, and taking into account the payments made to workers, suppliers, and subcontractors, as well as taxes paid to local, state, and federal agencies, it is highly unlikely that Ohbayashi sent any substantial sum out of the country.

Although there are a number of world class Japanese contractors operating in many regions of this country, their role on the American construction scene needs to be viewed in the proper light. Only one firm with Japanese origins appears on the *Engineering News Record* 1988 List of Top 400 Contractors. Kajima International, headquartered in Fort Lee, New Jersey appeared on several lists in this report. Kajima was ranked 19th in the Design-Construct category, 28th in the Top 50 General Building Contractors Division, and 74th in the overall listings.

The Ohbayashi project received more than its fair share of publicity and along with congressional furor over the exclusion of American contractors and engineering firms from participation on the Kansai Airport Project, may have been the final act that contributed to the Brooks-Murkowski Amendment.

Brooks-Murkowski

On December 21, 1987, the Brooks-Murkowski Amendment to a defense authorization bill was entered into the Congressional Record of the U.S. House of Representatives. Senator Frank Murkowski of Alaska and Representative Jack Brooks of Texas believed that one way to turn up the heat on U.S. contractor and engineering firm participation in Japanese construction projects would be to deny the Japanese access to any future U.S. public works jobs. Congress approved the bill and President Reagan signed Brooks-Murkowski into law. Under the provisions of the bill, the United States Trade Representative would determine which country denies fair market opportunities for products and services to the United States in procurement, or in bidding for public construction projects costing most than $500,000.

Japan was specifically mentioned as a country preventing access to its construction market and, as such, would be denied participation in any U.S. public works job. The ban applied to Japanese construction companies, Japanese architecture and engineering firms and any subcontractor or supplier where the owners are "citizen or national of a foreign country, or is controlled directly or indirectly by citizens or nationals of a foreign country."

Senator Murkowski, in order to make certain that the Japanese got the message, tacked an amendment onto the Airport and Airways Improvement act which made foreign participation in the $25 billion+ U.S. airport and airways improvement program contingent upon reciprocal opportunities for U.S. companies overseas.

This "knee jerk" reaction to Japanese construction activity in America is not shared by all persons familiar with international construction activity.

A More Modulated View

Mr. Charles T. Pinyan, Chief Editor for *International Construction Week* discussed the subject of global competitiveness in the construction industry when he spoke before the U.S. Department of Commerce Global Competitiveness Conference in Washington, D.C. on September 9, 1987.

Mr. Pinyan stated that although U.S. public works contracts awarded to foreign firms attracted a great deal of attention, they only amounted to $250 million in toto. (Of the total federal construction put into place in 1985, valued at $12.364 billion, this foreign portion of $250 million represents only 2 percent.)

Mr. Pinyan goes on to say that during 1985, all foreign contractors operating in the U.S. reported sales roughly amounting to $7.3 billion. He asks the following questions:

"Now what happened to that $7.3 billion foreign firms won in the U.S. in 1985? Did the money leave here, taking U.S. jobs with it? No! Much of it came from abroad to begin with and most of it stayed right here. Take a current New York City real estate development project as an example: Kumagai Gumi and New York real estate developer Zeckendorf each put up $15 million to form a joint venture to build a high-rise on my favorite parking lot in New York. The partnership received a $270 million loan from a consortium of Japanese and American banks, and spent $100 million to acquire the site. Kumagai Gumi has the construction management contract on a cost-plus-fee basis: $180 million for construction plus a $10 million fee. American architects and engineers were selected . . . that's another $10 million. Kumagai Gumi brought in some Japanese managers for the project, but most of that $180 million is going to American subcontractors, materials and equipment suppliers, and construction laborers. Net result? More construction jobs created, more materials and equipment sold, another attractive ratable for Mayor Koch, and I need to find another parking place.

A Look at The European Contractors on the U.S. Scene

There are many European contractors in the United States (see Table 1–1) and the value of their U.S. construction contracts exceeds those of the Japanese, but, in the main, they have avoided the notoriety that the Japanese seem to attract. Two large German firms have maintained very low profiles due to the way in which they have approached entry into the American market.

Philip Holzmann AG, the West German construction giant, was ranked No. 5 in the 1987 *ENR* Top 250 International Contractors list. Philip Holzmann owns the Jones Group, Inc. of Charlotte, North Carolina, whose 1985 sales volume in this country topped $1.6 billion. In 1987, 75% of Philip Holzmann's foreign sales volume came from their Jones Group subsidiary.

Another very large West German builder, Bilfinger + Berger Brau AG was ranked No. 19 in this same *ENR* Top International Contractors list. Bilfinger + Berger owns Fru-Con, a Ballwin, Missouri-based construction company operating nationwide with annual sales volume of $600 million in 1985 and $514 million in 1987. Together, Philip Holzmann and Bilfinger + Berger with combined U.S. sales contracts of $2.1 billion via their U.S. construction companies exceeded *all* Japanese construction work in this country during the same period of time.

According to the figures shown in Table 1–1, clearly German-owned construction firms gathered the highest percentage of foreign contractor construction awards in this country for the period 1980–1986. If three nations of the old British Empire, Great Britain, Canada, and Australia, were to be combined, their construction awards would put them in

Table 1-1. Foreign Contractor Penetration of the U.S. Market, 1982-1986 (expressed in millions).

Nationality	1982	1983	1984	1985	1986	1980-86	% of Total
German	2225.0	1355.6	1971.8	2004.5	2021.2	9578.1	33.8%
Japanese	114.1	359.2	671.1	1802.1	1976.9	4923.4	17.0%
British	1491.1	711.7	892.8	794.5	1400.9	5291.0	18.6%
French	418.4	555.5	595.4	1026.7	970.4	3566.4	12.5%
Australian	—	81.3	55.0	80.0	232.0	448.3	1.6%
Swiss	37.5	48.0	57.0	554.2	226.8	923.5	3.2%
Canadian	210.5	144.6	947.3	280.5	210.2	1793.1	6.3%
Dutch	61.7	25.2	65.3	159.0	190.4	501.6	1.7%
Italian	98.2	69.4	119.6	128.4	159.3	574.9	2.0%
Swedish	30.0	121.6	104.0	346.7	103.1	705.4	2.5%
Top 250	4798.2	3616.1	5638.7	7267.8	7554.2	28305.7	

Courtesy of *International Construction Week.*

second place with a grand total of $7.532 billion. When all European contractors, exclusive of the Germans, are lumped together, their U.S. market penetration would total $6.27 billion for this same period of time, exceeding the Japanese aggregate of $4.9 Billion.

Granted the Japanese contractors came on strong—from a meager 2% of construction contract share in 1982 to a 26% market share in 1986—it is equally clear that their U.S. market penetration is not overly significant when compared to that of other foreign-owned firms!

In 1987 Japanese construction companies operating in the United States were working on projects with a total value of about $3 billion, which, while a significant dollar amount, represented only about 1.7% of all U.S. construction exclusive of one and two family housing costs.

Table 1-2 contains a list of the 25 largest international contractors along with their record of total and foreign generated construction sales for the period 1981-1985. Of the American companies listed, M.W. Kellog conducted 92% of its business overseas in 1985. Number 2 ranked Parsons Group obtained $5 billion, or 58% of its annual sales from foreign markets. Bechtel Group, Inc., generated 49% of its $7.3 billion annual contract work in other countries, while Brown and Root relied upon markets outside of the United States for just about 53% of its total yearly volume. Lummus Crest, Inc., the other big American international contractor, performed 69% of its work away from home. Among the Japanese companies included in the Top 25 list, Chiyoda Chemical Engineering and Construction Company, Ltd. registered 69% of its revenue outside of Japan; although this is a high percentage, in actual volume it represented only about $800 million. JGC, another Japanese contractor on this Top 25 list, achieved 72% of its sales vol-

Table 1-2. Top International Contractors, 1981–85[a] (in millions of dollars).

1985 Rank	Company	Country	1981 Foreign	1981 Total	1982 Foreign	1982 Total	1983 Foreign	1983 Total	1984 Foreign	1984 Total	1985 Foreign	1985 Total
1	M.W. Kellogg Co.[b]	United States	$3,700.0	$6,700.0	$5,200.0	$8,200.0	$5,000.0	$8,500.0	$8,630.0	$10,855.0	$6,204.0	$6,757.0
2	Parsons Corp.	United States	3,654.0	10,244.0	3,074.4	4,319.7	4,160.7	6,627.7	3,010.2	7,514.7	5,027.1	8,620.0
3	Bechtel Group Inc.	United States	6,923.0	10,114.0	3,866.0	5,490.0	7,884.0	13,810.0	4,905.0	8,220.0	3,626.0	7,364.0
4	Brown & Root Inc.	United States	2,642.0	10,738.0	1,250.1	2,472.3	769.2	2,059.4	1,288.7	3,883.9	2,928.7	5,578.7
5	Lummus Crest Inc.	United States	2,693.0	4,600.0	2,392.0	3,252.0	1,720.0	2,460.0	2,300.0	3,200.0	2,420.0	3,500.0
6	Kumagal Gumi Co., Ltd.	Japan	216.9	2,899.8	550.3	3,115.3	737.1	3,283.5	850.2	3,227.7	2,175.8	4,692.2
7	Hyundai Engineering & Co., Ltd.	Korea	2,386.0	3,016.2	2,920.5	3,422.7	1,109.0	1,606.2	2,490.6	3,016.8	1,998.1	2,536.4
8	Philipp Holzmann AG	West Germany	3,726.0	5,390.0	3,223.7	4,864.4	1,755.1	3,657.1	1,931.3	3,291.3	1,948.2	3,232.5
9	SADE/SADELMI Group of Cos.	Italy	2,705.0	2,720.0	3,242.0	3,246.0	2,003.0	2,021.0	1,773.0	1,784.0	1,795.4	1,799.0
10	John Brown Engineers & Constructors	England	360.0	404.0	n.a.	n.a.	641.0	2,739.0	1,329.0	1,567.0	1,743.0	2,066.0
11	Davy Corp. plc	England	5,015.0	5,045.6	3,380.4	3,429.6	2,036.3	2,178.6	1,553.4	1,627.4	1,611.3	2,118.8
12	Foster Wheeler Corp.	United States	3,450.0	4,400.0	2,500.0	3,200.0	2,103.0	2,980.0	1,933.0	2,413.0	1,426.0	2,048.0
13	Mitsubishi Heavy Industries, Ltd.	Japan	801.6	n.a.	n.a.	n.a.	195.0	n.a.	720.0	3,800.0	1,425.0	4,410.0
14	Fluor Corp.	United States	4,200.0	10,600.0	4,097.3	6,507.1	1,193.1	2,605.1	1,527.8	8,353.3	1,392.2	5,127.4

Table 1-2. *(continued)*

1985 Rank	Company	Country	1981 Foreign	1981 Total	1982 Foreign	1982 Total	1983 Foreign	1983 Total	1984 Foreign	1984 Total	1985 Foreign	1985 Total
15	IMPRESIT SpA	Italy	2,052.0	2,757.0	1,631.6	1,914.7	1,652.8	1,805.2	860.0	1,210.4	1,345.9	1,632.4
16	Bouygues	France	n.a.	n.a.	560.0	2,122.0	1,100.0	2,320.0	1,000.0	2,700.0	1,250.0	3,050.0
17	Spie Batignolles	France	1,543.7	2,306.7	1,070.0	2,074.3	1,092.3	1,825.2	1,101.6	1,826.0	1,212.3	2,442.6
18	Italimpianti SpA	Italy	248.2	289.9	334.5	409.1	55.0	130.2	153.8	256.7	1,170.5	1,389.0
19	Societe Auxiliare d'Entreprises	France	990.0	2,180.0	1,505.0	2,808.0	1,446.0	2,487.0	660.0	1,454.0	1,164.0	2,501.0
20	Enka Construction & Industry Co., Inc.	Turkey	743.0	1,007.0	1,129.7	1,405.8	1,595.0	1,869.8	1,347.4	1,655.3	1,081.6	1,572.5
21	JGC Corp.	Japan	699.3	893.8	2,480.7	3,069.5	642.0	990.0	519.0	771.0	1,020.1	1,420.2
22	Daewoo Corp.	Korea	2,077.0	2,360.4	533.8	911.3	891.3	1,266.8	677.0	1,056.0	921.4	1,208.4
23	Saipem SpA	Italy	448.0	n.a.	304.2	n.a.	747.0	835.0	532.9	561.1	826.2	909.9
24	Chiyoda Chemical Engineering & Construction Co., Ltd.	Japan	711.1	1,433.6	1,118.0	1,391.0	991.0	1,218.0	350.0	616.0	801.0	1,161.0
25	Voest-Alpine AG	Austria	n.a.	n.a.	1,000.0	1,170.0	430.0	472.0	622.3	656.3	783.8	896.0

[a]Ranked in terms of foreign contracts won in 1985.

[b]In 1985 Kellogg Rust Inc. was split into its two operating units, M.W. Kellogg Co. and Rust International Corp. The contracts won by Kellogg Rust in the 1981–84 period are listed under M.W. Kellogg.

Note: Figures include prime contracts, shares of joint ventures, subcontracts, design-construct contracts and construction management contracts where the firm is exposed to financial liability similar to a general contractor (greater than $500,000). Design-construct contracts are based on the erected value of the plant, including installed equipment. Construction management contracts are based on the erected value of the project managed.

Source: Engineering News-Record and *International Construction Week.* Used with permission.

ume outside of its home country, but this only translated into slightly more than $1.1 billion. In 1987, the mix of domestic revenue to foreign sales by some of the large U.S. builders in Table 1–2 changed somewhat as the OPEC countrys' construction projects bottomed out and major infrastructure projects overseas were either scrapped or delayed. Foreign sales still accounted for a substantial part of the annual volume of these American builders:

Contractor	**% Foreign Work 1985**	**% Foreign Work 1987**	**Total 1987 Revenues**	**Total 1987 *Foreign* Revenues**
M.W. Kellog	92%	49.7%	$4.609 Billion	$2.295 Billion
Parsons Group	58%	28.3%	5.507 Billion	1.557 Billion
Bechtel Group	49%	41.7%	7.954 Billion	3.314 Billion
Brown and Root	52%	39.8%	2.985 Billion	1.189 Billion
Lummus Crest	69%	72 %	1.600 Billion	1.152 billion

It appears, in all respects, that U.S. contractors have gained much more from their foreign market penetration that their opposite numbers from Japan.

The automobile epitomizes the essence of America. It breathed life into the American Dream and gave mobility to the lives of those who pursued it. The way in which Detroit reacted to competition from Japanese auto manufacturers might just point the way for other industries, including the American construction industry. The malaise that existed in Detroit created a ready market for the small, well built machines from Japan. At first dismissing these imports as a flash in the pan, Ford, GM, and Chrysler were caught completely off guard by the oil crisis of the 1970s. Slowly they began to take stock of their companies, and no longer being able to dismiss the Japanese, began to learn from them.

In 1970, Nissan imported its first sports car, at first known as the *Fair Lady.* Savvy Japanese executives on the West Coast quickly changed the car's name to the factory model number, 240Z, and the rest is history. Not quite twenty years later, 6–7 million cars are being sold annually in this country and 2.3 million of them come from Japan. Add to this total another 700,000 "Japanese" cars produced in U.S. factories, and you have about 30% of all U.S. auto sales attributable to Japanese sources. But things have changed from those doom-and-gloom days of the 1970s and early 1980s. Ford now owns 25% of Mazda, GM has a 41% stake in Isuzu, and Chrysler has a 24% equity position in Mitsubishi Motors.

The globalization of one of America's basic industries has taken place very quietly under our very noses. Ford Motor Company announced that beginning in 1989, it will ship 10,000 of its new, sporty *Probe* models to Japan, and they have plans to sell at least 30,000 of these sleek automobiles annually there. Not to be outdone by Detroit; Honda Motors of Marysville, Ohio announced that they will export some 50,000 *Accords* to Japan.

Although the past several years have been difficult ones, the U.S. automobile industry is climbing out of the doldrums, in most cases much stronger than before, and resilient in much the same manner as the steel that has been tempered in the fires of their bustling plants.

Will American contractors embark on a learning curve cognizant of the theory that warns that those who ignore the past are doomed to repeat it? Will American builders act and react with Japanese builders in a cooperative mood, and will the joint ventures they create, some successful and others possibly unsuccessful, lead to a better understanding of each other's goals? The Japanese contractors are here to expand their horizons and their fortunes. They bring with them the riches from the Orient—discipline, technology, hard work, and the determination of an entire nation that we need to know much more about. Hopefully, the following pages will provide a parting in the Silk Curtain, permitting all construction and otherwise inquisitive Americans to peek through.

2

The Construction Industry in Japan

A LOOK AT THE COUNTRY

Imagine owning a construction company in a country slightly smaller than the state of California, where there are 515,999 other licensed contractors vying for work. That equates to about 3 1/2 contractors for every square mile! Now suppose that 15 of these 516,000 builders are among the 100 largest international construction companies, with combined annual sales volumes exceeding $54 billion, and they are all competing fiercely for business in this same small geographic area. This happens to be the picture of the construction industry in Japan and, as one can imagine, even in a country with a burgeoning industrial base and a gigantic trade surplus to spur on government investment in domestic construction projects, the need to maintain and increase yearly revenues must create tremendous pressures on these giant contractors to seek new work everywhere in the world. A look at the Japanese construction industry and its worldwide activities ought to be viewed through this perspective.

To begin to understand some of the unique differences between the domestic Japanese construction industry and our own, it is necessary to take a closer look at the Japanese psyche and at some of the history that has shaped and molded the culture of this island nation.

Demographics of Japan

Japan is an archipelago stretching some 1744 miles, composed of four main islands, Honshu, Hokkaido, Kyushu, and Shikoku, and 3918 lesser islands, containing in total about 143,500 square miles, roughly the same size as the state of Montana.

In 1986, the population of Japan was estimated to be about

121,500,000, representing a population density of 846 people per square mile. The United States, in contrast, with its 246,000,000 people spread over 50 states with a total land area of 3,540,940 square miles, averages only 69 people per square mile. Tokyo, by far the most densely populated Japanese city, has a current population of about 8.5 million, and continues to grow. Yokohama is in second place with a population of 3 million, and the city of Osaka ranks No. 3 with slightly more than 2.5 million inhabitants. In comparison, New York City, this country's most heavily populated urban area has 7.1 million people and is followed by second place Los Angeles with a little more than 3 million, according to the latest U.S. census count.

Japanese History

Japan is an ancient country. According to legend, it traces its origins back to the appearance of a sun goddess some 700 years before the birth of Jesus Christ. Recorded history in Japan began sometime in the Fifth Century A.D. and the feudal society ruled by emperors and shoguns (generals) at that time continued for about the next thirteen hundred years.

This island empire, inhabited by one race with its unique Eastern religion borrowed heavily from the Chinese, remained more or less completely isolated from the rest of the world until a Portuguese ship landed on its shores in the mid-1500s. This first contact by a Western power was followed by a series of visits by religious zealots anxious to convert those Japanese "heathens" to Christianity and Catholicism. All of these efforts were met with little or no success.

The first contact by the United States occurred when Admiral Matthew Perry sailed into Tokyo Bay in 1853, bringing with him many gifts for his Japanese hosts, including a perfectly functioning model of a steam locomotive and a working telegraph set. These two examples of western technology greatly impressed the Japanese dignitaries, and from that time forth Americans were viewed as people from whom a great many things could be learned.

The Age of Industry

Japan entered the Industrial Age in 1868 when the Emperor Meiji (pronounced *May-gee)* officially abolished feudalism and replaced the power of the shoguns with a parliamentary system. Japan started sending emissaries around the world looking for new ideas, and this Meiji Restoration, as the period was known, set the stage for the cultivation and nurturing of a domestic industrial base. It also set the stage for Japan's entrance into world adventurism and thoughts about expanding its influence.

Japan started to formulate ideas about imperialism, and anxious to expand its economic borders, waged war with China in 1894. It acquired the island of Taiwan and then went on to defeat Russia in the Russo-Japanese War of 1904. Further attacks were directed toward China and in 1931 Japan invaded Manchuria. Its boldest move of all came in 1941 when bolstered by an alliance with Nazi Germany, Japan attacked Pearl Harbor and consequently ended forever its claim to world power through military conquest.

The Phoenix

In ancient Egypt there existed a mythological bird called the Phoenix. The Phoenix, according to legend, lived for 500 to 600 years and then burned itself to death on a funeral pyre only to rise again out of its ashes, renewed, more beautiful than before, and eager to begin life all over again. Japan is certainly the economic Phoenix of the modern industrial world. She represents an unusual amalgamation of American style democracy and technology interwoven between the threads of Japan's old paternalistic-feudalistic society to create an unusual economic fabric whose patterns are closely observed throughout the rest of the industrialized world.

At the base of this new economic miracle lies a foundation of cultural values that play an important part in the creation of Japan's phenomenal growth. Without some understanding of several basic differences between our Western culture and their Eastern culture it will be difficult to comprehend the Japanese approach to business in general, and their approach to construction in particular.

The Island Mentality

Japan was an island empire in the truest sense of the word for many, many centuries. Divorced from outside contact and content with keeping it that way, a keen sense of close, human relationships developed in which people of the same race, and basically the same religion, grew to depend upon each other for survival. This collectivistic approach has transcended the centuries so that today most Japanese business decisions are collective, and approval by all parties prior to implementation is the order of the day in modern Japan.

Amae

Words can sometimes express cultural differences as well as anything else. Words can provide insight into the national character.

American slang might well have had its origins in our defiance of anything establishmentarian and our impatience with the restraints of formal language. German words, sometimes blockbusters, seem to mirror the Teutonic penchant for building one thing upon another in a methodical manner. For example, a boat with two propellers creates the German word *Zweischraubendampfer,* and the title of Under Secretary of State is transformed into *Unterstaatssekretär.* In Japan, the word *amae (Ah-my-ee)* is such a word. It tells us a great deal about the Japanese outlook on relationships. The literal translation of the word is "to presume upon the affections of someone close to you," but in its most commonly used form it is meant to be interpreted as "dependency."

Although dependency, at times, might be defined in Western culture as a sign of weakness or indicate the need to lean on someone for support, in Japan the word *amae* is used to convey the concept of, "We are all in this together and we need each other's help to survive and prosper."

When a Japanese company offers the concept of lifetime employment to a prospective employee, it is displaying *amae*—the employer extends an offer to join his company, his family, in a long term relationship. The employee, by accepting, implies that by joining this family, they will look after him all during his working days. *Amae* is displayed when a client decides to build a new plant and negotiates a vaguely worded construction contract with the same general contractor that he has been doing business with for the past 75 years. *Amae* is displayed when this general contractor, in turn, negotiates contracts with the same nucleus of subcontractors that he has been working with for just about the same period of time. *Amae* is again exhibited when both general contractor and subcontractor agree to perform extra work not specifically called for in the construction contract, but which would be essential in order to comply with the intent of the client's construction program.

A Japanese contractor related a story to the writer that illustrates the concept of *amae.* His company, one of the biggest in Japan, had a contract to build a new building for an American client in Tokyo a few years ago. The building, a cast-in-place concrete structure was scheduled to be completed in time for dedication ceremonies on Christmas Eve. There were problems with the foundation work and the building was coming out of the ground very slowly. By October, when the schedule appeared almost impossible to meet, the American representative in Tokyo suggested changing the dedication date to late January or

February. Since many of the executives from the States had to make elaborate plans to travel to Tokyo, a later date seemed to be in order. A meeting was held to discuss the job progress and the general contractor advised his client that he had met with all of his key subcontractors and they were reviewing the work that was needed to be done to get the project back on track so that the original schedule could be met.

The following week a three-shift, around the clock work schedule was implemented and the building was completed the morning of the day initially set aside for dedication. The bunting over the front door was being hung just as the last worker left the site. The writer asked this Japanese construction executive if he had a difficult time trying to collect an "extra" for all of this overtime work. There was a long pause that at first was interpreted as his not fully understanding the question. The reply came with a puzzled look on the Japanese builder's face, "Oh, we could not ask for any additional reimbursement, we gave our *word* that the building would be completed on time." When asked how much he had to pay his subcontractors for their accelerated efforts, there was the same pause and the same puzzled look, "Our subcontractors knew we had to do everything possible to meet our schedule and they were merely helping us out in this difficult situation." The writer said that this was a very noble concept, but one that would lead to bankruptcy in short order. The Japanese builder said that, of course this type of situation does not happen often. His client knew full well that neither his company nor his subcontractors made any money on this project, and in fact, they lost money, but there will be other work with this client in the future, and at that time, they all might be compensated in such a manner that their losses would be partially or completely repaid.

That is the essence of *amae*—We look out for each other and when we do, all of us prosper in the long term. Not a bad concept to pursue!

THE HISTORY OF CONSTRUCTION IN JAPAN

The Japanese construction industry can trace its origins back to the Tokugawa Era which existed from 1603 to 1868 when the forerunners of today's major construction companies such as Shimizu, founded in 1804, were in their embryonic stage. During the Meiji Period from 1868 to 1911, Japan began to look outward and began to embark on an era of modernization and industrialization. The government invested heavily in infrastructure and the Number 1 priority of the day was the construction of a national railway system to open up the country. Throughout this entire period of time, some 43 years, the government funneled 40–50% of its annual construction budget into the creation of the National Railway System, with river work, roads, ports, and agricultural projects assuming lesser priorities.

The Mid-1930s

Beginning in the early 1930s, Japan's construction investment philosophy took on military overtones and once again railway construction became a national priority. The role of contractors during this prewar period was confined to supplying just the labor required for the projects. The government planned the projects, provided all of the design work, and even furnished the equipment, supplies, and materials and tools needed for the construction. This was possibly the beginning of the close alliance between government and contractors that exists in Japan today.

Overseas construction in the mid-1930s was directed toward the building of military facilities in Japan's colonial sphere of influence, and their contractors moved into Korea, Manchuria, Formosa, and various other island outposts in the Pacific Rim.

The Postwar Period

After World War II, Japanese overseas construction took on a somewhat different role as these contractors began reconstruction work as part of a reparations program in those Asian nations ravaged during the war.

In 1948, due in part to the major role construction would play in the restoration of postwar Japan, and due in part, to pressures being mounted by the emerging construction giants, a Ministry of Construction was created to provide direction to this vital industry. At the time it was founded, the Ministry of Construction was given a threefold responsibility:

1. Initiate agricultural and civil engineering projects needed to create and increase existing food production capabilities.
2. Create flood control projects desperately needed since serious deforestation had taken place when huge amounts of lumber had been harvested for post war rebuilding efforts.
3. Administering the construction of U.S. armed forces facilities.

The influence of the U.S. military mission coupled with the emergence of a new era outlook helped to abolish some of the prewar conditions that existed in the Japanese construction industry. Before the war forced labor crews were used on several large construction projects in Japan. This system was known as the Kangokubeya System and immigrant workers from slum areas were pressed into service as laborers, working long hours under harsh labor bosses with no real recourse to their plight. In remote areas of Japan, such as the North, where labor shortages were prevalent, as many as 20,000 debt bondaged construction laborers were said to have been put to work, probably in much the same manner as we used immigrant labor in the early and mid-1800s to build the mighty transcontinental railroad system in this country.

The 1960s and 1970s

In the 1960s and 1970s construction in Japan reached extremely high levels of activity amounting to 15% of their Gross National Product in 1960 and 20% in 1970. Nationwide, 2.36 billion to 2.7 billion square feet of building area was being constructed annually. The resulting effect of all of this accelerated construction activity began to surface in the early 1970s when air pollution reached dangerous levels and water pollution became a serious threat. The major cities became congested as workers flocked to the new factories, which along with residential developments started to choke Japan's three major cities, Tokyo, Osaka, and Yokohama. When the oil crisis occurred in 1973 and again in 1979, the government started to take stock of national priorities and began to focus its concern on proper housing, and the rebuilding and upgrading of the country's infrastructure.

Existing water sources were being polluted and no new sources were being investigated. Because of the Japanese topography—steep mountains, quick flowing streams, and the ever present danger of earthquakes, typhoons, floods, landslides, and other natural disasters, untold loss of property and lives occurred every year. Japan, the emerging economic giant, was devoting all of its resources to building a strong industrial base and was disregarding an inadequate and dangerously overloaded infrastructure. Soon the politicians and the populace alike began to clamor for more government involvement in road, tunnel and utility construction, and a major portion of Japan's top contractor's efforts were channeled into a wide range of construction programs in the civil engineering field.

Table 2-1 contains a list of these major Japanese construction companies and their domestic and foreign revenue sources for the period 1981–1985. With only two exceptions, Kumagai Gumi and Mitsubishi Heavy Industries, Japan's Top Ten were devoting from 78–96% of their construction activity at home, primarily on the improvement of infrastructure.

The government in Japan, at all levels, tends to take a very active role in the orchestration of construction activity, and the relationship between contractor and government remains a close one.

Table 2-1. Top Japanese Contractors, 1981–85, Domestic Versus Foreign Sales (in millions of dollars).

Company	1981		1982		1983		1984		1985	
	Total	Foreign	Total	Foreign	Total	Foreign	Total	Foreign	Total	Foreign
Kajima Corp.	$4,494.0	$252.0	$4,352.8	$363.8	$3,840.1	$291.1	$3,787.0	$485.0	$4,953.4	$627.7 (12%)
Shimizu	4220.4	220.4	4,379.9	471.9	4,299.5	537.0	3,931.7	345.9	4,779.1	605.4 (12%)
Kumagai Gumi	2,899.8	216.9	3,115.3	550.3	3,283.5	737.1	3,227.7	850.2	4,692.2	2,175.8 (46%)
Taisei Corp.	4,521.0	485.0	4,067.0	481.0	4,834.0	453.0	4,213.0	247.0	4,641.3	266.5 (6%)
Mitsubishi Heavy Ind.	n.a.	801.6	n.a.	n.a.	n.a.	195.0	3,800.0	720.0	4,410.0	1,425.0 (22%)
Ohbayashi Corp.	n.a.	n.a.	n.a.	n.a.	3,378.8	375.4	3,122.5	359.9	4,280.0	377.8 (8%)
Takenaka Komuten Co.	n.a.	n.a.	3,043.8	311.4	3,301.3	304.4	3,011.8	191.5	3,787.0	189.0 (5%)
Hazami-Gumi	n.a.	n.a.	n.a.	66.9	1,841.5	346.5	1,855.0	319.1	2,266.3	499.8 (22%)
Fujita Corp.	1,929.0	216.0	n.a.	n.a.	n.a.	n.a.	1,611.7	81.7	1,980.0	83.0 (4%)
Sato Kogyo Co., Ltd.	1,397.7	123.6	n.a.	n.a.	1,172.9	134.2	1,219.6	113.6	1,913.2	121.1 (6%)

Note: Figures above listed in terms of total 1985 contracts. These figures include prime contracts, shares of joint ventures, subcontracts, design-construct contracts, and construction management contracts where the firm is exposed to financial liability similar to a general contractor (greater than $500,000). Design-construct contracts are based on the erected value of the plant, including installed equipment. Construction management contracts are based on the erected value of the project managed.

Source: Engineering News Record and *International Construction Week.* Used with permission.

PUBLIC VERSUS PRIVATE PROJECTS

Government projects have represented anywhere from 40 to 45% of Japan's total construction awards during the period from 1981 to 1985. This is considerably more than the 18–20% government share of construction work performed in the United States. Changes in government policy or priorities can have a profound effect on the sales volume and earnings of the large Japanese contractors and they have to be constantly on the prowl for work in other parts of the world.

When the Japanese government decides to open the tap on construction awards, they tend to do so on a grand scale. Toward the end of 1987, Japan announced that the country was to allocate $58.7 billion for public works projects in 1988. This was in addition to several monumental projects already underway or in the pipeline.

TUNNEL WORK

The mountainous nature of much of Japan, coupled with the separation of large land masses by water, has led their contractors to develop a great deal of tunnel construction expertise. Japanese technology in innovative tunneling techniques ranks among the world's finest. Shield tunneling machines including pressurized slurry face models, articulated machines for curved tunnel projects, enlargement shield excavators, and cocoon multi-shield excavators have been designed and built by several construction companies. Tunnel work is frequently required in concert with highway construction and the expansion of rail systems and utility extensions, and the value of these types of tunnel projects has been about $12.5 billion annually. A major tunnel project valued at $1.5 billion was started in 1987 as part of the Trans-Tokyo Bay highway system, and the Shinkanses (Bullet trains) had $9.375 billion set aside in 1987–88 for tunnel work required for some of their new routes. Subway and freeway projects during the same period required additional tunnel work valued at $6.225 billion. In March of 1988, the $5.3 billion Seikan Tunnel was opened for business, linking Japan's main island of Honshu with the island of Hokkaido to the north. Thus tunnel made headlines, as it is currently the world's longest undersea tunnel devoted solely to rail traffic. The tube is 24.8 miles in length; and it took 16½ years to build. This new Seikan Tunnel greatly surpasses the previous record holder, the Simplon, a 12.3 mile tunnel through the Alps, linking Italy to Switzerland.

URBAN DEVELOPMENT

In 1987 a $12.5 billion revitalization project was started in Yokohama, Japan's second largest city. The project, known as Minato Mirai 21 (New Port City Plan for the 21st Century), will consist of a rebuilt metropolitan district which will include a 5,000 seat convention center, exhibition halls, office buildings, shopping centers, hotels, parks, and museums.

For the past 120 years, Yokohama has been known as the Gateway to Japan, famed for its easily accessible waterfront and commercial shipping facilities. This waterfront will be radically changed when Minato Mirai 21 creates a new 460-acre center for international trade and convention purposes. Yokohama is strategically located just 18 miles from Tokyo, 30 minutes from Haneda Airport which handles domestic flights, and 90 minutes away from Narita International Airport. Minato Mirai 21 will build its own inner city arterial roads which will connect to the Tokyo-Yokohama expressway. The outdated railroad station will be revamped and the new business area created in this redevelopment project will be no further than 1300 feet from the new rail lines with access provided by ramps, bridges and walkways completely detached from the new road system.

Of the total of 460 acres in the redevelopment project, space will be allocated as follows:

Business, commercial, and residential areas	215 acres
Road and railway service	104 acres
Parks and open space	114 acres
Port related space	27 acres

It is anticipated that the daytime population of the city will be increased by 190,000, and with 3,000 new households to be created in Minato Mirai 21, a permanent population of 10,000 is to be expected.

The installation of a fiber optics communications system will link the city with the rest of Japan at the speed of light. A central heating and cooling plant will provide temperature control to all of the buildings and facilities in the complex. Massive construction activity is underway and visitors to Yokohama cannot miss the huge YES '89 YOKOHAMA EXOTIC SHOWCASE sign as the train pulls into the station (Figs. 2-1 and 2-2). This gigantic project (Fig. 2-3) is scheduled for completion in the year 2000.

In Tokyo, a three building government complex neared design completion in 1987 and contract awards were made in 1988 for this $900 million project. One of the structures in the project (Fig. 2-4) is slated to become Japan's tallest building, a 48 story, 797 feet tall gray granite-

Fig. 2-1 Yokohama Exotic Showcase—YES '89.

clad tower which will house the Governor's office. The other two buildings, known as Tower 1 and Tower 2 are to contain a total of 3.5 million square feet of space. When this new Metropolitan Government Center is complete in 1991, it will be home to 13,000 employees, and is expected to attract 45,000 visitors into a city that is already strained by overcrowding.

Across the bay from Tokyo and nearby Yokohama lies Chiba Prefecture, and an area known as Makuhari, a trade fair city that is under construction and will form the core of the Makuhari Metropolitan Center, scheduled for completion in 1989. A concerted effort is being put forth by the Japanese government to attract trade fairs from all over the world, and Makuhari has this goal in mind. The first full scale convention center in the project was started in 1987 and will soon be joined by exhibition halls, conference facilities, hotels, and 96 acres of parking for the 1 million square foot exhibitors arena that will be built. The entire Makuhari complex will encompass 1100 acres, employ 100,000 workers, and will house at least 26,000 of these new employees nearby. As of this writing there are more than 20 major government sponsored

(a)

(b)

Fig. 2-2. Massive construction effort underway in Yokohama. (a) Bridges, infrastructure, roadways proceeding at a rapid pace against the background of Yokohama. (b) Buildings are starting to take shape throughout the Minato Mirai 21. (c) A marine memorial park called the *Nippon Mara* Park, for the 1930 naval training ship permanently docked there is already attracting visitors to the area.

(c)

Fig. 2-3. The vision of Minato Mirai 21, the new port city for the 21st Century.

Fig. 2-4. Metropolitan Government Center, slated to be Japan's tallest building. Construction has already begun on the Metropolitan Government Center in the Shinjuku section of Tokyo. With typical Japanese precision, dump trucks move easily from staging area into the construction site and out again with little or no interruption to business in the immediate area.

and privately funded construction projects underway in the Tokyo Bay area as part of a concerted effort to stimulate the local economy and wean Japan away from its overwhelming dependency on world trade and export sales as a major source of revenue.

The role government plays in both Japanese and American construction activity can be readily discerned by scanning the figures in Table 2-2.

In Japan, construction is a greater contributor to the country's Gross National Product than in the United States, constituting 17% of their GNP in recent years versus just about 8.4% of America's GNP in the same period.

Table 2-2. Types of Construction Activity—U.S. vs Japan—1984.

Type of Construction	Japan (using 238 yen) Dollars	%	U.S. (current dollars) Dollars	%
Private housing	$60.5 billion	30.4%	$130 billion	42%
Government housing	3.4 billion	1.7%	1.8 billion	½%
Commercial-industrial	35.0 billion	17.6%	88 billion	28½%
Private civil engineering	22.7 billion	11.4%	31 billion	10%[a]
Government civil engineering	62.7 billion	31.5%	39 billion	12½%
Government—other	14.7 billion	7.4%	20.2 billion	6½%
Total	$199.0 billion	100.0%	$310.0 billion	100%

[a]Private civil engineering projects include railroad construction, pipelines, and private utility company construction.

Because government in Japan can impact greatly upon construction, and since construction activity contributes so significantly to GNP, this strong alliance between government and contractor will probably continue, and the effect of government fiscal policies may dictate when the major Japanese construction companies will have to become more aggressive in world markets.

OTHER GOVERNMENT ROLES THAT AFFECT CONSTRUCTION

Both federal, state, and local governments, through various tax concessions and incentives, can stimulate or retard construction activity in specialized or targeted areas. Americans have seen this phenomena occur in many parts of the United States as state governments wooed northern industries to the Sun Belt and northern states attempted to attract high-tech industries to replace those industries that moved south.

In Japan, where government assistance has taken on a slightly different role in targeting construction activity, more emphasis has been placed upon efforts to stimulate high-tech and research and development facilities.

HI-TECH AND INDUSTRIAL PARK DEVELOPMENT

Hokkaido is the northernmost island in the Japanese archipelago, located at about the same latitude as Portland, Maine. It has a population of about 5.7 million distributed over a land area of approximately 21 million acres. The capital city of Hokkaido is Sapporo, the city that gained fame as the home of the 1972 Winter Olympics. Industrial park development is not new to Japan as there are currently about 24,350 acres of land throughout the country developed in this manner. New park development is being encouraged by federal, state, and local governments to stimulate specific types of industrial park development for very specific purposes.

In the Ishikarai Bay New Port Area Industrial and Distribution Center, located on Ishikarai Bay about 9 1/2 miles from Sapporo, subsidies are being offered to create new high-tech companies, R&D facilities and computer related industries. Companies already engaged in these types of businesses in other parts of the country are being coaxed to relocate to this area and the state, or prefecture government, is offering cash payments to do so. The Ishikarai Bay New Port Area Center has made the following incentives available to industries willing to relocate there:

1. When factory investment exceeds $6.8 million, and more than 50 new employees are hired, cash payments of 6% of factory investment, up to a maximum of $1.37 million is being offered.
2. For high-tech companies of "small size," payments equal to 6% of factory investment up to a maximum of $680,000 will be made when plant investment exceeds $205,000, and more than 10 new employees are hired.
3. When R&D and computer programming companies locate in the park, payments equal to 10% of plant investment, up to a maximum payment of $684,000 will be made when plant investment exceeds $205,000 and more than 10 new employees are hired.

It is interesting to note that payments are pegged not only to capital investment but also to increased employment which in turn will stimulate growth in local consumer related businesses.

The Town of Ishikarai adds a further inducement to new plant owners when local citizens are hired to work in their plants. Over and above the state subsidies, the town government will pay factory owners a subsidy of about $411 for each worker hired beyond the first twenty employees, when the cost of the plant, exclusive of land, exceeds $342,000. To obtain this subsidy, the new employee must have been hired from the local labor force and the town will issue total payments to the company for as much as $8200 for this purpose.

At the Chitose City Industrial Park, some 23 miles inland from Sapporo, the city government is selling land at greatly discounted prices in order to attract advanced technology companies. Discounts amount to 10% of factory investment with a cap of $6.8 million, and Chitose also offers outright payments of $3400 per employee for all workers hired and remaining with the new company for one full year.

THE INDUSTRIAL RELOCATION PROMOTION LAW

In 1974, the federal government of Japan passed the Industrial Relocation Promotion Law and established the Japan Regional Development Corporation, empowered with the obligation to disperse the concentration of industrial activity and associated population growth away from the already overcrowded cities of Tokyo, Osaka, and Nagoya. The JRDC's other chartered purpose was to stimulate and renew the depressed economies of the coal mining regions, which had been plunged into recession when imported oil became Japan's dominant source of fuel.

JRDC was allowed to loan money from their initial capital account of $500 million, to industries willing to relocate to the less populated areas designated by the agency. JRDC was also charged with targeting 19 areas throughout the country for the development and construction of new industrial parks. To facilitate its mission, JRDC started to offer many types of incentives:

- Companies purchasing land would be required to post a 20% down payment; however, the balance of the money due could be paid back over ten years with interest rates ranging from 6.25% to 6.47%
- Plant and equipment loans would be made, not to exceed 40% of plant and equipment costs, with paybacks of up to 10 years, at interest rates of 6.25%
- To companies relocating to JRDC targeted areas, loans for operating expenses would be available with five year paybacks.
- Special loans were made available to firms relocating to certain specific areas and companies doing so would receive low interest loans based upon 80% of the value of the site they were vacating! These companies could also obtain a 3 year loan to offset half the cost to relocate and loans would also be available to cover 50% of the company's operating costs incurred during relocation.
- Preferential tax treatment would be afforded to all firms relocating, and in some special cases, depreciation allowances would be granted when plant investment exceeded $106,000.

TECHNOPOLIS

To sharpen Japan's edge in high-tech manufacturing, and to remain on the cutting edge of new technology, JRDC set about to create a concept they termed *technopolis*. The underlying goal of the technopolis is to create brand new cities throughout the country that would be devoted solely to advanced technology. JRDC was painting a picture of life in the technopolis as one of affluence and culture for all who worked and lived there. With cultural facilities in abundance and a network of new highways, airports, and bullet train routes to link all of these hubs together, a fresh new wave of construction activity would be created to pave the way for Japan's entrance into the 21st Century.

The prototype for this new concept evolved in 1987 when Nagaoko New Town was created in the center of Honshu, about 36 miles from a large seaport on the Japan Sea. In Nagaoko New Town, the Technological University of Nagaoka would train engineers to enter the local workforce and contribute to the success of the many R&D companies beginning to locate there. With an airport 36 miles away, a rail station 6 miles from town, and just an hour and twenty minute train ride to Tokyo, the anticipated population of 411,000 in the residential section of Nagaoko New Town is well under way. Industries planned for this technopolis are mechatronics (robotics), new materials manufacturing (exotic metals, fine ceramics, sophisticated resins), bio-industries, and medical care industries.

Toyama Yatso Technopolis, also located in Honshu, encompasses nearly 200 acres of land and, as of 1987, had as its core of high-tech companies Fujitsu, Kokusai Electric, Matsushita Electronics, and SMK a major manufacturer of electronic parts. Surrounding these industrial establishments are support facilities such as the Toyama Technological Development Foundation, High-Tech Industry Communication Center, Life Science Research Center, The Industrial Design Center, and two science and engineering universities with graduate schools.

More technopolises are on the drawing boards for other parts of Japan's four main islands.

A variation on the technopolis theme is the *science city* concept, where government and private industry work together to further scientific knowledge. There is the new Kansai Science City in the Osaka area, and about 60 kilometers north of Tokyo there is a well established research city called Tsukuba Science City (pronounced Scuba). Although just about a one hour bus ride from Tokyo, Tsukuba is a breath of fresh air, both literally and figuratively.

Arriving at the Transportation Center, a visitor is immediately struck by the amount of open space and greenery, which is in stark contrast to Tokyo's crowded conditions. Within a short walking distance from the bus stop there is a cluster of western style houses for sale, which by Japanese standards are quite extraordinary: two-story houses with lots of yard space and neat landscaped areas that give way to an interior with cabinetry-quality woodwork. Narrow, clear cedar tongue-and-groove siding is used on corridor walls and on a ceiling in what appears to be a living room area. Door frames and wooden window frames show a mastery of joinery skills; modern kitchens, bathrooms with polished granite vanity tops all catch the eye. And best of all, to the Japanese buyer, is space. These houses range in size from 1900 to 2800 square feet and sell for $300,000 to $400,000, excluding the land. This is quite a bargain to people moving to Tsukuba from Tokyo, where a 1,000 square foot house on 1800 square feet of land can sell for as much as $1.9 million.

There are many elements of Tsukuba Science City designed and planned to attract workers to move into this research town. (See Fig. 2-5.) Across from the Transportation Center and a short walk or bike ride away is the Seibo Department Store, a multi-storied building with a few fast food restaurants on the first floor along with an array of fresh and prepared foods in the grocery department that would please any palate. Prices are not too bad, by Tokyo standards: 3 tomatoes—$1.92; corn—$1.52 each; peaches—4 for $6.00; cantaloupes—$7.69 each; and steak—$7.69 for 12 ounces. There are a number of private and public research facilities in the City along with higher education centers. The Ministry of Construction maintains its Building Research Center in Tsukuba, and Toda Construction saw fit to locate its extensive Institute of Technology there several years ago. Among the research laboratories located in the parklike settings are:

Research Institute for High-Energy Physics
Tsukuba Space Center of the National Space Development Agency
National College of Library Science
Public Works Research Institute
Geographical Survey Institute
University of Tsukuba
Tsukuba Branch of National Science Museum
NTT Technology Center

The feeling of "tomorrowland" is heightened by the multitude of prerecorded messages that greet the visitor to Tsukuba City. As a driver begins to enter one of the many on-grade parking lots scattered around, a female voice reminds him to take a ticket from the dispenser. Riding the escalator to the upper floors in the Seibo Department store, another female voice announces the departments to be found on each floor just before reaching the floor. The many buses taking passengers to the various research centers announce, in Japanese and English, the next destination as the bus speeds up the uncrowded parkways that crisscross the city.

This self-contained village, complete with primary and secondary schools, hospitals, and stores of all kinds makes the pursuit of scientific research a much more pleasant task.

THE THIRD SECTOR APPROACH

Local business communities have joined with state and municipal governments, and, with the assistance of loans from the Japan Development Bank, have created what is called the Third Sector approach—a triad of local government, local businesses, and state or prefecture governments, jointly investing in projects that will enhance and upgrade the local community.

The City of Kawasaki had been advertising their Kanagawa Science Park as an "incubator for R&D firms looking to participate in the world market". This new science park is budgeted at $54.8 million and offers open type research centers, a venture capital incubator, a research institute and school, banks, hotels, and even convenience stores for its inhabitants. The long-term profit orientation of Japanese enterprises is clearly shown in the publicity being generated for the Kanagawa Science Park, where no profit is expected for at least 10 years.

A similar Third Sector approach has been taken at the Kurume Techno Research Park in Kurume City in Fukuoka Prefecture, where a research center is the cornerstone of the park. This center will contain a mechatronics research laboratory and a bio-technology research institute, both of which will be at the disposal of the venture capital businesses and software development companies locating in the park. Scheduled for completion in 1988 at a cost of $133 million, the park is not projected to turn a profit during the first 12 years of operation.

HOTEL CONSTRUCTION IN JAPAN

Japan has been experiencing a boom in the construction of luxury hotels, and while this activity has been going on for the last twenty years, it has recently taken on a new importance in the economy. Japan has begun to promote tourism as another way to diversify its economy that has been too dependent upon a manufacturing base. From 1982 to 1985 between 11,000 and 12,000 new hotel rooms have been added throughout the country, and the expanding Japanese economy has been bringing scores of businessmen to the country, along with an increasing number of visitors from all parts of the world.

In 1983 Tokyo Disneyland opened and millions of people poured into the city to visit this latest creation of American ingenuity. Attendance at Disneyland has brought visitors from all over the Pacific Basin, and it is now a major new source of revenue for the city of Tokyo.

More and more Japanese are changing their vacation habits and are coming to the cities to stay at these new luxury hotels that have been built as a focal point for many urban redevelopment schemes.

In 1982 the 1,000 room Takanawa Prince Hotel was completed in

(a)

(b)

(c)

(d)

(e)

Fig. 2-5. Tsukuba Science City. (a) The Transportation Center. (b) Looking across the plaza to Scibo Department Store. (c) NIT's office and laboratories. (d) Looking across the plaza at Tsukuba Government Center toward one of the many parks in the area. (e) There are still some construction cranes in the area as evidence that building continues at Tsukuba City.

downtown Tokyo and this was followed by the equally plush 761 room Akasaka Prince and the 363 Imperial Hotel Tower. Another hotel, the 377 room Yaesu Fujiya was recently completed. The Tokyo Hilton opened its 858 rooms in 1984, and the Japanese National Railway built two hotels with a total room count of 1262 right next to the All Nippon Airways 1,000 room hotel. The New Otani hotel with more than 2000 rooms, and situated on fifteen extremely valuable acres in downtown Tokyo appears to be the grandest new hotel built in the area (Fig 2-6).

HOUSING IN JAPAN

Immediately after World War II, the housing shortage in Japan was estimated at 4.2 million units, but until 1950 residential construction activity was given a back seat to industrial plant construction. Both housing and infrastructure fell far behind and remain so even today.

Fig. 2-6. The New Otani Hotel located in the Akasaka section of Tokyo. With landcosts in the district averaging $15,375 per square foot, it is difficult to place a price upon the magnificent 15 acre formal Japanese garden that is part of this hotel. The total value of the land on which the hotel is built is estimated at $9.5 billion.

Most of the population of Japan is centered around and in its major urban areas, and by the year 2000 it has been estimated that fully 70% of the entire country's population will be located in only six cities—Tokyo, Yokohama, Osaka, Nagoya, Kyoto, and Kobe. Tokyo is already home to 8½ million people and has become a magnet for Japan's increasingly migrating population. With land speculation fueling the fires, and a strong agriculture lobby intent upon keeping 25% of all habitable land in use as farm land, it is no wonder that available housing in Japan is both miniscule in size and astronomical in price.

There are still a great number of wooden multi-unit apartment houses in Tokyo with apartments averaging 1.4 rooms and having total floor areas of 125 square feet. Only 10% of these urban apartment dwellers have their own bathrooms, and the practice of sharing kitchens, washing spaces and bathrooms is commonplace. (One reason being given for the increase in hotel occupancy in Tokyo is rumored to be the need for young married couples to experience the joys of marriage in private, since they cannot very well do so in a shoe-box sized apartment.)

Apartments in Japan are still measured in terms of "mat" or "jo" size, this designation referring to the woven straw mat typically placed on the floor in traditional homes, and measuring about three feet by six feet. When an apartment is advertised for rent, the ad might read "Two tatami mat room" meaning a room that is six feet by six feet, or a "Ten Mat Apartment," very generous by Tokyo standards, measuring 180 square feet.

Small, modern apartments in downtown Tokyo can rent for as much as $3,000 per month, but outside the center city rentals are somewhat less. An apartment house just completed in 1987 and located 17 minutes by train from downtown Tokyo was advertising studio apartments with 413 square feet for about $1350 per month, and one and two bedroom units ranging in size up to 1040 square feet were renting for $2800 to $3600 per month.

One would expect that single family housing in this type of environment would be terribly expensive and it is. The average house in Tokyo containing slightly more than 958 square feet, situated on a 1646 square foot building lot sells for about $823,000. The average condominium in the city sells for $370,000.

Housing starts have increased dramatically since 1965 (see Table 2-3); they reached a peak of 1,860,000 in 1972, coincidentally the same year that housing starts peaked in the United States at 2,378,500 units.

The Ministry of Construction has implemented their fifth Five Year Plan for Housing Construction and 6.7 million units are scheduled for production between 1986 and 1990. This new housing program is designed to create housing which exceeds the minimum standards, as defined by the ministry. Minimum housing has been defined to be a

Table 2-3. Housing Starts—Japan versus United States.

Year	Japan	United States
1950	200,000	1,952,000
1955	260,000	1,646,000
1960	420,000	1,296,000
1965	850,000	1,509,700
1970	1,490,000	1,469,000
1975	1,430,000	1,171,000
1980	1,210,000	1,312,000
1985	1,240,000	1,650,000

Source: Ministry of Construction, Tokyo, White Paper 1986.

living unit with 538 square feet, containing three bedrooms and a combined living-dining room-kitchen, and suitable for occupancy by a family of four. A quick, hypothetical floor plan for this "minimum" house might show one bedroom is for Mom and Pop, measuring ten feet by twelve feet, one bedroom for aunt, uncle, grandma, or gramps, at eight feet by eight feet, and another bedroom of equal size for Junior, which would leave just about sixteen by eighteen feet for living room, dining room, and kitchen. As a means of comparison, the U.S. department of Housing and Urban Development (HUD) years ago developed minimum property standards for single occupancy elderly housing apartment units, and they established a minimum size of about 500 square feet as being adequate space for a single dweller.

The Japanese Ministry of Construction states that "average" housing standards ought to be 925 square feet and should contain three bedrooms, one living room, separate dining room, and a kitchen. This type and size residence is deemed to be suitable for a family of four.

VARIATION OF THE TWO FAMILY HOUSE THEME

Japan, not unlike the U.S., is an aging society where the lifespan of both men and women has increased to the point where almost 32 million people will be 65 years or older by the year 2020.

One of the cultural differences between the U.S. and Japan is dramatically highlighted by the way each country views its responsibility to house its elderly citizens. The "Empty Nester" is a term familiar to all Americans, and conjures up a picture of husband and wife, having shed the duties of raising a family, embarking on a new era of carefree living. The big house is sold, handyman chores ended, and Mr. and Mrs. M.T. Nestor move into a maintenance-free condo to spend their golden years playing golf and bridge, and window shopping at the mall.

This is not the way in Japan, where parents have traditionally lived with their children and their grandchildren when they have reached retirement age. This practice has given rise to a ready market for a variation on the two family house. With both families pooling their resources, usually the parents contribute the land that they had purchased years ago at half-way reasonable prices, and their son or daughter pays for the house, so that both families can live together but in separate quarters.

In typical Japanese fashion, the existence of two-family living has created a need for further study and hence the formation of the Two Generation Housing Research Institute, which after one year of intensive study concluded with the following profound observations:

1. A choice must be made whether son and wife will live with son's parents or with wife's parents.
2. Cooperative living must be done with mutual respect.
3. Both households must remain independent, emotionally and financially.
4. A new set of rules must be formed to create family guidelines.

Residential contractors have responded to the needs of this two family market and they sell a variety of housing styles and configurations specifically manufactured for this purpose. The Asahi Chemical Industry Company's Housing Division, a major producer of factory built homes, markets a series of two-family houses where one level is reserved for each family, with outside stairs for the upper level. Another configuration is a two-story affair but with a party wall down the center dividing the building, with ground level entry for both residences. Asahi markets the DUO model for the man of the family and his parents, with several models displayed in a blue colored brochure. The DUET model, marketed in a suitably colored pink pamphlet, is for the wife who has decided to share the residence with her family. Both DUO and DUET feature separate entrances, but aside from some models containing common utility rooms, each half of the building has separate facilities for each family.

LAND COSTS

With approximately 90% of Japan's population living on only 30% of the land, it is no wonder that real estate for both residential and commercial uses is so expensive in and around the major cities. A small parcel of land in the exclusive Ginza section of Tokyo recently sold for $22,000 per square foot, and a few years ago the site upon which a new hotel was to be built, measuring 8,700 square meters (approximately 2.15 acres), was valued at $1.8 billion.

In the United States, land costs account for 10–20% of the total cost of housing, whereas in Japan, land costs contribute 20–50% of the total cost of a home. The average price of a house in Japan represents 6.1 times annual income, while even in the high-priced New York marketplace it represents only 3.3 times annual income. Because the price of real estate in general is so high in Japan, it was estimated in 1988 that the total value of real estate in this small country exceeded the total value of all real estate in the United States! This fact was uncovered by the Maekawa Report in Japan, which was a study identifying housing and agriculture policies as major segments of the economy requiring restructuring. This survey is all the more remarkable when one takes into account that the land area in the United States is about 25 times greater than the land area of Japan. Apparently there is no end to these rapidly increasing land costs. In 1987, a survey conducted by the National Land Agency revealed that the price of residential land in and around Tokyo rose by a heady 57.4%, while commercial land prices increased by 76.1%! Because of these extraordinary land costs, the Japanese have come to be very innovative in their approach to many types of urban development projects.

KAMI MEGURA

The Kami Megura Housing Development was built in 1957 by the Japan Housing Corporation and consisted of three four-story buildings containing 68 condominiums. Each condo was very small, even by Japanese standards, with only 377 square feet per apartment, containing two rooms and a kitchen. With only a 10 amp electrical service in each apartment, the use of any new electrical appliance or equipment was impossible. The site upon which the project was built made poor use of the available land. Situated on 1 1/3 acres of what had become very valuable property, the footprint of the entire development covered only 3/4 acre, a 56% coverage factor.

Nippon Steel, seeing a potential in creating higher densities, investigated through its Housing Division several schemes to increase the number of units on the property which would generate enough profit to make the whole scheme practical. Since they were dealing with 68 individual owners, one of Nippon Steel's major concerns was the ability to obtain approval from the homeowners. Nippon wanted 100% agreement before starting the work and they knew this would be no easy task. Over the course of the next 2 1/2 years they visited some owners reluctant to give approval, but daily visits in some cases to review plans and construction sequencing finally paid off. Nippon Steel planned to create 98 units ranging in size from 517 square feet to 1044 square feet, increasing the entire complex by 140% to a total of 74,271 square feet.

They were also going to upgrade all incoming electrical services to 40 and 60 amps in all units.

It was Nippon's idea to swap each 377 square foot apartment for a new one with 549 square feet, and if any of the present owners wanted to upgrade further, they could pay the difference between the selling price of a new 549 square foot unit and the selling price of the larger one.

Nippon, with typical Japanese patience, finally obtained approval from all of the present owners, went ahead with the new construction, added 30 new apartments in the complex which they sold at a decent profit.

LAND FEVER

Beginning in 1971, land prices throughout Japan started to increase rapidly for all types of construction—residential, commercial and industrial. Land fever in Japan crops up from time to time, the last period extending from the late 1950s into the 1960s when land costs escalated due to the great demand for industrial sites. The period of rapidly increasing land prices in the 1970s, however, appeared to be a false demand brought on by land speculators using that old come on, "You'd better buy now even though prices are skyrocketing because they ain't making this stuff anymore." Parcels of land in some subdivisions began to appreciate as much as 100% in just one year's time. Sparked on by speculative fever, land prices increased by about 18% per year during the ten year period from 1960 to 1970.

Major industrial corporations and many large construction companies started to bank land with their surplus funds. Many other corporations borrowed heavily to buy tracts of land they felt were ripe for development, or would be in the path of some new proposed highway or bullet train route. By the mid 1970s, corporations had purchased almost 750,000 additional acres of land, bringing their total land holdings to 2.323 million acres. The Big Banks of Japan added fuel to the fire by liberally lending funds to buy speculative land; from 1972 to 1973, loans for this purpose had increased by 300%. By 1976 land prices in Japan had escalated 160% higher than they were just five years previously.

All good things usually come to an end, and the bottom fell out of the land boom when the oil crunch hit Japan in 1973. Corporations were faced with substantially higher energy costs and rapidly increasing costs of raw materials due to OPEC's stranglehold on oil supplies and prices. Business failures started to occur just as they did in the United States during that hectic period, and the companies in Japan that were saddled with huge interest and principal payments for land purchases had to either go into default or sell off their investments.

Many of the large corporations caught up in the land fever lost substantial sums of money when prices plummeted. Other corporations, such as Kajima Corporation, who had a corporate policy that basically said, "Stick to construction, that's our business," came out of this frenzied era in a more financially sound position than some of their fellow contractors who were caught up in the land fever.

PROFILING THE JAPANESE CONSTRUCTION COMPANY

By comparing two construction companies of almost equal size, some insight can be gained into the differences between Japanese and U.S. contractors.

Turner Construction Company, based in New York, has offices in 36 cities throughout the country and maintains offices in several foreign countries as well. In 1985 Turner reported completed sales volume of $2.149 billion. Toda Construction Company, Ltd., headquartered in Tokyo, with ten offices in Japan, also has offices in various countries throughout the Pacific Basin and has subsidiaries in the United States. In 1985 Toda reported $2.586 billion in sales. In 1985, Turner Construction was 84 years old while Toda Construction celebrated its 100th anniversary. Both companies are engaged in diversified areas of construction (see Table 2-4) with Turner's main source of sales volume coming from the commercial sector and Toda's coming from a combination of commercial and residential.

Let us take a look at some facts and figures gleaned from the annual reports of both companies. First of all let us compare financial data displayed in Table 2-5. Toda's gross earnings from construction are twice as great as Turner's, however Toda's General and Administrative overhead is just about twice as much as the U.S. company, not too surprising when it appears that there are about 40% more employees at Toda Construction than there are at Turner. Toda claims 5252 employees worldwide with 1467 administrative personnel, 2195 construction engineers, 945 civil engineers, 162 electrical engineers, 126

Table 2-4. Where Their Sales Came From, 1985.

Market	Turner	Toda
Commercial	61%	21%
Health care	13%	9%
Housing–hotels	13%	25%
Manufacturing	2%	18%
Educational	3%	11%
Public sector	1%	5%
Others	7%	11%
	100%	100%

Table 2-5. 1985 Financial Profile.

	Toda	Turner
Value of construction	$2.586 billion	$2.149 billion
Earnings from construction	164,901,000 (6.3%)	67,653,000 (3.15%)
Earnings from real estate	8,591,000 (0.3%)	1,813,000 (0.08%)
Other income	11,195,000	6,887,000
Gross earnings	173,492,000 (6.7%)	76,353,000 (3.5%)
General and administrative expense	130,259,000 (5%)	56,771,000 (2.6%)
Income before taxes	56,858,000 (2.2%)	19,582,000 (0.9%)
Net income after taxes	17,593,000 (0.6%)	11,068,000 (0.5%)

Income Before Taxes reported by Toda is just about 2½ times greater than Turner Construction, but after taxes (69% for Toda, 43% for Turner) Toda's net profit is just slightly higher.

Financial Statements—1985

	Toda	Turner
ASSETS		
Cash	$400,734,000	$60,102,000
Securities	45,212,000	
Receivables	444,192,000	165,750,000
Work in progress	911,094,000	357,649,000
Real estate	189,832,000	100,199,000
Land, bldg, equipment	69,899,000	19,994,000
Investments	244,260,000	7,938,000
Other assets	41,504,000	40,503,000
	$2,346,727,000	$752,135,000
LIABILITIES	$1,914,973	$686,492,000
SHAREHOLDER'S EQUITY		
Common stock	62,741,000	4,551,000
Paid-in capital	72,444,000	2,382,000
Retained earnings	280,988,000	67,129,000
Total equity	$431,754,000	$65,643,000
Dividends per share	$42.00	$2.64

mechanical engineers, and 357 engineers of various other disciplines. According to Turner's annual report is employs a total of only 2800 employees. The Japanese contractors like to staff their departments and construction projects in a much more generous manner than American contractors.

Japan's high rate of taxation is also evident in this analysis. Turner's income of $19.5 million, before taxes, is reduced by 42% to a net income after taxes of $11 million. Based upon Japan's rate of taxation at the time, Toda's $56.8 million before-tax figure was reduced by 69% to a net income after taxes of $17.59 million.

Toda appears from the figures in Table 2-6 to be a very cash rich company, having $445 million in cash and securities, compared to Turner's $60 million. Investment by the Japanese company greatly exceeds that of its American counterpart and by looking at the Shareholder's Equity portion of both balance sheets, there is no question as to which stockholder has the better deal.

PROFITS—JAPANESE VERSUS AMERICAN

Pretax profits of many of Japan's largest construction companies were, for a time, generally higher than those of U.S. builders. Table 2-6 reveals the pretax profits of five of the largest Japanese companies along with their sales revenues for the period 1981 to 1985. By contrast, a survey conducted by the Robert Morris Associates in 1985, from data collected from slightly more than 1000 contractors revealed the gross and net profit percentages shown in Table 2-7, as reported by these contractors with various sales volumes.

THE BANKING CONNECTION

By scanning the financial statements of many of Japan's major construction companies a characteristic common to most of them becomes apparent—their close ties to banking and financial institutions. Many Japanese construction companies have banks and other financial institutions as stockholders. These close ties to money sources and project financing contacts enhances their ability to compete in the world market.

Giant Shimizu includes among its stockholders the Dai-ichi Kangyo Bank, which had an asset base of $252.76 billion in 1987. The Dai-ichi Kangyo Bank is a commercial and long term credit bank that concentrates on industrial development. Dai-ichi Kangyo currently has seven offices in the United States, and if Shimizu America needed financing for a U.S. project, it probably would not be looking for long.

About 20% of Kajima's shareholders, as of 1985, read like the Who's Who of Japanese financial institutions:

Sumitomo Bank, Ltd.	4.7% stockholder
The Kyowa Bank, Ltd.	3.4% stockholder
Mitsui Bank, Ltd.	3.3% stockholder
Sumitomo Trust and Bank	3.3% stockholder
Mitsui Bank and Trust	2.4% stockholder
Sumitomo Life Insurance	2.2% stockholder
Nippon Life Insurance	2.4% stockholder
	21.7%

Takenaka Komuten Co., Ltd., another large construction company, with worldwide construction sales of $5.79 billion in 1986, lists Mitsu-

Table 2-6. Financial Profiles of Five Top Japanese Contractors (reported in millions of yen).[a]

	Total Revenues	Pretax Profits	Percentage of Sales
Shimizu Construction			
FY 1981	776,990	36,446	4.7%
1982	871,854	36,462	4.2
1983	964,131	36,672	3.8
1984	923,544	31,852	3.4
1985	1,052,293	28,370	2.6
KajimaCorp.[b]			
FY 1981	863,125	41,222	4.7
1982	942,592	38,621	4.1
1983	932,062	27,668	2.9
1984	913,628	27,108	2.9
1985	990,000	27,000	2.7
Taisei Corp.			
FY 1981	811,667	33,202	4.0
1982	845,461	33,465	3.9
1983	907,700	31,128	3.4
1984	968,322	28,350	2.9
1985	978,607	23,058	2.3
Kumagai Gumi Co., Ltd.[c]			
FY 1981	607,197	34,364	5.6
1982	631,417	34,509	5.4
1983	614,485	30,059	4.9
1984	709,863	32,559	4.6
1985	841,000	28,372	3.3
Ohbayashi Corp.			
FY 1981	681,594	24,667	3.6
1982	644,726	25,802	4.0
1983	689,883	25,918	3.7
1984	766,318	22,514	2.9
1985	797,786	20,415	2.5

[a]To convert yen to dollars use following exchange rates: FY 1981—Y227.52; FY 1982—Y249.64; FY 1983—Y236.33; FY 1984—Y244.19; FY 1985—Y221.09
[b]Fiscal year ending in November.
[c]Fiscal year ending in September.
Source: Japan Company Handbook

bishi Bank and Dai-ichi Mutual Life Insurance among its stockholders. Kumagai Gumi has powerful Sumitomo Bank as one of their owners.

Thumbnail sketches of the following eight construction companies with sales volumes ranging from $150 million to $424 million reveals the same pattern of banker-financial institution stockholder involvement.

Table 2-7. Gross Profit–Net Profit Percentages—General Contractors.

	0–$1 Million	$1–10 Million	$10–50 Million	Over $50 Million	Mean
Gross profit	26.7%	14.9%	9.4%	7.3%	14.6%
Operating expenses	22.4%	12.6%	8.5%	6.5%	12.6%
Net profit before taxes	5.7%	2.9%	1.6%	1.5%	2.9%

Courtesy Robert Morris Associates

Katsumura Construction Company, Ltd.—Tokyo. Founded: February 1947.

Business activity: building construction 62%
civil engineering 35%
real estate 4%

Number of employees: 652

Financial data (1985):

Sales: $150.49 million
Net Profit: $731,000
Gross Profit, %: 7.8%
Net Profit, %: 0.49%

Major Stockholders:

Mitsubishi Bank	4.5%
Mitsubishi Trust	2.2%
Employees	1.9%

Ando construction Company, Ltd.—Tokyo. Founded: March 1918.

Business activity: building construction 85%
civil engineering 13%
real estate 2%

Number of employees: 1,924

Financial data (1985):

Sales: $424.2 million
Net Profit: $2.605 million
Gross Profit, %: 9.32%
Net Profit, %: 0.61%

Major Stockholders:

Dai-Ichi Kangyo Bank	5.6%
Tokai Bank	4.9%
Chuo Trust and Banking	4.1%
Employees	3.4%
Meiji Mutual Life	2.9%

Matsui Construction Company, Ltd.—Tokyo. Founded: January 1939.
Business activity: building construction 95%
civil engineering 5%
Number of employees: 845
Financial data (1985):
Sales: $171.7 million
Net Profit: $945,378
Gross Profit, %: 8.20%
Net Profit, %: 0.55%
Major Stockholders:

Fuiji Bank	5.0%
Hokuriku Bank	5.0%
Yasuda Trust and Banking	4.2%
Employees association	4.1%
Asahi Glass	3.8%

Daisue Construction Company, Ltd.—Osaka. Founded: March 1947.
Business activity: building construction 75%
civil engineering 25%
Number of employees: 1,022
Financial data (1984):
Sales: $246.6 million
Net Profit: $663,865
Gross Profit, %: 6.84%
Net Profit, %: 0.27%
Major Stockholders:

Japan Securities Clearing	11.6%
Japan Securities Finance	5.4%
Toyo Real Estate	4.9%
Sanwa Bank	3.3%

Odakyu Construction Company, Ltd.—Tokyo. Founded: January 1939.
Business activity: building construction 58%
civil engineering 38%
real estate 2%
others 2%
Number of employees: 884
Financial data (1985):
Sales: $189.72 million
Net Profit: $630,000
Gross Profit, %: 6.21%
Net Profit, %: 0.33%
Major Stockholders:

Odakyu Electric Railway	44.7%
S. Nomura	6.0%
Odakyu Real Estate	2.7%
Employees Association	2.6%
Mitsui Trust	1.5%

Totetsu Kogyo Company, Ltd.—Tokyo. Founded: July 1943.
Business activity: civil engineering 24%
railroad works 33%
building construction 36%
others 7%
Number of employees: 1,121
Financial data (1984):
Sales: $245.7 million
Net Profit: $1.525 million
Gross Profit, %: 9.03%
Net Profit, %: 0.62%
Major Stockholders:

Nippon Life Insurance	5.2%
Fuji Bank	4.9%
Bank of Yokohama	3.75%
Yasuda Trust and Banking	3.0%
Sanwa Bank	2.3%

Taihei Kogyo Company, Ltd.—Tokyo. Founded: September 1946.
Business activity: civil engineering 11%
building construction 35%
machinery 22%
installation work 24%
electrical work 10%
Number of employees (1985): 5,882
Financial data (1985):
Sales: $393 million
Net Profit: $1.627 million
Gross Profit, %: 9.4%
Net Profit, %: 0.41%
Major Stockholders:

Nippon Steel	43.4%
Japan Securities Finance	3.5%
Employees	2.1%
Kyowa Bank	1.8%
Tokai Bank	1.7%
Yasuda Trust and Bank	1.6%

Daiho Construction Company, Ltd.—Tokyo. Founded: March 1949.
Business activity: civil engineering 86%
building construction 14%
Number of employees (1985): 1,464
Financial data (1985):
Sales: $301.15 million
Net Profit: $5.456 million
Gross Profit, %: 10.39%
Net Profit, %: 1.81%
Major Stockholders:

Sumitomo Bank	6.8%
Dai-Ichi Life Insurance	6.1%
Dai-Tokyo Fire Insurance	4.0%
Daiho Investment Assoc.	2.5%
Tokai Bank	2.0%

THE MINISTRY OF CONSTRUCTION

The alliance between government and business in Japan is so closely intertwined that it is often referred to as "Japan, Inc." and the Ministry of Construction with its many functions does little to dispel this concept.

The Ministry is composed of several bureaus such as Economic Affairs, City (involved in city planning, urban renewal, city infrastructure), River, Road, Housing, and it has several auxiliary organs, the Geographical Survey Institute, Public Works Research Institute, Building Works Research Institute, and the Construction College. Each year the Ministry prepares a White Paper known as the State of Nation's Building in which it analyzes the state of Japanese housing and public works construction and the government's policies as they relate to the current and long range policies relating to construction. In combination with the Ministry of International Trade and Industry (MITI), the Ministry of Construction has a great deal to say about the direction of Japanese construction activity. MITI can shift economic and industrial priorities to expand or contract certain segments of the industrial base; when it decided to accelerate various export industries in the 1960s and 1970s, it actually charted the course for construction priorities in the country.

The Ministry of Construction operates as an overseer of the construction industry in Japan, and is assisted in this endeavor by various laws relating to the construction industry.

The Ministry of Construction and MITI jointly authorized an independent private foundation to create the Japan Testing Center for Construction Materials in 1963 to establish testing standards for construction materials, building components, and equipment. The Ministry also created a joint venture with Osaka Prefecture to form the Japan General

Laboratory for Building Materials. The government's commitment to research was further displayed by attaching the three research institutes to the Ministry of Construction's sphere of responsibility.

The Geographical Survey Institute

Formerly known as the Land Survey Division of the Japanese Army General Staff Office, the Institute collects and analyzes earthquake data as well as pursuing its main function of preparing topographic, geodetic, and geographical surveys and maps. The Institute is involved in orogenesis studies (formation of mountains) and acts as a central repository for public works survey data. This agency conducts examinations for licensed surveyors and assistant surveyors along with the rest of its duties.

The Public Works Research Institute

This organization is one of the older government departments, having been established in 1921 to design various government civil engineering projects. The Public Works Research Institute maintains testing and research facilities for public works projects under the jurisdiction of the Ministry of Construction, but it is also available to local governments who need assistance in getting their public works projects off the ground.

The Building Research Institute

This organ of the Ministry acts as a consolidated research agency and is involved in research relating to housing construction, urban construction problems, and disaster prevention research. The Building Research Institute assists in the formulation of standard building specifications and acts in an advisory capacity when new building laws are being prepared. Their input was necessary when the Building Standard Law of Japan was formulated creating a nationwide building standards yardstick.

LAWS RELATING TO CONSTRUCTION IN JAPAN

The Construction Contractor's Law

Enacted in 1949 and revised several times since, this law contains three basic provisions:

1. Establishment of a licensing system for contractors.
2. Establishing the basic business and ethical guidelines for the industry.

3. Assigning a role to the federal government and the relationship with the construction industry.

The Construction Contractor's Law created, in turn, the Central Council on Contractors, an organization to bridge the gap between industry and government and act as an intermediary to relay industry problems to government and vice-versa.

THE LICENSING SYSTEM FOR JAPANESE CONTRACTORS

A rather elaborate system for licensing all but the very small carpenter–sole proprietorship type companies was established by the Construction Contractor Law in 1971. The 520,000 contractor licenses awarded as of 1986 are divided into categories of Prime Contractor and Subcontractor or Special Trades Contractor.

No license is required for those contractors or builders working on the following types of projects:

1. General construction work having a value that does not exceed 6 million yen ($46,154 figured at 130 yen to the dollar).
2. General wood-structure-type housing less than 1615 square feet in size.
3. General wood-type structures, other than housing, with a value less than 2 million yen ($15,385 at 130 yen).

Licensing Classifications

Prime or General Contractor licences are divided into two classifications based upon the relationship and degree of control that the general contractor will exert over the subcontractors that will be employed:

1. *Common* contractors are those who will be engaged in work valued at more than 10 million yen (about $77,000 at 130 yen) and will be performing this work with their own forces.
2. *Special* contractors are those contractors who will be allowed by law to subcontract work valued in excess of 10 million yen. These special contractors, who really could be called general contractors, are given the name "special" because the Ministry of Construction places upon them the "special" obligation to guide, support, and nurture the subcontractors that they will be employing.

The common contractor has no limit upon the total value of work he may pursue. He can subcontract work to other companies as long as the value of each segment of work is less than 10 million yen. If the work exceeds this amount, the common contractor must perform the work with his own forces. For instance, if a common contractor is working on a concrete foundation job valued at $2.5 million and the reinforcing steel portion of the work is estimated to be worth $250,000 (which exceeds 10 million yen), he must perform this work with his own crews. Conversely, if the re-bar work was estimated to be worth less than 10 million yen, it could be subbed out.

Categories of Licenses

Contractor licenses are issued for three-year periods, after which time they have to be renewed. Table 2-8 contains a list of the twenty-eight categories of licenses available as well as the number of common and special contractor permits issued as of March 1980. Category No. 5 contains the term *tobi,* which refers to a rigger, demolition, and/or scaffolding contractor.

The Criteria for Obtaining a License

In order to apply for a contractor's license the applicant must be able to meet four basic requirements:

1. The managing executive of the firm must have at least 5 years experience in general management control of his relevant work, or have approval of his past experience made by the Ministry of Construction.
2. There must be a regular staff engineer in every business office with background to meet one of three criteria:
   a. Mastery of his discipline while at high school and business experience of 5 years in relevant construction work, or three years experience in relevant work after leaving a junior college, university, or college.
   b. Business experience in relevant construction work for a period of ten years.
   c. Approval by the Ministry of Construction, and must have passed either the Engineering Test under the Construction Contractor Law, the Test of the Engineers Law, the Trade Skill Test of the Vocational Training Law, or the Test for First or Second Grade Architect of the Architect's Law.

A construction company having offices in two or more prefectures must obtain their license from the Ministry of Construction; a company with only one office will obtain its license from the governor of the prefecture in which it is located.

Table 2-8. Categories and Numbers of Types of Licensed Contractors in Japan.

License-Granted Work	Common Contractor	Special Contractor	Subtotal
1. General civil engineering work	95,572	10,737	106,309
2. General building work	200,707	12,471	213,178
3. Carpentry (rough and finish)	43,450	1,326	44,776
4. Plaster work	15,145	27	15,172
5. *Tobi*/earth/concrete work	61,935	3,194	65,129
6. Masonry	12,529	1,202	13,731
7. Roofing	11,193	933	12,126
8. Electrical work	30,450	1,443	31,893
9. Plumbing/piping/ air conditioning	48,280	2,079	50,359
10. Tile/concrete block/ brick	13,196	995	14,191
11. Structural steel	23,650	2,467	26,117
12. Reinforcement work	4,711	61	4,772
13. Pavement work	20,707	3,997	24,704
14. Dredging	4,731	1,314	6,045
15. Sheet metal work	7,215	15	7,230
16. Glass and glazing	4,313	3	4,316
17. Coating work	14,343	45	14,388
18. Waterproofing	3,739	20	3,759
19. Interior finish work	24,140	1,387	25,527
20. General equipment	10,924	1,055	11,979
21. Heat insulation	2,498	45	2,543
22. Electrical communication work	6,702	258	6,960
23. Landscaping	19,536	2,072	21,608
24. Well-drilling	2,603	67	2,670
25. Cabinet-making	12,285	21	12,306
26. Waterworks installation work	19,141	3,588	22,729
27. Fire protection	10,115	111	10,226
28. Garbage disposal facility work	722	201	923

Source: Ministry of Construction, March 1980

Registration Requirement for Public Works Projects

Unlike the system that prevails in the United States whereby any construction company that has not been blackballed and can obtain a bid bond is free to bid on government jobs, in Japan a contractor must be prequalified to bid. On the surface this seems like a desirable practice: the government agency soliciting bids can have some assurance that all bidders are well qualified and therefore contractor performance will be at an acceptable level no matter which one is low bidder; it assures contractors that some unqualified contractor might not submit a "low-ball" bid because his estimating staff was not experienced in pricing projects such as the one in question. Theoretically this is true, but on a more practical level, it has permitted the practice of collusive bidding, pre-bid consultation and price adjustment between the fraternity of large contractors on the invited bidders list.

The Contractor Rating and Qualification System

The Central Council on Construction Contracting Business was given the task of formulating basic business and ethical guidelines for contractors to establish a national, rational standard of tender so that credible contractors would be allowed to bid on public works projects. A contractor must achieve a certain rating to be eligible for inclusion on a government bidder's list. In order to obtain a rating, the contractor must supply certain information to the Council:

1. Annual sales volume of completed projects of a nature similar to the one being bid.
2. Number of employees in the company.
3. Owned capital (paid-in capital plus reserve capital).
4. Ratio of current assets to current liabilities.
5. Ratio of fixed assets to capital.
6. Ratio of net profit to total liabilities and net worth.
7. Number of years in business.
8. Work records and work history.
9. Construction machinery owned.
10. Safety record.
11. Record of labor relations.

Contractors are rated each year, and if there has been any evidence that they have "exhibited insincere behavior in implementing contracts" they can be disqualified or demoted on the ratings list. If a contractor is not satisfied with the rating he has been awarded, there is a review board where he may appeal.

Low Bidder Does Not Always Get the Job

On public projects, low bidder is not always the successful bidder. In most cases the government agency soliciting the bid has formulated a detailed estimate for the work. If a low bid falls well below the owner's estimate, it is deemed to be "dumping," an attempt to obtain work by submitting unrealistically low bids. Any low bid falling outside the parameters of what the government agency feels is "low" will be disqualified and the project awarded to the next low bidder, if his bid is within the range deemed acceptable by the government agency.

The Law Concerning the Improvement of Employment of Construction Workers

This Japanese law establishes a position known as Person Responsible for Employment Management, at each job site. PREMs are a combination of personnel manager and time-keeper on construction sites. They have a responsibility for recruiting new workers, and maintaining job site records for all employees. PREMs are required to advise workers on labor welfare laws and insurance matters, and must be responsible for adequate working environment. Although most large construction companies have trained Persons Responsible for Employment Management on their staff, small and medium size companies may not have such a person on their payroll. The government provides a three day seminar to train people for the role of PREM and they are held at various times in eight districts throughout the country. The education of a PREM is rather extensive; they learn about:

- The supply and demand aspect of the labor market.
- The Employment Improvement Law, public assistance programs, and related laws and ordinances.
- Labor-management organization and the manager's responsibilities.
- Methods for recruiting workers and laws related to the process.
- Hiring criteria—physical tests required, medical examination required.
- Explanation of the employment contract between worker and employer.
- Working hours in the industry, work breaks permitted, holidays to be observed.
- Educational and training programs available to construction workers, levels of qualification, and certification for various job positions.
- Safety programs and health education programs.
- When lodging is provided for workers, rules and regulations must be explained.

- Grievance procedures.
- Collective bargaining procedures.

The Labor Standards Law

This law, passed in 1947 and amended more than a dozen times since, has as its basic consideration respect for the worker and protection of the worker from employer abuses. It contains general provisions, sections on wages and hours, safety and health, women and minor workers, workers' compensation, inspection and violations procedures, and various rules of employment. A look at the first 12 Articles in the General Provisions Section provides insight into some of the provisions of the law.

Article 1	Working conditions must be such that they meet the need of the worker who should be allowed to "live life worthy of a human being."
Article 2	Working conditions are to be decided by worker and employer on an equal basis.
Article 3	No employer shall discriminate against any employee by virtue of his nationality, creed, or "social status."
Article 4	Women shall not be discriminated against.
Article 5	Workers shall not be forced to work against their will.
Article 6	Unless permitted by another law, no person shall "profit as vocation by intervening in the employment of others."
Article 7	Worker must be granted time off from work to vote, or respond to any call to public duty.
Article 8	The law applies to 17 groups of industries listed in this section.
Article 9	The definition of "worker" within the industries in Article 8 is explained.
Article 10	Definition of "employer" is established.
Article 11	The meaning and form of "wage" is defined.
Article 12	Average wage computation is spelled out for calculation of average wages.

The balance of the Labor Law contains 110 other articles in addition to various ordinances establishing terms and conditions of employment, under what conditions an employee can be fired, and a statement of annual paid holidays, among other matters. There is a separate section devoted to women and minors. No one under the age of 15 can be legally employed; however, children above the age of 12 may be employed in light-duty tasks if they have parental permission and a certification from their school master attesting to the fact that the child's employment has not hindered its school work. The ordinance pertaining to women in the work force and their working conditions lists six classes of jobs that are injurious to menstruation, and employers must provide women with extra recess time if they continue working in these tasks during that time of the month. Some employers provide dormitories for out-of-town workers, and another section of the law contains the Do's and Don't's of dormitory living.

The Building Standard Law of Japan

This law was established in 1950 and its purpose was to provide minimum standards for construction site work, construction equipment, and building usage. Because of Japan being in one of the most active earthquake zones in the world, standards of construction are an absolute necessity to avoid catastrophe in densely populated areas. The law provides for fundamental requirements for buildings in general and contains an enforcement section with related procedures. Some zoning codes are also included in the law. The Ministry of Construction has issued technical standards from time to time which are treated as supplements to the Building Standard Law. The Ministry issued a booklet entitled Designation of Non-Combustible Materials in 1970, followed by Designation of Semi-non-combustible Materials and Fire Retardant Materials in 1976. Another directive, Establishment of Technical Standards for Ensuring Structural Safety of Wood Frame Construction, was issued in 1982, and Establishment of Technical Standards for Ensuring Structural Safety of Log Construction.

THE TRAINING OF CONSTRUCTION WORKERS

A shortage of skilled workers has been developing in Japan even though increasing numbers of people are entering the workforce. The construction industry has not been viewed too favorably by young workers because of the industry's poor safety record in past years, and because many construction tasks are boring, unpleasant, or dangerous. The Japanese have taken steps to correct the image of the building business by stepping up efforts to create safe working environments, and developing robots to replace man when dangerous or hazardous tasks have to be accomplished.

The shortage of skilled craftsman is probably a sign of the times as young workers gravitate to more exotic jobs. During the summer of 1988, the shortage of carpenters became so acute in Tokyo that ads were placed offering craftsman double their normal wages, \$240 per day instead of the going rate of \$100–\$130 per day. In Japan today, the average age of the on-site construction worker is almost 44 years of age and new young workers need not only to be attracted to the industry, but trained to take their place alongside the older, more experienced tradesmen. Because there are no nationwide craft unions to provide apprentices and apprenticeship programs, craft training must be done by the contractors or by privately funded or government operated vocational training centers.

High School Training

High School education in Japan begins at about age 15 and consists of a three year program in either general studies or vocational training. Beyond the high school diploma, education continues in the form of junior college, technical college and universities offering 2, 3, and 4 year courses very similar to higher education in the U.S.

Vocational Training Schools

The Vocational Training Law of 1958 was enacted to encourage and regulate the growth of vocational training schools in both the private and public sector. A number of private schools operating within the guidelines established by the Vocational Training Law were created offering instruction in diesel mechanics, electronics, construction, hairdressing, and other familiar trade school programs.

The Hiroshima Cooperative Training Center for the Building Trades

This school was a byproduct of the Vocational Training Law. Located near the Fukui Construction Company in Hiroshima, it offers an excellent one-year training program consisting of general studies, science, and architectural and construction related subjects. The program is divided into two halves: the first six months are devoted to general and specialized classroom training; the second six months are spent in applying the practical skills. General subjects such as civics, physics, chemistry, practical foreign language, and physical training take up 250 hours, while specialized subjects such as architecture, construction methods, welding, safety, surveying, machinery, electricity, drawing, and construction laws and regulations comprise 410 hours of classroom instruction. Basic practical training takes 568 hours and includes the use of construction tools and machinery, basic mechanical work, basic construction tasks, scaffolding building, and safety instruction. Another 1,052 hours are spent on instruction in foundation work, framing techniques, and concrete work. School starts in April and does not end until the following March. Training is usually 8 hours per day, 6 days a week.

Tuition is partially paid by federal and state subsidies as well as from the Employment Promotion Projects Corporation in addition to an annual membership fee of 120,000 yen per student which is paid by the employers who send pupils to the center. According to officials at the Center, their graduates, after one year in the field, can be classified as 2nd Grade Skilled Workers, and after an additional two years will have learned enough to be classed as a First Grade Skilled worker. Just two years after graduation these graduates should be able to assume a position of supervisor in several basic construction trades, and after three years in the field, a trainee in architecture should be able to qualify as a Second Grade Architect.

CONCLUSIONS

The construction industry in Japan appears to operate not so much as a free agent but as another gear in the machinery known as Japan, Inc.

The major Japanese contractors are financially strong, well managed, and poised on the cutting edge of their industry's technology. They appear ready to take a giant step in any direction that they or their country feels is propitious.

Japan is one of the wealthiest nations in the world, but it is also a country where housing, infrastructure, and standards of living are not commensurate with its economic wealth, and this perception is shared by its citizens. Japan has recently been going through a reevaluation of its economic accomplishments and how it can best make use of its accumulated wealth.

The Maekawa Report, published in 1986 after a period of extensive retrospection, suggests that Japan's economic structure needs to be adjusted for the sake of achieving harmony in the international community. This report recommended that housing production and urban development proceed at an accelerated pace, and that consumer spending ought to be encouraged while working hours are decreased to allow the populace to have more time to enjoy the fruits of their labor. Japan needs to take a more active role in its contribution to the world's economy and assistance to developing nations, and it needs to open its doors to imports and encourage foreign companies to enter its market.

The inextricable link between government and industry will be the key to the direction the construction industry will take in Japan, and other parts of the world, during the last decade of the 20th Century.

3

The Functioning Construction Company

A Japanese construction company is first and foremost a Japanese *company,* and any attempt to understand how it functions needs to be preceded by an understanding of how a Japanese company functions.

THE JAPANESE CORPORATE ENVIRONMENT

The Japanese corporation is similar to U.S. corporations in many ways—in physical makeup, organizational charts, and accounting and reporting procedures. In other ways the two systems are as far apart as the distance between Kobe and Kalamazoo.

If a corporation were to be examined as a living, breathing entity, both Japanese and American companies would reveal the same basic skeletal structure, but the lifeblood of a corporation, its people—the executives, middle and lower echelon managers, technical staff, marketing and administrative groups, and the ways in which they relate to the company—are what sets the Japanese company apart from its American counterpart.

In Japan, the corporation offers more than just employment, it offers a home, a nurturing environment, and a lifetime commitment between employer and employee. The centuries old mores and customs within the country manifest themselves in unique corporate behavior that is strictly Japanese.

Pride in one's company is widespread, and working for certain well established construction companies imparts a certain degree of status upon the employee. One is not simply a project manager or an estimator, but rather one is a Kajima project manager, or a Toda estimator, or a Taisei engineer, or a Shimizu architect, with the company logo proudly displayed on the lapel of a business suit. The devotion to and pride in one's company is largely responsible for much of the Japanese management mystique. Managing is made much easier when those who are being managed are so dedicated to insuring that their company produces a quality product in the most productive manner, and within the alloted time frame.

One Sunday evening in Tokyo, the writer had dinner with an executive from Toda Construction Company at the Imperial Hotel, and after dinner they both walked through the Ginza section of the city discussing the nature of construction in the U.S. and Japan. At about 10:00 P.M. they found themselves near Toda's corporate headquarters building in the Chuo-ku section of Tokyo. As they walked past the building, it was dark except for the seventh floor which was ablaze with lights. The Toda manager said, "I guess our architects and engineers are working late again tonight because we have had a great deal of work to do lately." They were willingly working late "again" because they felt an obligation to their company to do so.

DIFFERENCES IN CORPORATE ENVIRONMENT—PHYSICAL AND PHILOSOPHICAL

There are many differences between the corporate environments of Japan and the U.S. It is apparent in the physical layout of a typical corporate office. Work areas are large open spaces filled with gray metal desks and chairs facing each other in long rows. Islands of desks and filing cabinets exist throughout the area. Floors are typically resilient flooring with carpet in the reception areas and some corridors. Task lighting is not common and everyone seems content working at their calculators, word processors, or computers under the normal level lighting provided by standard recessed troffers. Upon entering a Japanese office, its utilitarian atmosphere is readily discerned.

The differences between the organizational traits of the employees of Japanese and U.S. corporations is as pronounced as the contrast in office surroundings.

Organizational Trait	Japanese	American
Identification with company	Strong	Weak
Relationship with fellow workers	Togetherness—us	Lone wolf—me
Interaction with Managers	Open office mentality On daily basis	Private office mentality On "as needed" basis

Conflict between workers	To be avoided	To be brought to surface and dealt with
Decision-making	Bottom up By consensus Slow	Top down Individualistic Rapid
Corporate changes	Evolutionary	Revolutionary
Group suggestions offered	High	Low
Group suggestions—managers reactions	Positive	Ambivalent
Term of employment	Lifetime concept	Until something better comes along
Promotions	Seniority	Ability
Salary and bonus	Length of service Corporate profitability	Merit—contribution to corporate P&L
Corporate profit goals	Market share Long term	Short term
Dealing with poor performers	Transfer to another department	Dismissal if it persists

Corporate policies change with the times as social and environmental forces surrounding them change, and there are many experts in and outside of Japan that see some changes in the Japanese psyche. Many of the younger employees in the work force are not as passive and accepting as their fathers were, and they are beginning to become impatient with some of the old ways.

An engineer with one of the country's top construction firms, who is in his early thirties, said that he works in a group where one of the members has been given less responsibility because he is not as competant as the rest of the group. This person spends a great deal of time looking out the window near his desk, while others are busily occupied with their design work. (The expression "looking out the window" appears to be a figurative term that was used by several managers in several companies and was applied to those workers who were not so capable or could not be given as much responsibility as others and therefore were "given a seat by the window" to remove them from the field of action.) There appeared to be a certain amount of resentment in this engineer's voice as he said, "We all recognize that Hirohide wastes a great deal of time during the day while we are all very busy, and we really don't mind that. I do not think it is fair, however, that he is being paid the same salary as the other group members who really work quite hard."

A similar complaint was registered by another middle manager from one of Japan's construction companies a notch or two below the Big Six. He said that most of the executives and managers in his company work longer than the normal eight hour day and many employees feel obligated to wrap up loose ends after five o'clock. Secretaries would never think of giving a letter to their boss without having proofread it first, and they can be seen staying after work a few minutes or so to review their typing if they have not had a chance to do so during the day.

The employee being criticized had a habit of dawdling during most of the afternoon and allowing his work to pile up. After 5:00 P.M. he would get down to work and frequently spent two or three hours each evening completing his day's work just so he could catch the eye of his supervisor and impress him with his diligence. The criticism was being made not so much against this particular employee as it was against a company policy that seemed to reward quantity of working hours rather than quality of working hours. It was a remark made about a system that rewards both laggard and conscientious employee according to length of service, educational level, and number of hours worked, and not much else.

LIFETIME EMPLOYMENT

The concept of lifetime employment is still very much in effect in Japan, although it is beginning to be scrutinized a little more closely these days. It is a system closely tied to seniority and rewards longevity with annual wage increases and usually increases in job responsibility.

Japan is a chauvanistic society, where women are just now being accepted in many fields. One high level construction executive, a gentlemen in his late sixties, is tied closely to the old ways. He said that lifetime employment had even been extended to women in the workforce and his company had even hired some, even though there were other wonderful opportunities available to women on a part-time basis. "Why they can work in supermarkets nowadays from 10:00 A.M. until 2:00 P.M., four hours each day, and they can still be home in time to take care of the children and prepare dinner for their husbands."

The old ways are being questioned and looked at more objectively these days, which may be the first step in the long process of change. During a discussion about corporate policies in his company, one contractor drew a triangle to express his views about lifetime employment and the seniority system. The triangle (Fig. 3-1) was meant to represent the upper level of executives in his company some 29 to 30 years after they first joined the construction company. It was an accepted fact that a few had turned out to be exceptionally qualified, while others bordered on the incompetent. All would be assured of continued employ-

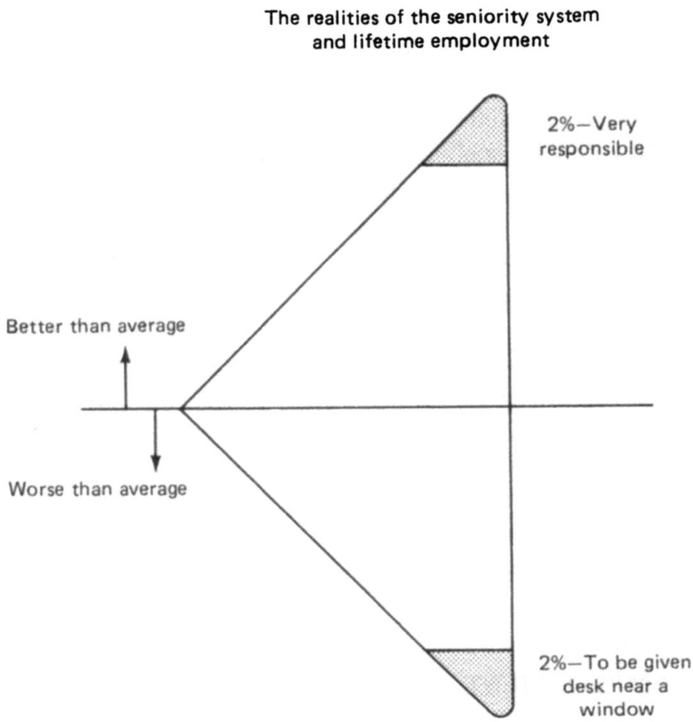

Fig. 3-1. The reality of lifetime employment—competents and incompetents. Looking at the class of 1959 who are now in their fifties, one or two might be outstanding and one or two might be incompetent, but ALL receive the same pay and ALL will have total job security.

ment because of the lifetime employment concept, and the differential in yearly remuneration between them would be small, even though their job responsibilities and authority might vary considerably.

THE JAPANESE EXECUTIVE

The role of the executive in the corporate structure in Japan is somewhat different from the typical American perception of a manager. The Japanese executive is more of a collectivist rather than the strong singularly driving force associated with managers in the United States.

Desirable Japanese leadership qualities are not the same as the American "take charge," strong willful type, but are based upon how well a manager can achieve harmony within the group and how well a manager can affect consensual agreement within the work group. The physical space occupied by Japanese managers mirrors the relationship they have with their employees. Top executives do have private offices, but they are used primarily for ceremonial purposes and for meetings with visitors from outside of the company. On a daily basis, these executives can be seen working in a large open office environment along with the rest of their staff. The boss's presence is highly visible since he is right in the middle, or at the far end, of the action where he can keep a sharp eye on office activity.

This contrasts sharply with the image of the imperious American executive, closeted in his private office, behind closed mahogany doors, firmly planted in his high-backed leather chair, and in full control of his antique wooden desk.

EXECUTIVE SALARIES AND AMENITIES

Salaries for most top Japanese executives are low by American standards. High ranking industrialists and executives command annual salaries and bonuses ranging from $125,000 to almost $400,000, although many owners of privately held corporations do make handsome salaries each year. Table 3-1 lists salary ranges for four top executive positions according to capitalization of the company. (Table 3-2 lists salary ranges for other positions) These salaries pale in comparison with those of many highly placed executives in some of America's large corporations.

In 1987, Navistar paid its head man $640,000 while USX (the old U.S. Steel Corporation) paid their CEO $1.36 million. Ford Motor Company's top executive received $3.37 million that year which was small change to Chrysler Corporation's payment plus stock-option considerations worth a total of $17.9 million when received by Lee Iacocca. Another difference between U.S. and Japanese executive salaries becomes apparent when corporate fortunes happen to take a turn for the worse. The Japanese executive will be one of the first to effect a reduction in his salary and/or bonus, the same as he would expect from one of his factory workers. He clearly wants to send a corporate message that the company is going through hard times and everyone has to tighten his or her belt a little further. He wants everyone to know that "we are all in this together, through good times and bad times alike." This is certainly not the same approach used by American CEOs when their companies fall upon hard times. Texaco Corporation, with a $4.4 billion loss in 1987, and in the midst of bankruptcy proceedings, boosted their Chief Executive's salary by 14% while suspending three dividend payments to stockholders. Mr. Iacocca's ill-timed yearly remuneration disclosure came at a time when he was asking Chrysler workers to tighten their belts.

Table 3-1. Yearly Remuneration of Company Presidents, Vice-Presidents, Directors in Japan (based upon range of company capitalization).

Position	Company Capitalized at $230,000	Company Capitalized at $7.7 Million	Company Capitalized at $77 Million
President			
Yearly Salary	$92,690	$131,586	$229,241
Bonuses	31,448	41,310	95,034
Total	$124,138	$172,896	$324,275
Vice-President			
Yearly Salary	$71,172	$95,172	$227,586
Bonuses	14,759	26,828	51,724
Total	$85,931	$122,000	$279,310
Executive Director			
Yearly Salary	$48,828	$71,172	$114,207
Bonuses	13,241	22,207	29,310
Total	$62,069	$93,379	$143,517
Director			
Yearly Salary	$44,690	$52,966	$91,862
Bonuses	11,931	20,000	40,552
Total	$56,621	$72,966	$132,414

Dollar amounts based upon 145 yen, 1987 average.
Source: Jetro Nippon Facts and Figures 1988, Nihon Keizai Shimbun 6/5/87

Table 3-2. Construction Staff Salary Levels.

Entry Level	Starting Salary Range	Year 2	Year 3	Year 4
High School Graduate	$925/month →			$1150–$1230/month
University Graduate	$1150–$1230/month →	$1308–1385/month		
Masters degree	$1308–$1385/month →		$1538/month	
PhD	$1538/month			
	Project Managers[a]			
10 years experience	$34,600 per year			
20 years experience	$54,000–$61,500 per year			
30 years experience	$77,000 per year			

[a]Figures quoted by executive of large construction company and include twice-a-year bonuses.

Japanese managers seem to be more sensitive to the feelings of their employees and perhaps the lack of sensitivity by American managers is contributory to the worker's lack of concern for company product, property, and fellow worker.

THE CORPORATE CHAIN OF COMMAND

A position on the corporate ladder in Japan carries with it a great deal of prestige and importance, and the hierarchy it creates within the corporation establishes a chain of command. In the very large construction companies, corporate titles, in descending order, are as follows:

Chairman of the Board of Directors
President
Executive Vice-President
Senior Managing Director
Managing Director
Director
General Manager
Deputy General Manager
Manager
Deputy Manager

Promotions are based upon age and seniority. At age 35, when the university graduate has been with the company for 12–14 years, he will be considered for his first managerial post, Deputy Manager. If his performance is very good over the next 10 years, he might advance to Manager or even deputy General Manager. Upon reaching the age of 45, with more than 20 years' seniority, the position of Department General Manager will be the next rung up the ladder, and as the promising executive approaches 50, further advancement into the upper levels of corporate management will be opened to him.

CORPORATE ORGANIZATIONAL STRUCTURE

The giant Japanese construction companies have large and complex organizational structures, as witness the multidepartmental makeup of Kumagai-Gumi, Kajima Corporation, Taisei, and Ohbayashi Corporation on the following pages and reflected in Figs. 3-2, 3-3, 3-4, 3-5. Smaller construction companies have organizational structures much like their U.S. counterparts—administration and finance, personnel, sales and marketing, project administration, estimating, and purchasing. It is in the big construction companies of Japan that their special approach to the building process is apparent.

A CLOSE LOOK AT OHBAYASHI CORPORATION

Ohbayashi Corporation, with 1987 revenues exceeding $5.8 billion, ranks among Japan's top five contractors. Ohbayashi has 17 subsidiaries, 8 affiliates, and maintains 59 business offices in Japan as well as 13 overseas branches. Total company employment is 11,000, including 6,000 architects and engineers. Ohbayashi Corporation is divided into five major divisions with 19 administrative, accounting, sales, finance, real estate, and other miscellaneous departments, including a total quality control department, all reporting directly to headquarters. A look into the World of Ohbayashi begins with a visit to their head office located at 3.2 chome, Kanda Tsukasa-cho in the Chiyoda section of Tokyo.

The forward-thinking philosophy of the corporation is readily expressed in its new teleconferencing room (Fig. 3-6). This 21st Century conference room electronically links the branch offices together with 768K bit per second telephone communications lines and sophisticated video equipment. Is there a need to discuss a problem just beginning to surface at the Osaka branch which needs assistance from Mr. Haga back at headquarters in Tokyo? And what about bringing in those two engineers from Nagoya who experienced the same kind of situation about two months ago? Touch a few switches in the teleconferencing room and the tiny red lights on the TV cameras placed strategically around the room begin to glow. Flip a couple of other switches and the blackened screens of the television sets situated around the oval conference table begin to light up and the two engineers from Nagoya are ready for the meeting to start. As Mr. Haga enters the room and takes a seat, the other television set comes to life and the Osaka people commence to present their problem. The discussion is lively, and possible solutions are passed back and forth until all agree on a solution. The audio-visual response is instantaneous and so clear that one must be reminded that the participants to this discussion are several hundred miles apart and not in the same room in Tokyo. The conference draws quickly to a close, thank-you's are exchanged, goodbyes said, TV sets go blank, red lights fade to black, and on to another task in the Tokyo office. Ohbayashi has five corporate divisions: building construction, heavy construction, technical development, engineering, and nuclear facilities. There are in addition 19 departments that provide support functions to the five divisions and headquarters. One of these support groups, the machinery department and machinery works, would be the dream of any American construction company.

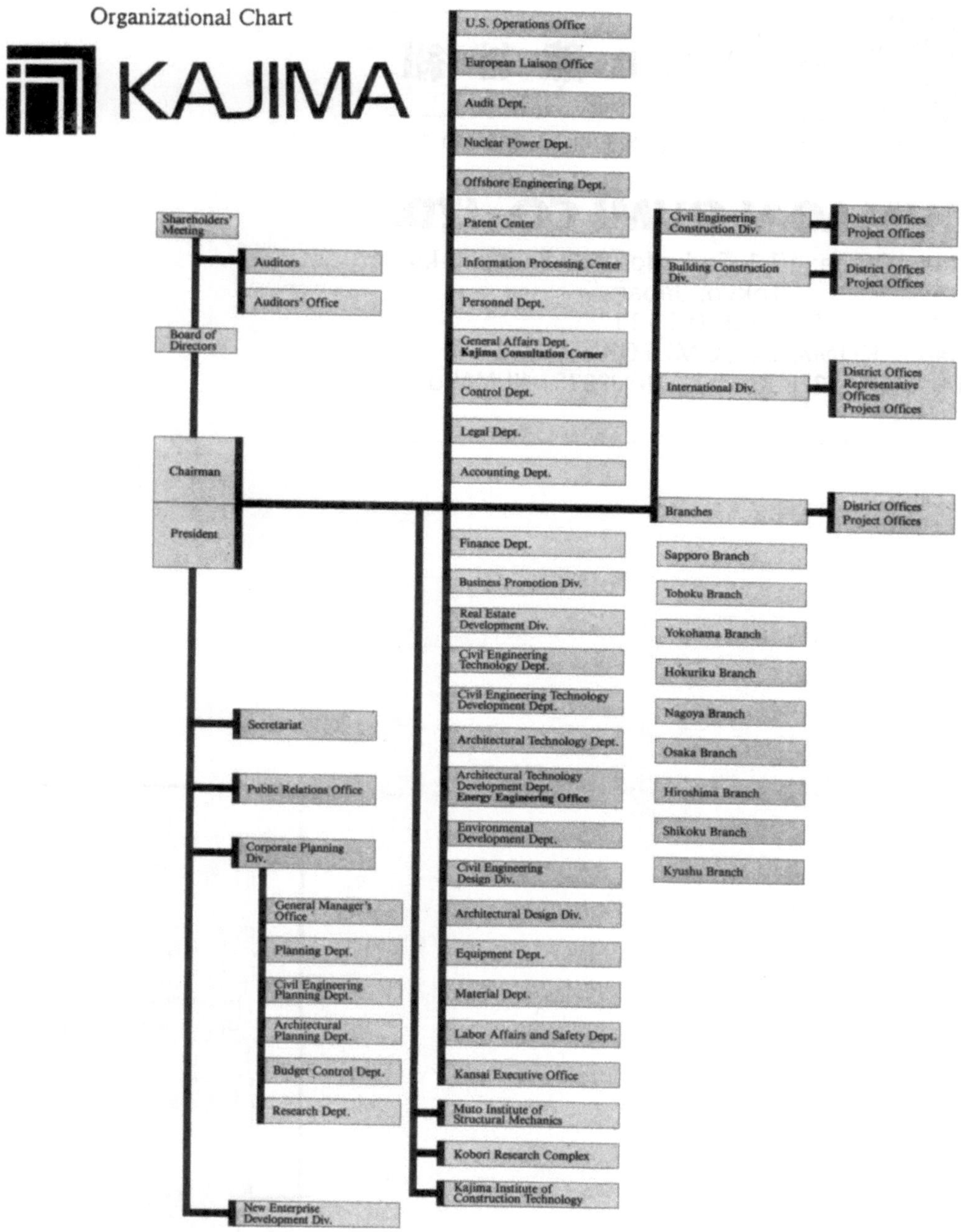

Fig. 3-2. Kajuna corporate organizational structure.

株式会社 熊谷組

本社●〒162 東京都新宿区津久戸町17-1 TEL03-260-2111

KUMAGAI GUMI CO., LTD.

Head Office: 17-1 Tsukudo-cho, Shinjuku-ku,
Tokyo, Japan
Telephone Tokyo 260-2111
Cable: KUMAGAIGUMI TOKYO
Telex: 232-2228 KUMAKU J· J26751 KUMAKU J

お問い合せは——————

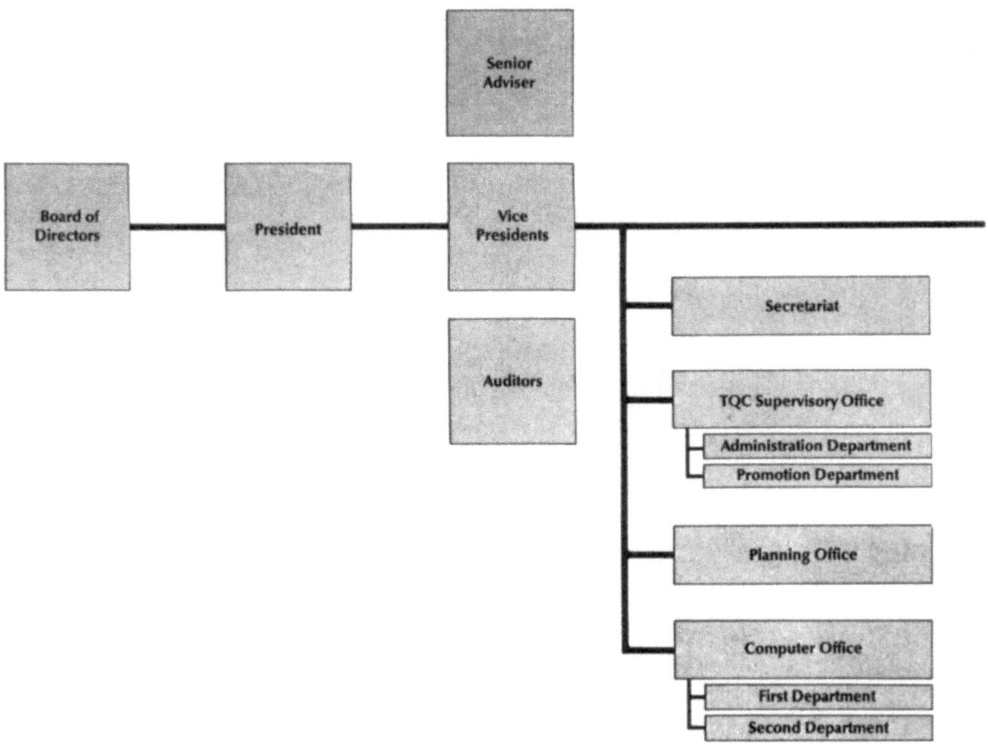

Fig. 3-3. Kumagai Gumi organization chart.

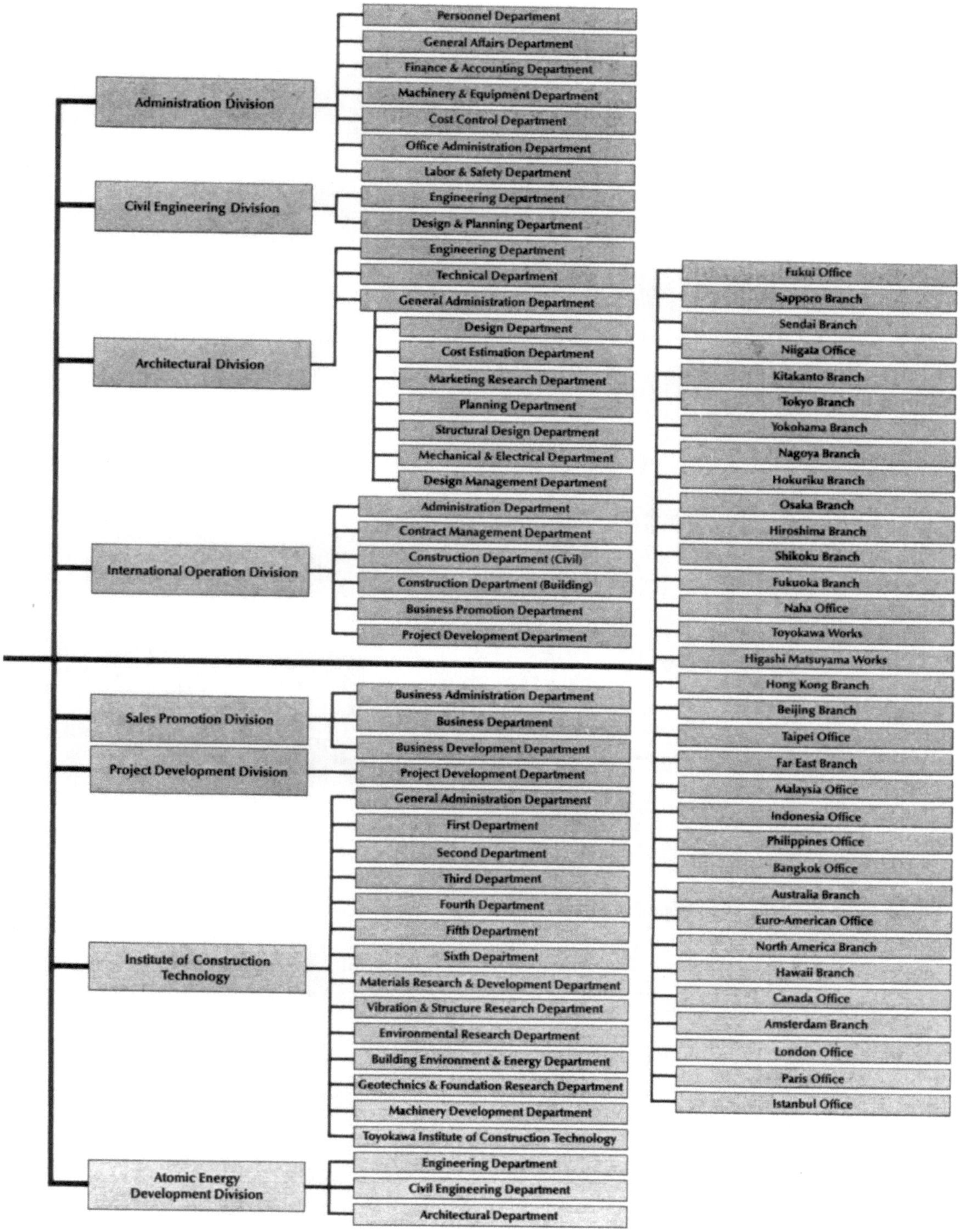

Administration Division
Personnel Department
General Affairs Department
Finance & Accounting Department
Machinery & Equipment Department
Cost Control Department
Office Administration Department
Labor & Safety Department
Civil Engineering Division
Engineering Department
Design & Planning Department
Architectural Division
Engineering Department
Technical Department
General Administration Department
Design Department
Cost Estimation Department
Marketing Research Department
Planning Department
Structural Design Department
Mechanical & Electrical Department
Design Management Department
International Operation Division
Administration Department
Contract Management Department
Construction Department (Civil)
Construction Department (Building)
Business Promotion Department
Project Development Department
Sales Promotion Division
Business Administration Department
Business Department
Business Development Department
Project Development Division
Project Development Department
Institute of Construction Technology
General Administration Department
First Department
Second Department
Third Department
Fourth Department
Fifth Department
Sixth Department
Materials Research & Development Department
Vibration & Structure Research Department
Environmental Research Department
Building Environment & Energy Department
Geotechnics & Foundation Research Department
Machinery Development Department
Toyokawa Institute of Construction Technology
Atomic Energy Development Division
Engineering Department
Civil Engineering Department
Architectural Department
Fukui Office
Sapporo Branch
Sendai Branch
Niigata Office
Kitakanto Branch
Tokyo Branch
Yokohama Branch
Nagoya Branch
Hokuriku Branch
Osaka Branch
Hiroshima Branch
Shikoku Branch
Fukuoka Branch
Naha Office
Toyokawa Works
Higashi Matsuyama Works
Hong Kong Branch
Beijing Branch
Taipei Office
Far East Branch
Malaysia Office
Indonesia Office
Philippines Office
Bangkok Office
Australia Branch
Euro-American Office
North America Branch
Hawaii Branch
Canada Office
Amsterdam Branch
London Office
Paris Office
Istanbul Office

Fig. 3-4. Taisei Corporation organization chart.

The Machinery Department

With approximately 2,000 construction sites functioning at any one time, the need for construction equipment of the right kind in proper working order when needed is a foregone conclusion. Ohbayashi deals with this problem by maintaining machinery depots in Hiroshima, Kyushu, Shikoku, and Nagoya, with main works operations in Osaka and three other depots, one in Sapporo, one in Tohoku, and one at the head office reporting to another main depot in Tokyo.

The machinery department is staffed by 100 mechanical engineers, 40 electrical engineers, 250 mechanics, 60 electricians, and an additional support staff of 50 people. This department is charged with the following responsibilities:

1. Design, produce, improve, and maintain the equipment necessary to run Ohbayashi's construction operations effectively.
2. Incorporate mechatronics and robotics into construction equipment whenever possible.
3. Act as the corporation's purchasing department for all new equipment purchases.
4. Stockpile and inventory all equipment. (As of FY 1987, Ohbayashi's equipment had a book value of about $220,000,000.)
5. Provide equipment for specific projects along with accurate estimates for its use and operation so that these costs can be incorporated into the Estimating Department data base.
6. Train mechanics and electricians in the care and repair of equipment.

The machinery works department works closely with Ohbayashi's technical research institute to develop specialized equipment that the company needs in order to operate effectively.

A giant electric bucket excavator was designed and built that can cut 2370 cubic yards of earth per hour out of a hillside. The electrically operated conveyor attached to the excavator can load waiting trucks rapidly, quietly, and with relatively little pollution due to its electric power operation.

Diaphragm wall construction is used with great frequency in Japan, because of the nature of their soils and the need to excavate foundation walls at zero lot line tolerances. The works department designed and built the Super Hydrofraise, a computer controlled deep excavator that has the capability of digging for walls up to ten feet thick and as deep as 550 feet while maintaining vertical tolerances no more than one millimeter in a thousand. If bedrock is reached during the deep exca-

OHBAYASHI CORPORATE ORGANIZATION

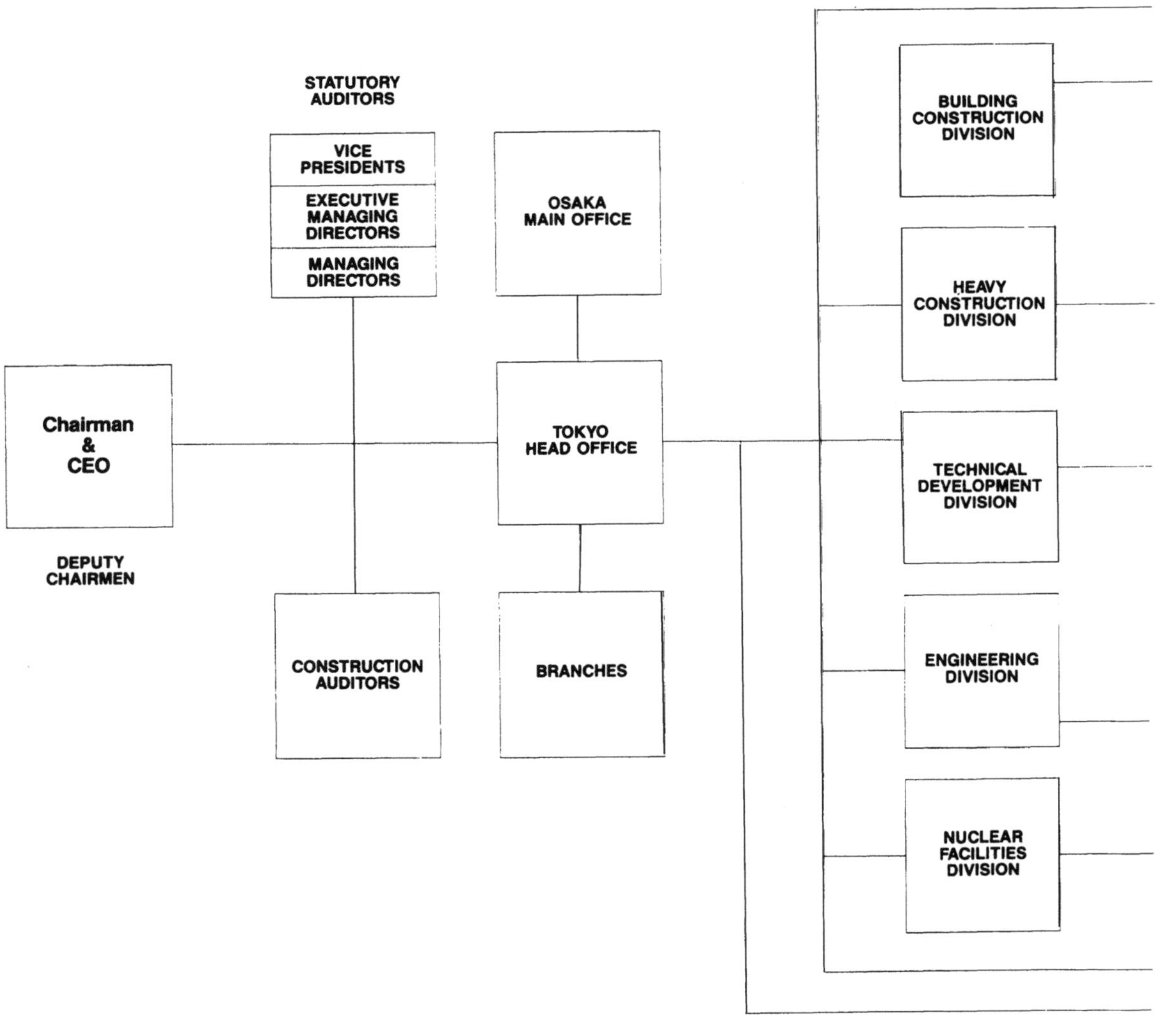

Fig. 3-5. Ohbayashi's corporate divisions.

- Audit Dept.
- Integrated Management Planning Dept.
- Total Quality Control Dept.
- Executive Secretariat
- Public Affairs Dept.
- General & Administration Dept.
- Personnel Dept.
- Occupational Safety Dept.
- Finance Dept.
- Accounting Dept.
- Machinery Dept.→Machinery Works
- Materials Dept.→Materials & Equipment Depot
- Real Estate Sales Dept.
- Development Administration Dept.
- Development Sales Dept. No.s 1 & 2
- Overseas Business Dept.
- Overseas Sales Dept.
- Overseas Operations Dept.

- **Civil Engineering Administration Dept.** Administrates, does market research, and manages construction.
- **Sales Dept. No.s 1—19.** Perform all sales and marketing activities.
- **Design Dept.** Does the necessary design and engineering for each heavy construction project.
- **Safety Administration Dept.** Conceives and administrates all safety oriented programs in the Heavy Construction Division.
- **Specialty Construction Dept.** Takes care of projects or portions of projects that require special skills and/or technology.
- **Sales Administration Dept.** Administrates sales department-generated projects, preparing reports & statistics. Charged with the long-range development of new building construction business.
- **Civil Engineering Dept. No.s 1—5** Oversee all heavy construction projects currently in progress.
- **Overseas Civil Engineering Dept. No.s 1 & 2** Supervises all international civil engineering projects.
- **Overseas Estimate Preparation Dept.** Prepares all estimates for international civil engineering bids.

TECHNICAL RESEARCH INSTITUTE

- **Administration Dept.** Administrates the entire Technical Research Institute physical plant.
- **Project Administration Dept.** Takes care of administrative details for overall R&D activities.
- **Research Dept. No.s 1—17** Put together as necessary to conduct the research and development projects approved by management.

- **Planning & Inspection Dept.** Plans forward-looking development projects, and inspects ongoing work to make sure it is up to standard.
- **Sales Dept.** Maintains constant communications with major clients and informs of major new projects.
- **Technology Dept.** Controls and develops technologies concerning nuclear power and nuclear power facilities.

EDP Center

- **Design Administration Dept.** Administrates both design and construction. Applies for both building & development permits.
- **Design Dept. No.s 1—8** Perform all necessary architectural functions.
- **Facilities Design Dept. No.s 1 & 2** Designs all necessary facilities for execution of each construction project.
- **Design & Construction Inspection Dept.** Ascertains regularly that all design & construction are up to designed standard.
- **Sales Administration Dept.** Takes care of administrative details generated by the sales force.
- **Project Planning Dept.** Charged with the long-range development of new building construction business.
- **Development Project Dept. No.s 1—4** Look after specific development projects.
- **Development Presentation Dept.** Conceives and prepares presentations to potential clients of large development projects.
- **Sales Dept. No.s 1—11** Carries out day-to-day sales and information gathering activities.
- **Tenant Search Dept.** Finds tenants for commercial buildings whether Ohbayashi- or client-owned.
- **Construction Administration Dept.** Oversees all construction projects currently in progress.
- **Safety Administration Dept.** Conceives and administrates safety programs at all building sites.
- **Specialty Construction Dept.** Performs all jobs that require special technology and skills.
- **Estimate Preparation Dept.** Prepares all estimates for any job Ohbayashi bids on
- **Construction Dept. No.s 1—6** Executes all construction jobs in the building construction division.
- **Site Support Dept.** Plans and estimates all on-site temporary structures. Provides on-site technological assistance as necessary.
- **Facilities Planning Dept.** Control all facilities on construction sites. Control of facility construction budgets. Development of facility-related technology.
- **Facilities Dept.** Estimating costs, preparing bids, and other such work, both domestic and overseas.
- **Facilities Construction Dept.** Preparation of facility construction plans, management and control of construction process.
- **Housing Dept.** Takes care of housing design & construction.
- **High-density Housing Dept.** Does high-density housing design & construction.
- **Overseas Construction Dept. No.s 1—3** Oversees international projects.
- **Service Center** Administrates technical development & property rights acquisition.

- **Project Administration Dept.** Takes care of the administration of all ongoing projects.
- **Patenting Dept.** Charged with patenting as much of the results of Ohbayashi R&D work as possible.
- **Civil Engineering Technology Dept. No.s 1 & 2** Works on developing and improving civil engineering technology.
- **Offshore Development Dept.** Develops technology for use in offshore structures.
- **Research & Development Groups** Assigned projects as required, especially cooperative jobs done with government or other companies. There are currently 19 R&D Groups.

- **Planning Dept.** Identifies directions and aligns Ohbayashi in those directions with carefully planned research and development projects.
- **Bio-environmental Project Dept.** Works on refuse and bio-sewerage treatment processes.
- **Intelligent Facilities Dept.** Lend intelligent properties to current buildings and design more efficient intelligent capabilities for the future.
- **Special Space Frame Structures Dept.** A group concerned primarily with the creation of large, covered spaces such as stadiums.
- **MENSHIN Base Isolation System Dept.** Works on the further development of Ohbayashi's revolutionary base isolation system.
- **Materials Handling & Factory Automation Dept.** Develops new systems for handling large quantities of bulk materials, and new systems for automating factories.
- **Hi-Tech Facilities Dept.** Responsible mainly for the development and improvement of clean room technology.
- **Space Development Project Dept.** Working on construction in space as well as colonization of the lunar surface.

- Sales Branches
- Site Offices

(a)

(b)

(c)

(d)

Fig. 3-6. The teleconferencing room at Ohbayaski corporate headquaters. (a) The group assembles in Tokyo. (b,c) The branch offices are alerted. (d) A discussion with the Nagoya office begins.

vating operation, the Super Hydrofraise has the ability to cut a keyway into the rock.

Machinery works is the repository for all of the shield tunneling machines used extensively by Ohbayashi in their civil engineering projects. One official at the machinery department said that their huge Tokyo storage yard has been criss-crossed with underground tunnels created when new boring equipment was undergoing tests or when existing equipment was being upgraded and put through its paces.

The Building Construction Division

This is the largest division in the corporation, accounting for about 70% of Ohbayashi's 1987 sales volume. The building construction division incorporates the following departments:

- Design. Design, design administration, facilities design department, and a design and construction inspection department
- Sales. Sales promotion, sales administration, tenant search department.
- Planning. Project planning, development project department, presentation department.
- Construction administration department.
- Safety administration department.
- Estimating department.
- Facilities planning and construction department.
- Housing department, including a high density housing department.
- Specialty construction department.
- Overseas construction department.
- Service center.

The design department of this division is further divided into a series of subdepartments:

1. Interior design, special projects, plans, and proposals.
2. Local area development, education and research and related buildings, office buildings.
3. Commercial buildings, sports facilities, office buildings.
4. Medical facilities, hotels, office buildings.
5. Welfare facilities, high density housing, office buildings.
6. Factories, warehouses, nuclear power facilities.
7. Develop working drawings and new technology for projects originating in design departments 1, 2, and 3.

8. Develop working drawings for projects emanating from design departments 4 and 5.

The Heavy Construction Division

This division of the company is composed of a civil engineering administration and sales department, safety administration, specialty construction department, overseas civil engineering department with its own estimating department, and five other civil engineering groups to handle the $1.74 billion sales volume that flowed through this division in 1987.

Heavy Construction is very proud of the world's records that they hold:

- Southeast Asia's largest land reclamation project—5362 acres in Singapore.
- The world's longest tunnel—Seikan Tunnel, 53.85 kilometers.
- The world's longest undersea shield method tunnel—a 3.7 kilometer gas supply tunnel.

This division holds various Japanese construction records: it has constructed 35% of the country's gas storage tanks, the largest train station in the country—Tokyo Station—and the country's longest railway tunnel—the 10.9 km Kanetsu Tunnel.

The Engineering Division

Within this division there are several other groups or departments. A bio-environmental department has been working on the treatment of refuse and bio-sewage treatment projects utilizing microbes to break down solid waste. There is a department for intelligent building design and another department devoted to high-tech facilities construction including clean room work. A materials handling and factory automation department works with problems relating to handling bulk materials and automated production facilities construction.

Seismic construction techniques play a major role in building design in Japan and the menshin (base isolation) department is kept quite busy. The special space frame department has been developing a unique skylight dome structure for sports arenas. This dome has the capability of opening and closing as weather conditions dictate. The space development project department, in line with the company's long range business approach, is investigating technologies that will have future application when applied to space exploration.

The Nuclear Facilities Division

Since 30% of Japan's energy sources are nuclear, the ability to construct nuclear power plants is of prime concern to any large construction company in that country. Ohbayashi's nuclear facilities division began work on prestressed concrete containment vessels in 1967 and put them into actual use when they built the Tsuruga Nuclear Power Plant No. 2, which went onstream in 1987. This division constructed Japan's first fast breeder reactor, known as Monju, a 280 MWe prototype fast breeder reactor owned by the Power Reactor and Nuclear Fuel Development Corporation of Japan.

The Technical Development Division

Ohbayashi's technical research institute, located in Kiyose, falls under the jurisdiction of this division, which includes an offshore development department, two civil engineering technology groups and a Research and Development department. With a 1988 R&D expenditure budget of $63.8 million and a projected budget for 1989 in excess of $66 million, it is quite obvious that the Ohbayashi Corporation is fully committed to the development of cutting edge construction technology.

WHAT IS IT LIKE TO WORK FOR A COMPANY LIKE OHBAYASHI?

A recently retired veteran of the company, Mr. Takashi Akaogi, spoke warmly about the company where he had spent most of his working life. The company was an extension of his family. He spoke about receiving a beautiful wooden sake cup after having been employed for 12 years. After 22 years he received 120 shares of company stock, and after 32 years Ohbayashi presented him with a travel voucher worth 100,000 yen.

A much younger member of the Ohbayashi Corporation, an eight year veteran, talks about how he will progress up the corporate ladder as his seniority increases. He says that his friends in the company do not resent having to wait 20 years in order to become a manager because, "That's the way it is in Japan and there is a spirit of cooperation that exists between all employees."

Seniority not only affects advancement and salary increases within the company, but it also has an impact on an important company fringe benefit—subsidized housing. With the exorbitant land costs in Tokyo and the resulting effect on housing, all but the very wealthy have to live outside the city. Younger employees often have to travel one to two hours to reach their offices.

Most companies in Japan, especially the very large ones, recognize that the cost of buying or renting a dwelling will consume a major portion of an employee's salary. Ohbayashi, like other big contractors, subsidizes the cost of apartment rentals for their employees. Mr. Yoshira, married just two months and with the company eight years, lives in a small three room apartment about one hour by train from the office. The apartment measures 62 square meters (about 667 square feet) and costs 60,000 yen per month ($462). The company pays one-half of Mr. Yoshira's rent; however, if Mr. Yoshira should move shortly because of a growing family, the company may not pay him more than $231 per month even though his newer, bigger apartment would cost more.

Employees with more seniority are afforded more liberal conditions. For instance, someone with the company for 20 or 25 years would probably be able to rent an apartment for 100,000 yen per month, for which the company would contribute 50,000 yen. Salaries are paid monthly and the company augments these wages with a generous bonus plan. Bonuses are calculated on the basis of five times monthly salary, and since salary increases with seniority, the dollar amount of the yearly bonus does as well. The bonuses are distributed twice yearly, in July and in December. Someone with a monthly salary of $2000 will receive a $5000 bonus in July and another $5000 bonus in December.

Other Company Benefits

Ohbayashi provides its employees with low cost vacation accomodations. The company makes arrangments with local hotels in resort areas and subsidizes the cost of the employee's room. The only costs that the worker will incur will be for meals and any entertainment costs. The company provides banking facilities that yield higher interest rates than commercial banks. These company banks will also lend employees money at less than market interest rates.

The company has a pension fund which is payable upon retirement. Ohbayashi's retirement age is set at 60, but if one is a Director, the retirement age is 65–70 years of age. The company's pension plan is meant to be supplemented by employee savings and the federal government's pension plan. Since retirement age at the company is 60 and the government pension plan does not begin to issue monthly payments until age 65, there is a five year gap that needs to be addressed.

Health insurance is provided by the government, with all costs being paid except for a reasonable deductible.

Job related travel and entertainment expenses are reimbursed by Ohbayashi, and when a manager takes a client out for a day of golfing, it will be one of those rare times when the average corporate employee can enjoy playing the game. With country club memberships costing $100,000 on the low end and $1,000,000 on the high end, except at

Tokyo's Kogaei Country Club, where an individual membership costs $3.5 million in 1988, golf is a rich man's sport in Japan. Because the sport is so popular and so expensive (an average round will cost 30,000 yen—$231) the driving range becomes the only golfing outlet for many Japanese. Because of the extremely high cost of land in and around Tokyo, it is not uncommon to see a golf driving range enclosed in netting on the rooftop of a downtown mid-rise office building.

Training at Ohbayashi

As in most large Japanese construction companies, on-the-job-training is an important part of a newcomer's orientation program at Ohbayashi. New university graduates commence their job training immediately. For new entrants, the company provides a dormitory where all single men live together. This housing arrangement encourages the spirit of being classmates. New trainees, whether they are engineers or liberal arts graduates, spend some time at a construction site. No matter what the employee's position within the company ultimately ends up to be, he or she is first and foremost in the construction industry and must have knowledge of the company's product. One of the first assignments is always at a construction site, and a new worker might be stationed in the field for one or two years before being brought back into company headquarters. After returning to an office environment, there will be a series of work assignments in different departments so that a good overall picture of the company's functions can be realized. At some point in the training program, possibly three to five years from the time of first hiring, a permanent position within the company will be determined, based upon a decision deemed best for employee and employer. At this juncture in the training program, a new type of training course will be introduced, one that will prepare the aspiring manager for the responsibilities that lie ahead. Courses will be offered that focus on the skills of proper communication and the art of working harmoniously and effectively with both senior and junior employees.

KAJIMA—A $7.6 BILLION GIANT

Quality contractors such as Ohbayashi might share similar organizational charts with other top flight contractors in Japan, but Kajima Corporation seems to have created its own special footprint. It has built such noteworthy projects as the Saudi Arabia Royal State Palace, the Grand Hotel in West Berlin, the massive Shanghai Exhibition and Hotel Complex on the mainland of China, and civil engineering projects of staggering proportions in many parts of the world. But unlike possibly any other construction company, Kajima has contributed to the cultural world, through several of its affiliates, with the same intensity that it contributes to the physical world.

Under the guiding hand of Dr. Morinosuke Kajima, former chairman of the corporation, Kajima has expanded its company horizons beyond construction and engineering and has contributed to the cultural life of Japan and the world through several philanthropic organizations and non-construction related business ventures:

The Kajima Foundation	Established at the request of Dr. Kajima to provide support and encouragement to scientific research for both Japanese and non-Japanese programs.
Kajima Institute Publishing Company, Ltd.	This company publishes engineering and architecture books as well as general interest magazines and *The Dalton Report,* a construction related newsbulletin. The Institute also produces educational video cassettes.
Yaesu Book Center Inc.	In 1978, Kajima opened Japan's largest bookstore, located a few blocks from Tokyo Station. The Yaesu Book Center stocks more than 1 million books on five floors. Magazines from around the world abound, and traditional Japanese woodblock prints and watercolors round out the offerings in this spectacular store.
Kajimavision Productions Co., Ltd.	This corporation produces cultural, educational, and documentary films. This recent addition to the Kajima Group has already won high praise in the film industry, as well as earning numerous honors for its productions.
The Kajima Institute for International Peace	The Institute serves as a research center for the study of various international issues and topics affecting world peace.
The Kajima Foundation for the Arts	The Kajima Foundation for the Arts contributes to the promotion of the arts and artists around the world, particularly in the field of painting.

Kajima's Corporate Philosophy

Dr. Taiji Sayo, Executive Vice-President of Kajima Corporation, speaking from the corporate conference room at their new headquarters building near the New Otani and Akasaka Prince hotels said, "The primary concern of a construction contract is cost. A general manager or a project manager must stay within the budget. But if quality is violated, or the schedule is not met, or if the client is dissatisfied, then the construction process has failed."

Kajima is deeply concerned with customer satisfaction when building in Japan as well as in markets outside of the country. Dr. Sayo is very clear in his statement of corporate philosophy: "The final objective in any construction project is owner satisfaction." This philosophy works well in Japan, where Kajima has established long term relationships with scores of owners and strives to perform at their absolute best. If the client is pleased when the project has been completed, and the builder has sustained some losses, the assurance of future business will allow a recapture of lost profits.

Unfortunately, this type of reasonable attitude is not shared by many clients outside of Kajima's nucleus of steady clients, but Kajima, rather than compromise their principles, will stick to their corporate policies even if it means losing money on a project. Client satisfaction includes delivery of a building on time, and rigid adherence to the construction schedule is almost a point of honor with Kajima Corporation. According to Dr. Sayo, several of Kajima's construction contracts in Southeast Asia contain liquidated damages clauses. If for reasons beyond the company's control the construction schedule slips and the project completion date is in jeopardy, Kajima will initiate an accelerated pace that might cost them far more than the penalty imposed by the liquidated damages clause, because completion on schedule is a key ingredient in the formula for client satisfaction.

Dr. Sayo said that the philosophy that Kajima and other major Japanese contractors follow should be recognized and appreciated by the rest of the world, but he doubts that this will be so within the foreseeable future. It is becoming more difficult to maintain these policies in light of increased competition, according to Dr. Sayo. Korean contractors are good builders and tough competitors, but Dr. Sayo says that if a client demands top quality or sophisticated design capabilities in Southeast Asia they must look to Japan.

Kajima's concern for the client is matched by a concern for the employees. Housing is a serious problem for most working class people in and around the major cities in Japan. With Tokyo land prices ranging from $15,000–$35,000 per square foot in the most fashionable parts of the city, housing is prohibitive for all but the very wealthy. Even a one hour train ride away from Tokyo will unearth land prices ranging from $200 to $300 per square foot, so that a small parcel of 1500 square feet will cost between $300,000 and $400,000. A typical, tiny, two story house with a total floor area of 100 square meters (1076 square feet) situated on a postal stamp site will more than likely cost $700,000 in the 1–1 1/2 hour commuting radius from central Tokyo.

Kajima realizes that in order to attract top quality people, it must provide some form of housing assistance. The company has built a number of apartment houses which they rent to employees at very low rates, as low as $77 per month, but these apartments are old-fashioned and very small, and the younger employees are not too happy living in them. So Kajima is in the process of building new, more spacious and more modern apartments for their people and has set aside $77,000,000, exclusive of land costs, to do so. By Western standards these new apartments are still undersized, ranging from 645 to 753 square feet, but by Japanese standards these three room rental units are more than adequate.

Kajima's Other Fringe Benefits

Stock Purchases. When an employee buys company stock, the corporation contributes an additional 10% toward the purchase; an employee who purchases ten shares of corporate stock would actually be issued eleven shares.

Low Interest Loans, High Interest Savings. The company "bank" will lend money at below market interest rates. To encourage employees to become home owners, the company will provide loans for as much as 20 million yen (about $154,000) for that purpose, at below market interest rates. When money is deposited with the company, it earns higher yields than commercial bank offerings.

Assistance in Daily Life. One Kajima manager said, "We try to assist our people in their daily lives." This assistance takes the form of subsidized lunches in the company cafeteria, where a decent lunch costing 500 yen can be purchased for half price.

There is a company association, distinct from the enterprise union. Each employee contributes 500 yen per month (about $3.86) to the association and when a company employee gets married he or she receives a 10,000 yen check. If a parent should die, the association sends flowers in the name of the company and Shoichi Kajima, company president. If there is a personal tragedy and emergency money is needed, the association will provide rapid, low cost loans.

Company Retirement Benefits. Retirement at Kajima Corporation is mandatory at ago 60. Bearing in mind that government pensions do not begin until age 65, Kajima offers the newly retired the option of receiving a lump sum payment or regular monthly checks. If the lump sum payment is elected, it is computed on the basis of fifty times the monthly salary at retirement. If a manager's monthly salary at retirement was $50,000 per year, or $4167 per month, he would expect to receive a lump sum payment of about $208,000.

GOVERNMENT BENEFITS AND THE CONSTRUCTION COMPANY EMPLOYEE

The Japanese federal government and to a lesser degree the prefecture governments offer a number of assistance programs. The benefits that follow were obtained from the Ministry of Labor, as of 1987, and the conversion to U.S. dollars is based upon a rate of 150 yen.

Pension Insurance

All people living in Japan having worked more than 25 years between the ages of 20 and 60 are eligible to receive a pension once they reach the age of 65. As of 1987, the yearly pension for a single retiree is about $4177, and double that amount is provided for a married couple. The Old Age Plan, as it is known, contains a disability feature similar to the one in the social security system in the U.S. The government sponsored pension plan in Japan is funded through employer and employee contributions, and as of 1987 the total premium of 12.4% was shared equally by company and worker.

National Health Insurance

All companies except those with less than 5 employees participate in this insurance plan, although corporations with more than 300 employees can form their own insurance society. Health insurance premiums were about 8.3% of earnings in 1987 and the cost of the premiums is usually shared equally by employer and employee. Coverage includes examinations, drugs, medical treatment, surgical and hospital costs, nursing care, and transportation costs. An unusual maternity benefit is included in the plan for women who have been actively employed before becoming pregnant. A new mother will receive 60% of her daily earnings for a period commencing 42 days before she delivers and for 56 days after she has her child. A mother also receives a lump sum grant amounting to 50% of her monthly earning, with a minimum payment of 100,000 yen ($1333). She also receives a lump sum nursing grant of 2,000 yen.

Workman's Compensation

Similar to the program in the U.S., workman's compensation is provided for all workers injured on the job, or suffering from an occupational disease, or being killed while at work. Along with providing the injured or disabled worker with a monthly check, lump sum payments can be arranged.

Unemployment Insurance

Unemployment insurance is available for all workers in the private sector in Japan, except for those workers employed by a company having less than five employees. Basic coverage is computed on the basis of an employee's total wages for the year with a minimum benefit of $21.40 per day up to a maximum benefit of $81.42 per day. A sickness allowance is provided for those unemployed workers who cannot seek work because of a temporary illness. Elderly workers are given special consideration because the government recognizes the fact that it is more difficult for older workers to find jobs, and the many retirees at age 60 need to find employment until their federal pension plan payments begin at age 65. Unemployment benefits to workers who qualify as senior citizens are extended from 50 to 150 days while they seek work. If they do find employment, only to be laid off shortly thereafter, they may be eligible to receive a lump sum payment.

The Children's Allowance Plan

Another unusual feature in the Japanese government benefit plan has to do with an allowance program applying to all families with children not yet graduated from the compulsory education program. An allowance is paid to those families having a second child under the age of 4, and another allowance is paid to each family for their third child until age 9, and for all subsequent children until they reach the age of six. A flat monthly sum of 2,500 yen is paid for the second child and for all others under the age of six. This program is funded 70% by the companies participating in the various other federal programs, and the federal government contributes 20% while local and prefecture governments each contribute 5%.

THE ROLE OF THE SUBCONTRACTOR IN THE CONSTRUCTION PROCESS

Most large Japanese general contractors do not directly employ skilled workers on their construction sites; they subcontract nearly all of the work to be performed (see Table 3-3), including daily cleanup. Although many U.S. contractors maintain their own concrete foundation crews and generally rough and finish carpenters on their payroll, this is not a common practice among Japanese builders.

A major difference between subcontracting in Japan versus the U.S. is that the Japanese general contractor will more than likely purchase the necessary materials and equipment and hire subcontractors for installation labor only.

Table 3-3. How Work Is Subcontracted.

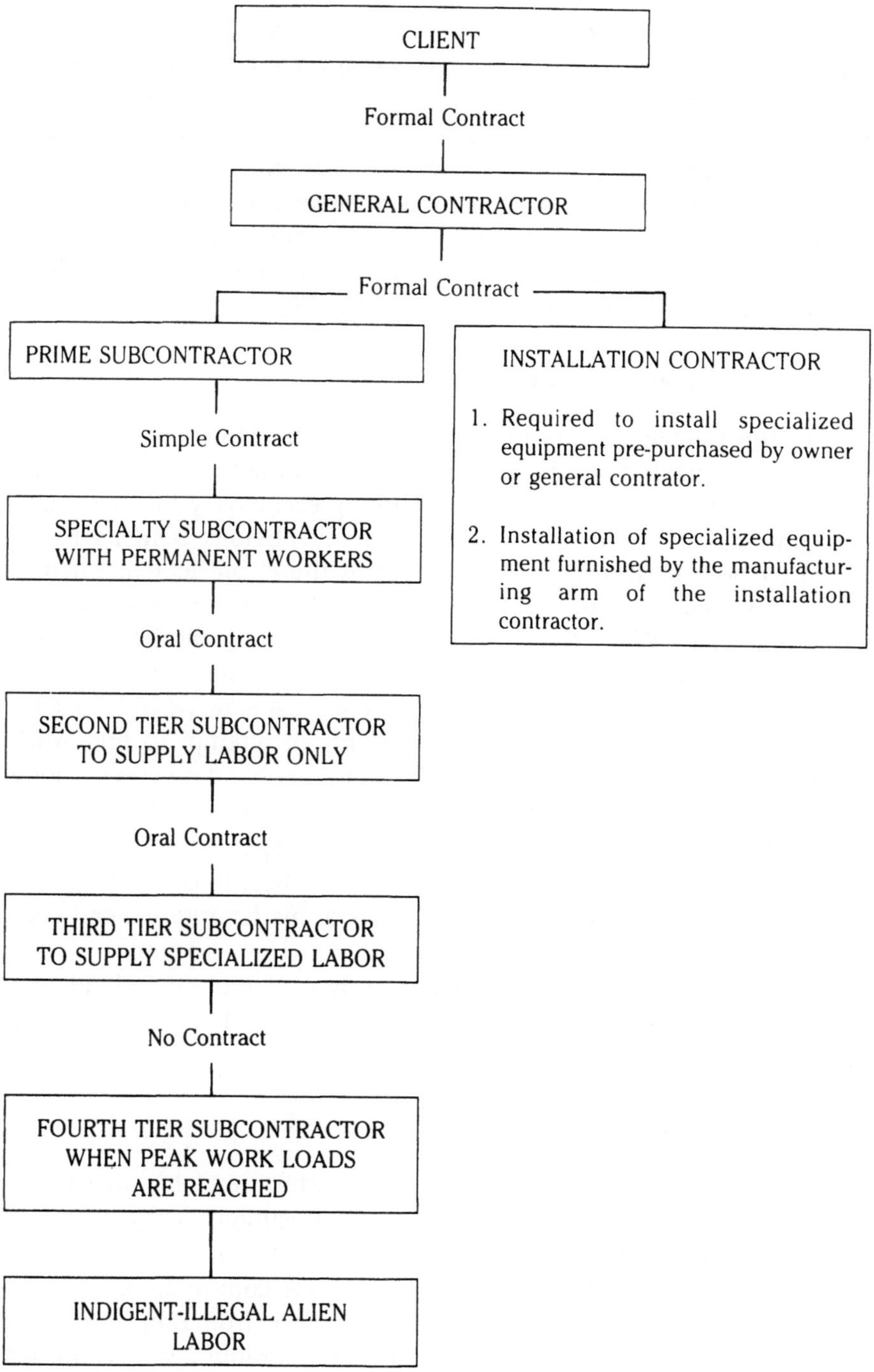

The Installation Subcontractor

The term *installation contractor* denotes a company that will furnish equipment and also install it, and is usually confined to companies manufacturing specialized equipment requiring highly skilled installation crews. Mitsubishi, a major manufacturer of elevators, is an installation contractor, since they not only furnish the elevator, but also supply crews for installation. Hitachi, a manufacturer of complex heating and cooling equipment in Japan, would also qualify as an installation contractor since they would supply, install, and connect much of the sophisticated HVAC equipment that they sell.

The Prime Subcontractors

Most of the established prime subcontractors in Japan have had long standing relationships with certain general contractors and have become almost captive subcontractors, working exclusively for these general contractors. There are, of course, many obvious benefits to be gained by both parties in situations such as these, but as construction activity fluctuates up and down, the subcontractor's work crews are affected radically. When the general contractor has a great deal of work, the subcontractor will have to be able to supply increased work crews, but when the general contractor is experiencing a slow period the subcontractor will have to reduce his workforce or attempt to find work with other contractors or with owners. This rubber-banding effect among prime subcontractors has created a whole system of second, third, and fourth tier subcontractors ready, willing, and able to jump into the breach when called upon. By using lower-tier subcontractors, the prime sub can more easily adapt to the demands of his favorite general contractor.

There are many problems with this system of multi-tiered subcontractors and they start at the bottom rung on the ladder and work their way up.

The Lower Tier Subcontractors

The third, fourth, or fifth layer of subcontractors tend to have lower skill levels and are lacking in basic business acumen. They offer low wages, little or no employee benefits, and quite often ignore workman's compensation laws and other required employer commitments.

Japan is experiencing an influx of illegal workers from other countries in the Pacific Rim where daily wage rates are extremely low. Attracted by the relatively high wage rates in Japan, illegal workers from the Phillipines, Pakistan, Bangladesh, Thailand, China, and Korea have begun to invade the country. The Labor Ministry estimates that fully 43%

of these foreign workers find jobs in the construction industry. With little or no skills, willing to work for low wages in construction jobs that are too hazardous, strenuous, or unpleasant for Japanese workers, these aliens tend to gravitate to the bottom tier subcontractors. The Japanese seem to condone these actions because it has become difficult to attract new workers to the construction industry, and when younger people are offered jobs by contractors, they will turn away from the very types of jobs that the illegal workers will readily fill.

The Ministry of Construction is concerned about the use of lower tier subcontractors and frowns upon the entire multi-layered subcontracting system as now practiced in Japan. One official in the Ministry said that he would favor the American system of subcontracting where a general contractor issues a contract to one subcontractor for a specific phase of work, and that subcontractor performs the work with his own crews.

The Ministry of Construction is concerned about the plight of subcontractors in general, and has gone so far as to publish a series of guidelines to be used to establish the relationship between general contractor and subcontractor. Called "Guidelines for Rationalization of the Relationship Between Contractor and Subcontractor," they contain the following advice:

For general contractor's use when evaluating a subcontractor:

1. Is the subcontractor capable of executing the work at hand?
2. Investigate the way in which the subcontractors manage their workers and how they handle safety matters.
3. Determine if the subcontractor considers the welfare of the workers he employs.
4. Evaluate the subcontractor's ability to conduct business transactions.

General contractors need to monitor subcontractors to insure that:

1. They have satisfactory work record.
2. They they are technically trained to perform their tasks properly.
3. That they have sufficient manpower to perform adequately.
4. They have proper tools and equipment to do the job.
5. They have sound financial status.
6. They can provide proper job-site supervision.
7. They are complying with all laws and regulations.
8. They maintain good safety practices and have a good safety record.
9. If the subcontractors have to provide temporary lodging or meals for their workers while away from home, that adequate provisions have been made.
10. If lower tier subcontractors are to be used, they will be paid fair wages and paid promptly.

General contractors are also concerned about the welfare of their subcontractors, recognizing the fact that they are dependent upon them to a great extent. General contractors will assist their subcontractors in many ways. They will provide them with safety training programs and help them establish quality control programs and QC procedures. General contractors will offer their subs accounting and financial assistance and advice and they will help them update and upgrade their estimating systems.

The entire concept of contractor-subcontractor relationship is much different in Japan and is evident in the manner in which GC and sub will negotiate a contract between themselves.

NIHON JOHNSON CONTROL

Mr. Paul R. Marting, a Staff Manager in the Marketing Department of Nihon Johnson Controls Company, Ltd., in Tokyo has had experience working with general contractors in Japan and the U.S., and is able to compare methods of bidding and bid evaluation as practiced in both countries.

According to Paul, the method of obtaining bid information and preparing proposals for building automation systems is slightly different in each country. Once bids are received by the Japanese client or general contractor, the method by which these proposals are analyzed is quite different as well.

In the United States, in the design stage of a project, the electrical engineer will most likely contact one or several building automation systems subcontractors and request some assistance in the design of the controls systems. A company such as Johnson Controls will respond with hardware specifications and information regarding the sequencing of various control operations. At times, this engineering input will work to the contributing BAS subcontractor's advantage inasmuch as systems might be designed around captive equipment, slightly limiting competition.

In Japan, the control subcontractor will receive hardware specifications in the bid package which describe the level of quality, or occassionally, a part of a specific manufacturer. The project design engineer will also furnish the bidder with detailed layout drawings showing the number and location of sensors, controllers, and so forth. When a BAS contractor has been selected, these detailed drawings will form the basis for a shop drawing submission.

When a proposal is prepared by Nihon Johnson Controls, the bid contains a detailed breakdown of all materials by quantity and price. Other BAS contractors will be submitting the same detailed analysis of materials, but usually labor costs are treated as a lump sum dollar amount in the proposal.

All of these BAS proposals will be reviewed by the client or general contractor to determine if each subcontractor has the same quantities, more or less, and that the quality levels of major pieces of equipment are equal and meet the client's requirements. If there are questions as to quantity clarifications or quality levels, the building automation systems subcontractor will be requested to adjust his proposal accordingly.

Paul said that his company recently submitted a controls proposal to a major Japanese contractor, and one item in the list of materials was a CO_2 monitor which the GC said was five times the cost of other bidders. Johnson Control explained that they interpreted the quality level of this piece of equipment in such a way that the higher price was warranted. The general contractor said that the quality level exceeded the owner's needs and the price should be adjusted accordingly. The adjustment was made and now JCI's bid could be fairly reviewed.

Although negotiations in Japan are extensive and thorough, they are not the cut-throat kind of haggling so common in the U.S. Contractors in Japan do not endlessly "shop" prices but approach negotiations in a methodical and very professional way. The bid price submitted by Nihon Johnson Control and their competitors is usually a "list price" that would be the same whether submitted to an owner, a general contractor or a prime subcontractor. Nihon Johnson Control will then offer discounts from list, depending upon their past relationships with the contractor or owner and other negotiating factors in play at the time.

Paul was quick to add that this professional approach to bid analysis and bid award was confined to Japanese contractors and Japanese clients. Working with other Southeast Asian contractors can be just as frenetic as negotiating with many U.S. contractors, and after a slight pause he added, "and maybe more so."

CORPORATE PLANNING IN THE CONSTRUCTION INDUSTRY

As part of their long range outlook, many of Japan's major contractors formulate five year management plans in which they establish corporate goals and priorities for the near term. Although some corporations limit their planning to three year lookaheads, the five year plan is the more prevalent. Japanese contractors are not content to leave many important factors to chance and by forecasting goals and formalizing corporate philosophy into concrete corporate action seems to give them an additional level of comfort. Some of these corporate plans are called "rolling plans" which mean that they are subjected to review and revision each year, but all of them depict a company and an industry determined to strive for improvement in the years ahead.

The management plans set forth below have been condensed to reflect their most salient points.

Sumitomo Construction Company—"Second Long Term Management Plan"—1982–1986

Goals:	Increase market share.
	Improve company's performance.
Strategy:	Increase amount of private sector orders.
	Reinforce commitment to technological development.
	Develop total quality control system.
	Increase share of overseas work.

Taisei Corporation—Management Five Year Plan—1981–1985.

Goals:	Restructure management.
	Extend market share, increase amount of overseas orders.
Strategy:	15% increase in annual sales volume.
	Establish housing project strategies.
	Continue technological development.
	Increase corporate efficiency.
	Adopt a rolling plan method of corporate planning.

Fujita Corporation—"The Fifth Plan"—1982–1986.

Goals:	Develop turnkey capabilities.
	Develop innovative management procedures.
Strategy:	Apply value engineering concept to every department in the corporation.
	Adopt a rolling plan of corporate planning.

Hazama-Gumi Ltd.—"New Three Year Plan"—1981–1984.

Goals:	Apply efficient, positive management techniques.
	Develop powerful, exclusive subcontractors base.
Strategy:	Increase productivity and profits.
	Increase annual sales by 10%.
	Develop new technologies in the energy related field.
	Increase real estate project volume.
	Develop more overseas work.
	Improve fiscal position.
	Amalgamate subcontractors and improve labor conditions of subcontractors.

Toda Construction Co.—"New Five Year Management Plan"—1982–1986.

Goals:	Establish Management Grand Vision for the Second Century (Toda was 100 years old in 1983).
	Enhance civil engineering department.
	Increase urban development work.

Strategy: Increase productivity of administrative departments.
Change over from value engineering concept to total quality control concept.

Shimizu Construction Company—Management Five Year Plan—1982–1986.

Goals: Formation of flexible management to cope with environmental changes.
Develop EC (engineering-contractor) capability.
Increase annual orders by 12%.
Win the Deming Prize (construction's most prestigious award in Japan).

Strategy: Develop matrix approach incorporating building, civil engineering, technology, financing, overseas development as part of corporate strategy.
Full scale development of total quality control systems.

Mitsui Construction Co.—"A Management Three Year Plan"—1980–1983.

Goals: Increase management efficiency and increase financial strength.

Strategy: Develop ability to handle more energy related orders.
Increase amount of sales received from Mitsui business group affiliates.

Of the eleven corporate management plans reviewed by Professor Yasuyuki Hippo when he conducted a survey of the construction industry in Japan for the Asian Productivity Organization, several priorities seem to be common to most of the companies surveyed:

1. A desire to increase market share, both domestically and in overseas markets.
2. Development or enhancement of a total quality control program.
3. Enhancement of profits, combined with a desire to improve management.
4. More efficient utilization of manpower (in recognition of the aging population combined with the lure of more attractive job opportunities) to combat the possibility of a labor shortage in future years.
5. Commitment to increased technological efforts to improve the construction process.

The corporate goals of the top twenty contractors in Japan appear to be focusing on increasing profits while turning more toward an engineering-construction concept, particularly in overseas markets. Their management remains firmly commited to increasing the quality of the end product.

4

Marketing–Domestically and Overseas

Marketing is more than merely selling. To be effective, it is necessary to be creative. There is an old story about a cannery in Alaska that was offered several truckloads of salmon at cut-rate prices. These salmon, for some reason, had very whitish meat and since most people were accustomed to buying pink salmon, the fisherman was having a difficult time selling his catch. The owner of this particular cannery was very enterprising and he bought the entire shipment at rock bottom prices. Several months later the fisherman stopped by to inquire about the sale of the off-color fish. The cannery owner said, "Great—in fact fantastic! We had labels for the can printed—'Sockeye Salmon—Guaranteed Not To Turn Pink in the Can.' " Now that's marketing!

It is not enough to be low bidder in today's high stakes international construction market. Other, equally important factors play a major role in contractor selection. The Japanese contractors seem to have been preparing for this day:

- The money they have been pouring into research and development is starting to pay off in the development of new products and innovative designs which set them apart from the competition.
- Specialty equipment and specialized construction techniques make them ideal candidates as joint venture partners in complex projects.
- The close alliance between government and industry and the concept of one united front—"Japan, Incorporated" can cut through red tape in a hurry, while throwing up roadblocks for outsiders.
- The relationship between construction companies and banks and other financial institutions in Japan makes the availability of project funding much easier and more rapid.

Armed with all of these tools of the trade, the large Japanese contractor has become a formidable marketer of construction services.

RESEARCH AND DEVELOPMENT AS A MARKETING TOOL

Within the country, long term relationships between contractor and client are very valuable to the contractor who has such a relationship but must be very frustrating to the competition. One method of wooing a client away from a long standing relationship is to come up with a design so innovative that it cannot be dismissed by the property owner. Land costs in Tokyo are extremely high, and when a new project is being considered for, say, a 1/4 acre parcel, the cost to acquire the land could be as much as $27 million. Obviously maximum site utilization is a must. Companies like Mitsui Construction wrestle with ways to design earthquake resisting buildings with the slimmest beams and columns to maximize interior rentable or salable space. A new design which creates considerably more interior area will not be so readily cast aside by a client, and allows the "outsider" contractor entry into a potential relationship.

Mitsui Construction developed a super beam structural system which allowed them to build a 36 story apartment house in Dojima, Osaka, with a Y shaped floor plan for optimum sunlight exposure while keeping both wings of the building column free. This design was made possible by their development of a 46-foot-long PRC-synthetic super beam which is able to span to the tie beams.

Taisei Corporation has established a "full turnkey" system for hotel and restaurant construction where the owner can truly turn the front door key and open a building that is fully equipped to function immediately.

For projects requiring agricultural expertise during the design stage, Taisei Corporation offers the services of in-house biologists and agronomists who were brought on board for an irrigation project in Nigeria. Taisei felt that specialists were needed who understood how plants grow when design-construction services were being offered.

Let us look at some of the other products of R&D that help eliminate the competition:

- Carbon fiber reinforced precast concrete panels which are so lightweight they can be "ganged" and lifted into place.
- Super fiber reinforced concrete pipe and other concrete products imparting very high strengths to concrete while reducing weight.
- The development of prestressed concrete structures for underground LPG and LNG tanks as well as for nuclear power plants.

- Earthquake resistant structures beginning at the foundation level with *menshin*–base isolation technology and progressing upward with precast concrete wall panels creating two- and three-dimensional shear planes.
- Shield tunneling techniques and equipment combined with computer assisted monitoring systems.
- Zero tolerance lot line foundation wall construction using diaphragm walls and slurry technology.
- Abrasive high speed water jet demolition techniques for use in congested urban areas or where excessive shock or vibration must be avoided.
- Computer and software developments to enhance every aspect of the construction process from scheduling concrete pours to estimating the cost to lift structural steel beams into place.

Japanese tunneling expertise and sophisticated equipment development has paid off for many of their contractors. Kumagai Gumi developed shield tunneling technology in 1956 and were pioneers in this new method of boring. They have obtained a number of contracts throughout the world based upon their experience and quite recently won a major contract in China, thereby gaining a foothold in this potentially dynamic market. Kajima, another leader in tunnel work, teamed up with a local contractor in Washington, D.C. to submit a proposal on a 440-foot subway tunnel and subway station for the Metropolitan Area Transit Authority. Although they were the apparent low bidder, they were not awarded the contract in March of 1988 because the then existing Brooks-Murkowski Amendment prohibited an award on government work to Japanese contractors. The Japanese presence on the tunnel under the English Channel has already been firmly established. Kawasaki Heavy Industries of Tokyo designed the mammoth tunnel boring machines for this huge project and has formed a joint venture agreement with Robbins Company of Kent, Washington and Markham and Company of Great Britain to produce the massive 1,300 ton machines costing as much as $42 million a pair.

JOINT VENTURES—A VALUABLE MARKETING CONCEPT

The joint venture concept has been popular in Japan for a number of years. Some experts claim that its popularity is due to the existence of the *dango* system whereby pieces of large projects are divided among certain preselected contractors in a "share the wealth" scheme.

The joint venture has taken on added significance recently and, with the many advantages accruing from this method of contracting, is likely to remain and grow.

- Small companies can form joint ventures and participate in large construction projects that they could not otherwise consider.
- Less sophisticated companies can team up with more experienced ones and obtain a great deal of on-the-job-training.
- Joint ventures spread risk.
- Joint ventures permit the transfer of technology.
- A foreign corporation can enter a local market with some level of comfort by joint venturing with a local established contractor.
- A joint venture may result in better control over costs if one of the members is a local builder with established subcontractor relationships and proven supply sources.
- Joint ventures allow contractors to participate in those phases of construction where they have the most experience and expertise.
- Lastly, the concept of joint ventures seems to have taken on a more important role as worldwide construction contracts have begun to recede in many parts of the globe.

The international construction market is in a state of flux. Construction contracts have declined by 30–40% over the past 5–7 years, and the OPEC countries are no longer the source of multibillion dollar projects. The twin city construction of Jubail and Yanbu in Saudi Arabia at a combined cost of $13 billion is probably a thing of the past.

While these mega-projects were in their heyday, they marshalled contractors and subcontractors from all corners of the globe. Builders from South Korea, Taiwan, Yugoslavia, Turkey, Pakistan, and points east were able to work side by side with experienced contractors and received valuable on-the-job training in the process. They all returned home with a great deal of construction experience under their belt by working on these complex projects. Although the home countries of these newly trained contractors might have had to solicit bids from more technically equipped contractors in the past, these local builders had acquired the ability to participate in sophisticated projects in their home countries without too much assistance from outside contractors.

Many of these developing or newly developed nations are no longer inviting foreign contractors; rather, they are taking steps to restrict outsiders from bidding on their domestic projects. The joint venture approach permits them to encourage foreign participation in government projects while not appearing to be so protectionist as to exclude foreigners entirely.

Saudi Arabia recently raised the price of admission to their construction projects, requiring foreign contractors interested in bidding on government projects to share the work on a 50–50 basis with a Saudi partner, most probably a cousin or nephew of King Fahd. The price of a ticket to a Saudi construction project can be expensive, as several U.S. contractors have found out. The Saudis are famous for finding reasons, real or imagined, to withhold substantial sums of money until minor disputes are resolved. American companies claim to have a total of $1 billion withheld from their contract payments because of disputes.

Joint venture alliances which might have gone begging in recent times may become as valuable as stock options, as various partnerships are formed, and the list of available participants on the international construction scene begins to dwindle. Fluor Daniel, who had previously worked with Ohbayashi on Toyota Motor Corporation's Georgetown, Kentucky assembly plant recently concluded a joint venture agreement with Ohbayashi to build 25–30 projects in the U.S., Japan, and a few other countries. If the relationship works successfully, these two world class contractors might not be interested in any other relationships with joint venture partners.

Tishman Realty and Construction Company of Manhattan signed an agreement with Aoki Corp. of Tokyo in 1988 to develop a 42 story twin-tower office complex in Osaka. As part of the deal, Aoki will assist Tishman in obtaining a Japanese contractor's license and will also help the New York builder to explore participation in the Kansai International Airport project, which is not far from Osaka.

Shimizu Construction Co., Ltd. is departing from its conventional role of engineer-builder as it forms a joint venture with a Texas based aerospace contractor, Bell and Trotti. Both companies are planning to develop human habitation modules for use in space exploration, and Shimizu feels that this agreement will give them an opportunity to increase their expertise in robotics. Shimizu also had space research program tie-ins with another U.S. contractor, Parsons, and now Shimizu has expanded the relationship between the two companies. Parsons and Shimizu, early in 1988, announced that they would enter into a basic agreement to cooperate in technical and marketing matters, primarily centering around design and engineering work. Parsons said that it would plan to open a branch office in Tokyo and utilize its relationship with Shimizu to obtain marketing information in Japan and Southeast Asia. Parsons also made it known that it would take steps to obtain a contractor's license in Japan, undoubtedly calling upon Shimizu for assistance in doing so.

THE MARKETING LINK BETWEEN GOVERNMENT AND INDUSTRY

It has been called an embarrassment of riches: Japan's annual trade surplus with the United States has been running at about $83 billion, and the assets of their big international banks have reached the trillion dollar level. What to do with all this money? How to disburse it to appease adverse foreign reaction while still being able to reap some benefits from any largesse? Aid to developing countries is a program that will win universal approval in the world community.

Japan is now committing $10 billion per year to developing countries and promises to double that commitment over the next several years. Loans granted by the Official Development Assistance (ODA) are carrying out this program. It has been reported that 30% of ODA's loans have been officially tied to Japanese goods and services and as much as 60% finds its way back to Japanese companies by some accounts: almost all of these funds are being disbursed in Asia, Japan's back yard. The link between the Japanese government and their construction companies provides Japanese builders with a leg up on many of their competitors, enhancing their ability to obtain work. Just knowing that funds might become available to finance a construction project in a developing nation may be enough for a Japanese contractor to offer to provide a feasibility study at little or no cost to the client.

The Overseas Construction Promotion Fund can supply these Japanese contractors with low interest loans to cover the cost of any of these pre-bid studies. When Japanese contractors let it be known that a project in a developing nation could receive funds from Japan's Overseas Economic Cooperation Fund as well as Japan International Cooperation Agency funds, or monies from ODA, the chances of their landing a construction contract will increase dramatically.

The Overseas Economic Cooperation Fund (OECF) recently made possible a $77 million loan to three Buddhist pilgrim centers in India. This project was estimated to cost at least $200 million and includes the construction of religious buildings, lodging facilities, and improvements to transportation and communication systems, in addition to upgrading the local water supply. At about the same time, the Overseas Economic Cooperation Fund lent millions of dollars to a utility company in Manila, which was probably one of the reasons why New Japan Engineering Consultants, Inc. of Osaka was awarded a contract for consulting services for this $290 million, 300 MW coal fired powerplant.

When an international tender was issued for the Hong Kong Subway project in 1973, it was a foregone conclusion that the ability to provide some form of financial assistance would greatly increase a bidder's chances of obtaining the contract award. The Overseas Construction

Association of Japan, a private organization, appealed to the Japanese government who, in turn, talked to the Export Import Bank of Japan. EXIM came to the rescue by offering a deferred-payment loan to Hong Kong, and a joint venture agreement, which included Japanese contractors, secured the deal.

This amalgamation of government and contractor is not only recognized by the Japanese as a means of bolstering their construction industry. The Italian government saw the potential for improving their contractor's chances of winning contracts by being able to offer some type of financial packages. Italy extended developing nations more than $1 billion in aid in 1987 and 1988, and this ability to tie financial aid into the bid package increased the sales volume of several Italian contractors. According to *Engineering News Record,* in 1987 more than half of the respondents in a poll of Italian contractors said that their contract bid requirements included major financing clauses, and they successfully obtained $3.5 billion out of a possible $7.3 billion worth of contract-finance work that year.

A senior member of Kajima's International Division said that the conventional way of doing business in the construction industry is no longer the new way. Construction nowadays must move into the business of creating projects. Top Japanese contractors can no longer respond to market conditions; they must find ways to create them. In other words, the time has come to stop reacting and start acting. Japanese contractors are increasingly looking to Japanese trading companies for information and to Japanese banks for financial assistance.

No longer can a contractor merely look at a set of construction plans and specifications and determine how he will estimate the cost of the project. Builders now need all the construction and engineering expertise their companies can muster, and they must delve into the project's feasibility, its financial requirements, its return on investment, its risks, its marketability, its rewards, and proceed only if all or most elements look favorable.

The Asahan project in Indonesia is a case study that demonstrates how all of these factors can come together and create a project.

THE ASAHAN PROJECT—A CLASSIC CASE OF CREATING A PROJECT

In 1967, the Japanese engineering consulting firm Nippon Keoi saw an opportunity to garner a substantial contract by suggesting that a hydroelectric power plant would be feasible on the Asahan River in North Sumatra, Indonesia. This hydroelectric power source could be used to provide electrification to the surrounding area, but more importantly, could provide power for an aluminum smelting plant and therefore boost the economy of the area and the entire country. One of the key ingredients in the smelting of aluminum is cheap power, since aluminum production requires 15,000 Wh of electricity to produce one ton of metal.

At the urging of Nippon Keoi, the Japanese government extended a loan to the government of Indonesia so that a feasibility study could be conducted by none other than Nippon Keoi. When the survey showed that a venture like this was indeed practical, Nippon Keoi contacted Sumitomo Aluminum Smelting, Nippon Light Metals, and Showa Aluminum, who all expressed interest in the project. The manufacturing of aluminum in Japan was becoming more and more expensive. Japanese smelters needed more power stations so that they could expand production, but power plant construction had been severely curtailed because of environmental issues and the potential for increased air pollution. The Japanese aluminum smelters saw the potential in this Indonesian project. This opportunity to come to the aid of a developing nation while helping oneself was not lost on the rest of the world. Kaiser Aluminum and Alcoa were attracted to this project and conducted their own feasibility study. The Indonesian government, realizing that a golden opportunity had been placed upon their doorstep, decided to invite international tenders in January 1972 on the power plant construction, the smelting operation, and the basic infrastructure that would be needed to support these two large projects. Sumitomo, Nippon Light Metals, and Showa decided to join up with Mitsubishi Chemical Industries, Mitsui Aluminum, Alcoa, and Kaiser to form a seven-company consortium, since the entire package would be extremely expensive.

The projected costs for the power plant and the smelter appeared to be so high that the consortium asked the government of Indonesia to fund the hydroelectric plant, and when the government refused to do so, all negotiations collapsed. On bid day in July 1972 not one proposal was received. Both Kaiser and Alcoa finally backed out of the consortium in 1974, and the five Japanese aluminum smelters decided to go it alone and worked diligently to present an acceptable proposal to the Indonesian government. After months of hard work a proposal was submitted and accepted by Indonesia, and in December 1974 a master agreement was signed. The Indonesian government was willing to finance a certain portion of the combined project, but expected the aluminum smelting companies to participate in the financing package. Loans were also being solicited from foreign banks at the same time. By the time a firm agreement was reached, the status of the costs and methods of financing took on the following characteristics, including a new name—Nippon Asahan Aluminum JV:

Cost of the Project	
Power Stations	123 billion yen
Aluminum Smelters	240 billion yen
Infrastructure	48 billion yen
Total Project Cost	411 billion yen ($3.15 billion, 1988 U.S. dollars)

Financing: Equity—22%, Loans—77.8%.

Equity	
Indonesian Government	22.775 billion yen (25%)
Nippon Asahan Aluminum	68.325 billion yen (75%)
Total equity	91.1 billion yen ($700 million, 1988 U.S. dollars)

Loans	
G.O.I. (Indonesian Government loan)	31.99 billion yen
OECF (Overseas Economic Cooperation Fund)	61.55 billion yen
EXIM joint loan (Bank of Japan and Japanese Commercial banks)	226.36 billion yen
Total loans	319.9 billion yen ($2.46 billion, 1988 U.S. dollars)

A final agreement was signed in July 1975 in Tokyo and the joint venture combine known as Nippon Asahan Aluminum was formed to act as an investment company. In January of the following year, P.T. Indonesia Asahan Aluminum, a venture formed between Nippon Asahan and the government of Indonesia was formalized, and construction began in mid-1976. Japanese contractors and equipment suppliers participated in various aspects of the project using local labor and local contractors wherever possible.

Aluminum ingot production began in 1982, and the entire project was completed in 1984. When exports of aluminum, primarily to Japan began, the faltering Japanese aluminum industry was bolstered and returned almost $300 million to Indonesia. Ports, roads and infrastructure continued to be added as the effects of this huge complex ripple through the area. Japanese contractors will undoubtedly find that further participation in Indonesian government projects will be welcomed with open arms.

THE OVERSEAS CONSTRUCTION ASSOCIATION OF JAPAN

Working along with contractors and governments of various countries, an organization known as OCAJI, The Overseas Construction Association of Japan, Inc., is a very effective marketing tool for Japanese contractors. OCAJI was formed in 1955 to promote cooperation among Japanese contractors interested in participating in construction projects outside of the country. OCAJI's mission included fostering goodwill and economic cooperation between member contractors and the foreign communities in which they would be operating. OCAJI has a membership list that reads like the Who's Who of Japanese contractors:

Ando Construction Company
Aoki Construction Corporation
Asanuma Gumi Co., Ltd.
Chizaki Kogyo Co., Ltd.
Daiho Construction Co., Ltd.
Dai Nippon Construction
Daito Kogyo Co., Ltd.
Fudo Construction Co., Ltd.
Fujita Corporation
Hazami-Gumi Ltd.
Ikeda Construction Co., Ltd.
Itogumi Construction Co., Ltd.
JDC Corporation
Kajima Corporation
Kandenko Co., Ltd.
Kinki Electrical Construction Co., Inc.
Kitano Construction Corp.
Konoike Construction Co., Ltd.
Kubota Construction Co., Ltd.
Kumagai Gumi Co., Ltd.
Maeda Construction Co., Ltd.
Matsumura-Gumi Corporation
Matsuo Construction Co., Ltd.
Mitsubishi Construction Co., Ltd.
Mitsui Construction Co., Ltd.
Mitsui Harbour and Urban Construction Co., Ltd.
Morimoto Gumi Construction Company
Nakanogumi Corporation
Nippon Hodo Co., Ltd.
The Nippon Road Co., Ltd.

Nishimatsu Construction Co., Ltd.
Nissan Construction Co., Ltd.
Ohbayashi Corporation
Ohmoto Gumi Co., Ltd.
Okumura Corporation
Penta-Ocean Construction Co., Ltd.
P.S. Concrete Co., Ltd.
Rinkai Construction Co., Ltd.
Saeki Kensetsu Kogyo Co., Ltd.
Sato Kogy Co., Ltd.
Shimizu Construction Co., Ltd.
Shiraishi Corporation
Sumitomo Construction Co., Ltd.
Taisei Corporation
Takenaka Komuten Co., Ltd.
Tekken Construction Co., Ltd.
Toa Harbor Works Co., Ltd.
Toda Construction Co., Ltd.
Tobishima Corporation
Tokai Kogyo Co., Ltd.
Toko Electrical Construction Co., Ltd.
Tokura Construction Co., Ltd.
Tokyo Construction Co., Ltd.
Toyo Construction Co., Ltd.
Wakachiku Construction Co., Ltd.
The Zenitaka Corporation

OCAJI states its aims and objections clearly:

1. Maintain close governmental relations on behalf of its members and present proposals that reflect the common interest of the entire construction industry with regards to overseas construction work.
2. Promote technical cooperation with foreign counterparts through activities of international federations or similar organizations.
3. Study market potential of overseas construction projects and collect and disseminate information to assist members (Table 4-1).
4. Organize overseas construction survey missions or study teams from among member contractors interested in fact finding or feasibility studies on prospective developing projects.
5. Operate a data bank to provide background information and data when requested by fellow members.
6. Provide educational opportunities for member contractors on overseas construction projects by holding seminars and meetings.

Table 4-1

THE OVERSEAS CONSTRUCTION ASSOCIATION OF JAPAN, INC.

OVERSEAS CONSTRUCTION CONTRACTS AWARDED IN THE LAST FIVE YEARS BY REGIONS(1982-1986)

(UNIT:¥ 1 million)

YEAR	Region	Corporation registered under Japanese Law			Japanese Corporation in J.V. with Local Partner			TOTAL		
		No.	Amount	%	No.	Amount	%	No.	Amount	%
1982	S.W. & S.E. ASIA	250	600,470	77.90	140	95,957	63.60	390	696,427	75.60
	MIDDLE EAST	73	97,251	12.60	7	21,073	14.00	80	118,324	12.90
	AFRICA	16	9,594	1.20	0	0	0	16	9,594	1.00
	NORTH AMERICA	9	13,919	1.80	47	19,347	12.80	56	33,266	3.60
	C & S.AMERICA	11	27,726	3.60	42	5,733	3.80	53	33,459	3.60
	WESTERN EUROPE	4	25	0.00	13	8,203	5.40	17	8,228	0.90
	EASTERN EUROPE	1	13,550	1.80	0	0	0	1	13,550	1.50
	OCEANIA & OTHERS	6	8,127	1.10	5	581	0.40	11	8,708	0.90
	TOTAL	370	770,662	100.00	254	150,894	100.00	624	921,556	100.00
1983	S.W. & S.E. ASIA	260	583,409	68.80	195	92,311	55.80	455	675,720	66.60
	MIDDLE EAST	43	149,532	17.60	5	13,739	8.30	48	163,271	16.10
	AFRICA	8	10,683	1.30	0	0	0	8	10,683	1.10
	NORTH AMERICA	14	41,744	4.90	65	41,506	25.10	79	83,250	8.20
	C & S.AMERICA	17	9,790	1.20	57	4,867	2.90	74	14,657	1.50
	WESTERN EUROPE	3	837	0.10	8	6,394	3.90	11	7,231	0.70
	EASTERN EUROPE	0	0	0	0	0	0	0	0	0
	OCEANIA & OTHERS	14	52,425	6.20	4	6,763	4.10	18	59,188	5.80
	TOTAL	359	848,420	100.00	334	165,580	100.00	693	1,014,000	100.00
1984	S.W. & S.E. ASIA	315	436,015	59.20	120	65,715	33.20	435	501,730	53.70
	MIDDLE EAST	42	70,415	9.50	3	5,323	2.70	45	75,738	8.10
	AFRICA	5	3,378	0.50	2	24	0.00	7	3,402	0.40
	NORTH AMERICA	30	95,931	13.00	161	111,191	56.10	191	207,122	22.10
	C & S.AMERICA	8	16,389	2.20	81	6,032	3.00	89	22,421	2.40
	WESTERN EUROPE	3	77	0.00	11	9,463	4.80	14	9,540	1.00
	EASTERN EUROPE	0	0	0	0	0	0	0	0	0
	OCEANIA & OTHERS	32	114,754	15.60	3	320	0.20	35	115,074	12.30
	TOTAL	435	736,959	100.00	381	198,068	100.00	816	935,027	100.00
1985	S.W. & S.E. ASIA	270	329,721	38.40	162	66,454	39.00	432	396,175	38.50
	MIDDLE EAST	56	54,619	6.40	1	656	0.40	57	55,275	5.40
	AFRICA	12	19,980	2.30	1	2,795	1.60	13	22,775	2.20
	NORTH AMERICA	50	209,842	24.40	67	58,696	34.40	117	268,538	26.10
	C & S.AMERICA	16	21,838	2.50	54	29,760	17.50	70	51,598	5.00
	WESTERN EUROPE	7	7,926	0.90	25	8,878	5.20	32	16,804	1.60
	EASTERN EUROPE	1	16,370	1.90	0	0	0	1	16,370	1.60
	OCEANIA & OTHERS	34	198,921	23.20	15	3,192	1.90	49	202,113	19.60
	TOTAL	446	859,217	100.00	325	170,431	100.00	771	1,029,648	100.00
1986	S.W. & S.E. ASIA	287	295,718	44.70	220	61,216	30.50	507	356,934	41.40
	MIDDLE EAST	41	34,589	5.20	1	1,016	0.50	42	35,605	4.10
	AFRICA	14	8,152	1.20	0	0	0.00	14	8,152	1.00
	NORTH AMERICA	84	260,261	39.40	94	96,075	47.80	178	356,336	41.30
	C & S.AMERICA	16	5,712	0.90	57	19,210	9.60	73	24,922	2.90
	WESTERN EUROPE	16	33,051	5.00	54	22,817	11.40	70	55,868	6.50
	EASTERN EUROPE	0	0	0.00	1	15	0.00	1	15	0.00
	OCEANIA & OTHERS	21	23,573	3.60	4	480	0.20	25	24,053	2.80
	TOTAL	479	661,056	100.00	431	200,829	100.00	910	861,885	100.00

OCAJI is very active in promoting conferences that bring Japanese contractors together with builders from other parts of the world. This free and open get together might just be one of OCAJI's more important contributions to overseas marketing. The first Japan-Turkey Construction Industry Conference was held at Tokyo's New Otani Hotel in 1987 and its sole purpose was to foster closer ties between the Turkish construction industry, which has grown rapidly in the Middle East, and the Japanese construction industry. In meetings like this Japanese builders can keep their hand on the pulse of middle east construction activity. Twenty seven members of OCAJI mingled with 25 Turkish builders and Mr. Nurettin Kocak, President of the Association of Turkish Contractors, agreed to actively exchange information about building activity in Turkey and surrounding third world nations.

The Tenth Anniversary General Assembly of the Japan-Korea Construction Cooperation Council was held in Yokohama in November 1987 and was attended by 35 Korean delegates and 50 Japanese delegates, during which time the subject of international cooperation and technical relationships were discussed.

OCAJI attended the International Contracting Symposium held at the Kunlun Hotel in Beijing, China in October 1987 at which Mr. Yasuo Satomi, president of Taisei Corporation, delivered a key speech.

Japanese contractors traveled to Europe to attend the 2nd General Meeting of the Confederation of International Contractors held in West Berlin in May 1987. Mr. Hirohide Nakamura, Executive Director of OCAJI and Mr. Sadao Asahi, Manager of International Activities are two of the driving forces behind the organization. During an interview at their downtown Tokyo offices in the No. 3 Yaesu building, they responded to the following questions:

Q: Where do you think your prime markets will be in the 21st Century?

A: As far as the prime markets for the new century are concerned, no significant changes will take place. It will depend upon what becomes of the marketing potential in the Middle East after the ending of the Gulf War. There is a high probability to believe that the Middle East Market would appear again as a booming one as it was once before.

Q: Do you anticipate that China and the USSR will be potentially large markets?

A: Yes, we do anticipate both countries have high potentiality as large markets. However, accessibility or profitability involved is still in question. In the China mainland, for instance, they have their own large human resources mobilizable for construction and all foreign contractors may face more and more intensified competition if the market becomes known as an attractive one.

Q: What do you think about Japanese construction company participation in Central and South American markets and in Europe?

A: Considering geographical and other factors involved in Central and South America, we do not foresee any large expansion of the marketing share which still remains at a relatively small percentage of 2 to 5. In Europe, possibility for participation is varied by countries depending upon each national policy, protectionism or regulation. We are expecting some encouraging results from reducing regulations within the EC territory.

Q: How do Japanese contractors conduct business in other foreign countries? Is most of their work design-build or is it competitive bid work?
A: Combined or mixed by both types. A new pattern in the current years is featured by BOT or BOTT, which means "Build," "Own," "Operate," "Training," and "Transfer." The contractor doing the job in this manner is no longer substantially the "contractor." He stands on behalf of the client as an entrepreneur.

Q: Is the ability to provide construction and long-term financing a key ingredient?
A: Yes, financing at a relatively low interest rate is the decisive factor to award the contract to any contractor both technically and financially capable.

Q: Is the government helpful in obtaining work in foreign countries?
A: No specific direct support or aid to the construction industry by the government. [*Author's note:* The answer might have hinged upon semantics. Direct government support is not prevalent, but *indirect* support certainly is.]

Q: How do you market your services in other countries? Through branch offices in the country or directly from Japan?
A: Marketing activities are carried out by joint effort of local representatives in the countries and head office personnel involved.

OCAJI has been obtaining government subsidies in order to finance fact finding teams that are sent to all parts of Asia, the Middle East, Africa, and South America to keep abreast of international construction activities.

In the last thirty years, OCAJI has conducted in-depth studies of various development projects and has performed market research in more than 60 countries. OCAJI maintains overseas offices in Burma, the Philippines, Thailand, Egypt, and most recently in London, England.

Is this a portent of what is to come?

THE JAPANESE AND THE EUROPEAN MARKET

In 1992 the European Community will have been formed, and all tariffs and trade impediments will be dropped between the twelve nations that make up the EC. West Germany, France, Spain, Portugal, The United Kingdom, The Netherlands, Belgium, Denmark, Luxembourg, Ireland, Italy, and Greece will essentially become one large marketplace.

As far as construction is concerned, beginning with 1992, standardization of building codes, materials specifications, and construction means and methods will be phased in to create a barrier-free construction market of substantial size. In 1987 the total value of all construction contracts in these twelve countries was just about $380 billion, and this does not include the $10 billion English Channel Tunnel construction which was begun on December 1, 1987. The service portions of this tunnel, which were started from both the French side and the British side, are scheduled to meet in 1990. When this channel tunnel is fully complete and operational it will link Folkestone, England with Calais, France via three parallel tubes, each 52 kilometers long. One tube will act as a service tunnel and the remaining two tubes will provide train service between Great Britain and the Continent. The Eurotunnel trains will shuttle cars, trucks, and other commercial vehicles back and forth as well as carrying conventional freight and passengers.

Along with the Kawasaki Heavy Industries–Robbins Company–Markham joint venture effort on this tunnel job, another well known Japanese equipment manufacturer, Komatsu, is providing a special piece of equipment for the project. Komatsu will be installing a seal on the tunneling machine to prevent water outside of the machine from entering the shield. Komatsu will install a series of narrow steel brushes on the rear edge of the shield acting as a gasket in much the same manner that a brush-type door sweep keeps cold air from entering under an exterior door.

WAS THE U.S. MARKET SIMPLY A LABORATORY?

Several Japanese construction companies set up shop in the U.S. in the early 1970s because their long time Japanese manufacturing clients were beginning to build factories and distribution plants in America. The Japanese contractors came to the U.S. to act as consultants and construction managers to watch out for their client's interests in a foreign country. While waiting to be called upon, Japanese contractors would have ample time to study the American construction scene, possibly enter into a joint venture or two, or simply make an investment in real estate for future use. All the while the Japanese builders would be observing, taking notes, and filing them away. If this was a well conceived, long term approach to obtaining recognition prior to market share so that future participation in construction projects could be made easier, then the European market ought to brace itself for further Japanese construction activity.

JAPANESE CONTRACTORS ON THE EUROPEAN SCENE

Shimizu Construction Company has a London branch office, Shimizu UK Ltd., and also owns a company called SC Properties in England. Shimizu Europe BV has its headquarters in The Netherlands and branch offices in Düsseldorf, Germany and Madrid, Spain. The Düsseldorf office was opened in 1974 and Shimizu brought new cooling tower technology to West Germany in 1975. It also worked with West Germany's Polensky and Zollner to introduce prestressed concrete bridge spanning technology to the area. Taisei opened Taisei Europe Ltd. in London and introduced its shield tunneling engineering system in Belgium back in 1980.

Kumagai Gumi has an office at 33 Rue Galilee in Paris and Kumagai Gumi U.K. Ltd. operates out of 8 St. James Square in the heart of London. Kumagai Gumi has been very active on the London construction scene, converting a building on the Thames, with traditional English architecture, into a 200 room luxury hotel, the Millbank. They have rehabilitated a building containing 6 floors and 376,000 square feet known as Saint Martins for their client Nomura Properties. Kumagai Gumi purchased Distillers House, a distinguished 1774 multi-storied building in the West End of London and set about to completely renovate the interior of this 75,000 square foot structure into a state-of-the-art intelligent building while maintaining the integrity of the building's historic facade. Kumagai Gumi recently acquired a site in the prestigious section of London known as Bishopsgate, and plans to develop an office complex there. Kajima has a London office located at 40 George Street, and Kajima GmbH is headquartered in Frankfurt, West Germany. Kajima has had a presence in Germany dating back to 1981 when it built the Hotel Merkur in Leipzig. In 1984, Kajima constructed the Hotel Bellevue in Dresden, Germany and followed up with the construction of the Grand Hotel in Berlin in 1987. That same year, Kajima opened Kajima Europe BV with headquarters in Amsterdam.

Japanese construction volume in Europe in 1986 was only $600 million, but by 1987 it had risen to $1.6 billion. The pattern of construction activity seems to parallel the investments by Japanese corporations, which rose from $500 million in 1981 to $1.75 billion in 1985. As Japanese investments in the EC rise, if the Japanese construction company's policy of Look-Observe-Note-Learn-Act continues to be operative, then further involvement in local construction activity is certain to follow.

JAPANESE INVESTMENTS IN EC COUNTRIES

Japanese investments in the twelve European Community countries accelerated during the 1980s, probably for a number of reasons. Japan would like to reduce or prevent the "bashing" they suffered in the U.S. and would hope to contribute to the economies of their European trading partners by producing some of their products locally. In this way some otherwise unfavorable trade balances might be reduced or eliminated and healthier economic ties between Japan and the EC might be created. The increased value of the yen has made foreign expansion more attractive to the Japanese, and because of the distances involved, a strong marketing effort could be enhanced by having parts manufacturing and assembly plants closer to the marketplace.

Investments in the United Kingdom

Nissan Motor Company began producing passenger cars in the plant they built in 1984 and Honda's participation in Rover's new Sterling Automobile has been established. Yamazaki Mazak Corporation, a manufacturer of lathes and machining centers, opened a new plant several years ago, and Matsushuta, Hitachi, and Sharp have all built plants in England to produce VCRs and CDs. Ricoh and Brother are making facsimile machines and copiers in England and Sony has substantially increased its investment in Great Britain, doubling its production of color TVs. Sumitomo Rubber Company has taken a more direct approach: it purchased world famous Dunlop Tire, insuring the 2,900 employees working in the factory continued employment.

Investments in West Germany

In 1985, Japanese investments in West Germany accounted for 16% of all foreign investment in that country. Forty-five Japanese firms have opened manufacturing plants in West Germany, including Sony, Victor, Toshiba, Hitachi, Sanyo, and Matsushita. Canon has been making copiers in West Germany since 1972 and was recently joined by Konishiroku and Minolta. Tokyo Electric is producing electronic typewriters in that country and TDK is manufacturing audio and video tapes in Germany.

Investments in France

As of 1986, thirty-three Japanese corporations were operating in France. The machine tool manufacturer Toyoda Machine Works has joint-ventured with the French company Ernault, and Mori Seiki and Mitsui Seika have launched their own machine tool production facilities in

France in 1987. Sony and Akai produce VCRs in France and Pioneer Electronics and Aiwa make CD players in their own French factories. Toshiba produces copiers and microwave ovens in France and Canon turns out printers and electronic typewriters while Matsushita manufactures high-fidelity tuners in local French plants.

Investments in Spain

Japan has been drawn to Spain because that country has recently liberalized foreign investment restrictions and Nissan Motor Company's direct participation in Motor Iberica has been duplicated by Suzuki Motor Company's investment in Avello, and Honda Motor Company's investment in a motorcycle production plant. Kubota is joint venturing with Motor Iberica and Marubeni to produce farm tractors in Spain. The electronics firms of Fujitsu, Sharp, Sanyo, Matsushita, and Sony all have production facilities in Spain.

Investments in Italy

Yoshida Industries produces fasteners in Italy. Honda makes motorcycles there and Nissan produces motor cars in that country. Japan's Torsy Industries makes artificial leather in Italy. In 1986, Seiko and Italy's Tecnocompy formed TECD-IS, a new corporation to produce liquid crystal panels, and Hitachi Construction Machine Company and Fiat-Allis, the largest construction equipment maker in Europe, teamed up to produce hydraulic excavators.

Investments in Belgium

Eighteen Japanese manufacturers maintain operations in Belgium. Honda produces automobile instrument panels for the Austin Rover Group in Great Britain, and Honda Benelux makes mopeds in Belgium. Toyo Sodo Company purchased an equity position in Eurogenetics Co., a manufacturer of medical equipment in Belgium.

Investments in Ireland

Nippon Electric Company and Fujitsu operate plants in Ireland which produce integrated circuits. Yamanouchi Pharmaceuticals and Toshiba and Sord, both business computer manufacturers, have plants in this country. Ireland's efforts to attract advanced technologies firms have netted them 10 Japanese manufacturers.

Investments in Portugal

Six manufacturers from Japan opened operations in Portugal, and in 1986, Yazaki Corporation formed a joint-venture known as Yazaki-Saltano to produce wire harnesses. A new plant was built for this venture and eventual employment is projected at 2,000 workers.

Investments in Denmark

At the end of 1986 there were only two Japanese manufacturers operating in Denmark, but the availability of skilled, highly productive labor makes this country a very attractive place for the Japanese to expand their operations.

Investments in Greece

A Japanese steelmaker by the unlikely name of Hellenic Steel is producing cold-rolled sheet steel in Greece. Hitachi Karmel makes color television sets and car radios in Greece, and Yoshida Industries manufactures sliding fasteners in that country.

THE BATTLE OF BRITAIN

The next Battle of Britain will not be fought in the skies but will be waged on the ground, possibly by American, European, and Japanese contractors. The construction market in England and particularly London opened when Great Britain deregulated its stock market and foreign investment firms flocked to London, spurring an unprecedented demand for office space. In the two-year period 1987–1988, it was estimated that at least 10.8 million square feet of new office space would be constructed in London. Usually such a major influx of new office construction creates secondary markets for everything from coffee and donut shops, or rather tea and crumpet stores, to office supply companies and electronic data processing hardware and software firms, and this delayed flurry of other construction needs should create even more work for local builders.

American architects, engineers and contractors are following on the heels of Shearson Lehman, Goldman Sachs, Merrill Lynch, and Salomon Brothers to garner multi-million dollar contracts for projects such as the $920 million Broadgate Development complex, the seventeen building Chelsea Harbor project, and the massive $1.3 billion Canary Wharf job. The presence of U.S. designers, engineers, and builders has elicited very favorable comments from the usually reserved British. English developers find that Americans have a knack for producing efficient and cost effective buildings. U.S. designers are impressing the British with their speed and quality of work; U.S. builders are impressing the British with their ability to produce buildings quickly and economically.

At some point the Americans will square off with the Japanese as competition heats up for many of these lucrative contracts. This might be the first neutral testing ground for both American and Japanese contractors, and the results will provide fodder for future discussions and analysis. The competition will be formidable because the Japanese contractors have been working on developing unique marketing strategies for many years. At least two companies have created sound approaches to increase their sales volumes. These will be discussed next.

KUMAGAI GUMI AND THE FINANCIAL APPROACH

Kumagai Gumi has developed a market strategy that seems to be paying off in all hemispheres. They are bringing substantial sums of money and new financing sources into the projects they want to pursue. In New York City, they teamed up with William Zeckendorf, Jr. and invested $1.2 billion through their KC Land Corporation in Manhattan real estate projects. As of 1988, Kumagai Gumi was the largest Japanese real estate investor in that city. On the West Coast, this Japanese contractor has joined with another developer in Seattle, Washington to create a $225 million office complex called Gateway Tower.

Halfway around the world in Australia, Kumagai Gumi has invested approximately $750 million in projects with a total value in excess of $2.25 billion, and has plans to double that investment. These Australian projects unveil a planned method of operating, and reveal the gambling nature of this huge Japanese construction company. Entering into a competition to build the Sydney Harbour Tunnel, Kumagai Gumi formed a consortium with an Australian developer and conducted a feasibility study to prove that the tunnel would make economic sense. This study lent a great deal of credibility to Kumagai Gumi's proposal, which was finally accepted by the government of New South Wales. The $260 million project, currently underway, was awarded to Kumagai Gumi without competitive bidding. In other Australian ventures, Kumagai Gumi looks for high-risk projects that are not popular with local builders. High risk projects are usually frowned upon by local bankers, so Kumagai can arrange financial packaging by calling in its favorite Japanese banker, Sumitomo.

Kumagai Gumi may be using the Australian market to develop, test, and analyze a new market strategy—to assume the role of developer on complex projects involving the three things the company does best: provide extensive civil engineering expertise; provide extensive tunneling experience; bring financing sources into a construction program.

Armed with these three potent weapons, plus a few more, Kumagai Gumi has rapidly advanced in its quest to capture a significant portion of the market Down Under. The financing Kumagai Gumi has brought to its Australian projects has, in itself, been an innovative approach. The company will sign a loan agreement as guarantor, and this is backed up by a letter of credit from a major Japanese bank. With Kumagai's guarantee, and a bank such as Sanwa backing the letter of credit, it is relatively easy for the joint venture company in which Kumagai Gumi has participation to obtain local financing. Once everything is put into place, Kumagai Gumi, if it is interested in obtaining a construction contract or a Construction Manager contract is certainly in an advantageous position to do so, as it has done elsewhere, including New York City. This formula is a blend of superb market strategy. First, the Japanese builder-developer decides to become involved in a high-risk project that is not looked upon favorably by local developers or builders. Therefore Kumagai Gumi cannot be accused of coming into a local market area and stealing lucrative contracts from established companies. Second, in order to "spread the wealth," it enters into joint venture agreements with local participants who would not have been able to initiate the deal on their own. Third, with guarantees and letters of credit from major banks in Japan, the risk venture is secure enough that local banks will willingly loan funds for construction.

Kumagai Gumi is probably picking up substantial fees at every bend in the road, and if the project succeeds, local builders, developers, suppliers, and banks are happy. If the project fails, Kumagai Gumi will have to make good on some portion of its guarantee, but with all various financing, finder's, developer, and construction services fees it has garnered to that date, the effects of a failure are not so earth shattering, if, in fact any money is even lost. Kumagai Gumi is honing this strategy to a razor sharp edge in Australia and is probably teaching the real estate magnates in that area a thing or two before it moves on to more fertile feeding grounds, possibly the lucrative European Community Market of the 1990s.

Although Kumagai Gumi might have cornered the market on creative financing techniques to increase its yearly volume, there are five other Japanese construction companies of slightly smaller size but with financing sources no less impressive than Kumagai Gumi's that will not let one of their associates steal the show.

KAJIMA CORPORATION AND ITS SPECIAL MARKETING TOOL

Kajima Corporation has developed a unique marketing tool which will allow them to offer turnkey manufacturing facilities to major manufacturers around the world. Their brand new Process Mechanical Division has been providing design, project management, and installation services to industrial clients in several parts of the United States. When a new manufacturing plant is being designed and built there is the monumental task of coordinating the construction of the plant with the manufacturing equipment that the owner will be purchasing and installing in this new facility.

From such simple problems as providing properly sized reinforced roof openings for owner's equipment that will penetrate the roof, to the problem of insuring that properly sized utilities will be installed in close proximity to the machines that need them, coordination problems can be monumental—and costly. On large projects hundreds of thousands of dollars can be wasted if the placement of manufacturing machinery is not properly coordinated as the construction of the plant progresses. When construction is complete and equipment is being installed and brought on line, delays can be equally expensive as startups are postponed and newly hired workers stand around waiting for equipment to begin functioning.

Enter the Process Mechanical Division of Kajima Corporation. This new Division had its baptism under fire when Kajima recently built the 2.7 million square foot manufacturing plant for Mazda Motor Manufacturing (U.S.A.) Corporation in Flat Rock, Michigan. The whole construction process began when Kajima had to move 1 million cubic yards of earth during the excavating phase of the project, which commenced in 1985. Next came the concrete subcontractor's crews, who placed about 150,000 cubic yards of concrete so that the 18,000 tons of structural steel framing could be erected.

As construction of the base building progressed, engineers from the Process Mechanical Division met in a series of meetings with owner's representatives, equipment manufacturers, and equipment vendors. Some of the equipment being purchased was manufactured in the United States while some was being manufactured in Japan. This situation required coordination between Japanese specifications and requirements and American specifications and requirements, and it also created a need for translation from Japanese to English and vice-versa.

The information obtained by the Process Mechanical Division was passed along to Kajima's Design and Construction Division to insure that proper roof openings were created as the roof steel was being erected and that roof curbs were coordinated with roofing installations. Close attention was paid to items to be embedded in concrete slabs and pits so that the manufacturing equipment could be accurately placed and secured when it arrived at the plant. It was a monumental task to insure that the correct size electrical conduits and feeders were being brought to the area where the companion piece of machinery would ultimately be placed. As production equipment began arriving at the job site, the Process Mechanical Division made arrangements for installation and actually installed more than 68,000 tons of this equipment in the factory. On September 1, 1987, the Mazda plant was completed, and according to Japanese tradition, it was "On Time and Within Budget."

Kajima has since become involved in two other automobile manufacturing projects, one is the Bloomington-Normal Illinois plant for Chrysler and Mitsubishi's Diamond Star Motors Joint Venture operation. The other project is in Lafayette, Indiana, where Kajima is constructing a 2.25 million square foot plant for Subaru-Isuzu Automotive Inc. These three automobile factories have caught the eye of General Motors and Ford as well as other non-automobile manufacturing U.S. industries, who see the economies and efficiencies in working with a contractor such as Kajima who has the responsibility and resources to carry out this complicated assignment.

HOW LONG TERM CLIENT RELATIONSHIPS BECOME MARKETING TOOLS

Obviously a long term relationship between builder and client connotes an absence of a competitor, but when another contractor knocks on the owner's door, the information gleaned from prior projects becomes a valuable asset when there is a need to fend off the competition.

Mr. Yukihiko Kimura, Corporate Secretary and Assistant Treasurer of Toda America, Inc., interviewed in his New York office, said that his company enjoys long term relationships with a number of Japanese industrial corporations and because client and contractor have been doing business together for years, negotiated work presents many advantages for both parties. Because the contractor has constructed similar buildings for his client, he can assemble budget estimates for new projects more rapidly and with much more accuracy, since he is so familiar with the owner's space requirements, levels of finishes, environmental requirements, electrical loads, and heating and cooling preferences. The contractor is also familiar with the client's priorities, so that if budget estimates are approaching the owner's construction budget, reductions can be made in those areas that are not considered critical.

Mr. Kimura said that Toda and other major general contractors in Japan maintain staffs of engineers and architects who can prepare schematic drawings for new projects very quickly primarily because of their knowledge of client needs and priorities. Since design criteria and budget estimates can be assembled in the quickest possible time, the entire construction cycle—design schematics, preliminary budget proposal, final design and pricing, contract preparation, and construction start—can be substantially shortened.

The ability to provide a rapid and comprehensive program that meets all of the client's requirements when it is first presented will place any potential competitor at a distinct disadvantage.

According to Mr. Kimura, Japanese clients are notorious for their demands for on-time completion of their office buildings or factories. In many cases, an industrial client will plan to have production lines operational within 30–40 days after the building is substantially complete, and will accept nothing less than strict adherence to schedule. Very fast track construction can be implemented when a general contractor is so familiar with his client's requirements that drawing preparation and on-site construction work can begin in short order.

Long term relationships aside, this "inside information" is what makes it very difficult for another construction firm to take business away from the previous builder.

Marketing requires innovation, skill, insight into client needs, perseverance, and probably a little luck. The Japanese construction companies seem to have marshalled many of these forces in a relatively short period of time, and although they didn't invent the wheel, they certainly have learned how to make it spin.

5

Estimating, Bidding, Cost and Quality Control

In a country where a great deal of construction work is negotiated, and where emphasis is placed upon quality and on-time project completion, accurate construction cost data continues to be of vital importance to Japanese contractors.

The need to build a complete and concise database for estimating purposes not only satisfies the Japanese penchant for precision and knowledge, but permits them to prepare competitive bids when required domestically, in other parts of Southeast Asia, or in any other part of the world in which they operate. The method by which construction estimates are prepared in Japan is very similar to the system employed in the United States. In fact, several Japanese contractors freely admit having patterned their estimating departments on those of American contractors.

Estimating data in Japan originates from basically the same sources as it does in the U.S.:

- From field generated cost reports.
- From subcontractor and vendor cost information.
- From previously completed projects.
- From estimating manuals published by private and government organizations.

Since most large general contractors in Japan perform little or no work with their own forces, daily reports submitted by their subcontractors provide most of the jobsite-generated cost data.

Because of the close relationship between general contractor and subcontractor, estimating data and current unit costs flow freely back and forth between the two, and in many cases the general contractor will actually prepare the cost estimate for his subcontractor, and once it has been reviewed and modified, or refined by the subcontractor, this estimate will form the basis for their contract. When subcontractor estimates and bid proposals are reviewed by the general contractor, any additional or current cost information developed will find its way back to the general contractor's database, where the information is then augmented by costs developed by the contractor's purchasing department.

THE ESTIMATING PROCESS

Many Japanese contractors have computerized estimating equipment including such accessories as light pencils and digitizers similar to those in use in the United States. However, one of Japan's largest general contractors was observed using ordinary desktop calculators to estimate projects the old fashioned way. Staff members could be seen poring over drawings, checking off materials such as doors, door frames, windows, and louvers, using colored pencils for each item. The totals were entered on tally sheets, which were passed on to other department members sitting behind rows of plain gray, linoleum topped desks, busily writing in unit prices and extending both quantities and prices to complete the various portions of the estimate. When asked if a more sophisticated estimating system was being considered in this construction office, the answer was, "We do have several digitizers on order, but we mainly perform take-off and pricing functions manually."

The approach to the estimating process is just another one of those typical Japanese anomalies; one major contractor still uses the old, laborious, and time consuming method of estimate preparation, while another contractor will go to incredible lengths to automate its estimates.

A CASE STUDY OF AN EXCAVATION ESTIMATE

A general contractor described how he had prepared an estimate for a recent bid involving the construction of a small dam for a public agency, so that he could illustrate how thoroughly his company assembles cost data.

When the invitation to bid was received at the company headquarters, it was decided that a branch office nearer to the actual construction site would be able to accumulate much of the required cost information more accurately than the home office, since a great deal of site investigation and local cost data would be needed.

Substantial quantities of earth would have to be moved and this might require searching out potential cut or fill sites near the proposed construction area. There appeared to be considerable amounts of rock that would have to be blasted and possibly rock crushing equipment would be needed at the site and a nearby dump area located if it became necessary to dispose of any rock off site.

It was planned that the completed estimate would be a combined effort of local supplier and subcontractor input plus database information, guidance, and assistance from the home office.

The question of rock removal was addressed first and the branch office was requested to obtain samples of the rock that would eventually have to be blasted and send these samples to the company's research technology institute for analysis and evaluation. After their engineers analyzed the rock samples to determine composition and hardness, a profile of the area thought to contain rock was prepared at the branch office and the various sections of existing rock were also sent along to the institute. The company engineers calculated and plotted the number, size, and depth of holes to be drilled in the rock formation in order to maximize the removal time and minimize the cost of drilling and blasting. Based upon their evaluation of the rock samples, the engineers at the company's research facility determined the type and size of explosive charge required, and all of this information was forwarded to the central estimating department for pricing.

With the data in hand, the branch staff could now discuss the drilling and blasting operations with local subcontractors so that the most competitive price could be obtained. They began to review each blasting subcontractor's bid proposal. How many holes are you planning to drill? 350? We feel that 325 will be sufficient, please explain why you figured to drill 25 more. How long do you think the drilling operation will take? Based upon your hourly rate for a drilling rig and an operator what have you estimated your drilling costs to be? What type and size explosive charges are you planning to use? Our engineers feel that most of the rock is metamorphic with a modulus of rupture of 1300 lbsf/in^2 and your charges do not have to be that large—do you agree? And so the questioning went.

Each subcontractor's proposal was put under a microscope in order to confirm that they had the correct scope or they needed to revise their bid based upon discussions with the general contractor. This was standard operating procedure for all subcontractors submitting bids for this project.

Meanwhile, back at company headquarters, the estimating department prepared a complete list of excavating equipment needed on the site during and after blasting. The estimators analyzed the types of equipment best suited to the job; bucket size and earth moving capacities were considered. Depending upon the type and size of the equipment selected, a calculation was made of the number of cubic yards of earth that would probably be moved per hour and per day. These calculations formed the basis for the estimated cost and also generated time of completion and therefore scheduling information as well. After an equipment list was compiled, rental costs of the equipment were determined, but not by leafing through the green book. The estimating department determined how much they were willing to pay per hour for many pieces of equipment based upon their knowledge of the capital cost of the equipment, the estimated operating costs per hour including fuel and lubrication costs, and even including a factor for depreciation. The productivity figures developed by the estimating department and the hourly costs for the equipment compiled by the estimators were discussed in the same detailed manner as the rock drilling and blasting operations, as subcontractors were interviewed.

What kind of earth moving equipment do you plan to use? Wouldn't a Kobelco 9452 with a six yard bucket be more economical? How many yards of earth do you think you can move per hour? Information was traded back and forth as general contractor and subcontractor went through almost every phase of the excavation operation. When these sessions were over, both parties had formed a much better game plan for the performance of the work, and their budgets and estimates were revised accordingly. The general contractor's role during all of these sessions was one of teacher and advisor, and costs were lowered not by arbitrary methods but by rethinking how various parts of the work could be performed in a more economical fashion.

If it had become necessary to set up a concrete batch plant on the construction site, the general contractor would have either transported one he already owned to the site, or leased one just for the project, utilizing local labor or subcontractors to operate the plant. The estimating and purchasing departments would already have solicited prices for sand, cement, and aggregate, and the capacity of the plant would be determined. Depending upon the construction schedule and the rated output of the batch plant, extended work days might be necessary, and therefore premium pay will have been included in the estimate.

Intricate concrete forms were needed, and the company's engineering department was ready to engineer a forming system using standard forms, and where necessary, to design special forms just for this project.

Vacations, holidays and weather days were factored into the equation after the project schedule was roughed out. In Japan, July is the rainy season and some parts of the country are subject to typhoons. Any schedule for outdoor work performed during certain parts of the summer must take these severe weather conditions into account. This construction project took place in a remote area of the country, so the general contractor had to "import" labor from towns and cities some distance from the work site. Dormitories were required to be constructed to house the workers, and the estimating department had to arrive at the cost to set up, maintain, and dismantle these temporary facilities. The subcontractors will be responsible for providing this temporary housing for their workers at the site, and the general contractor's duties included insuring that these facilities were maintained in accordance with the laws governing worker's dormitories.

Japanese estimates include provisions for general requirements or general conditions, but they are referred to as "Common Temporary Work" and "Site Expenses." Together these two categories form their general conditions and include items such as field supervisory salaries, telephone, temporary utilities, field office supplies, and so forth. The Ministry of Construction publishes a list of Standard Preliminary or Preparatory Work Expenses, and the items contained in this government publication form the basis for most general conditions cost estimates. The general conditions in this case included the cost of one of the engineers from the company's research institute, since it was determined that he would be needed during certain critical phases of the construction work.

All of the information developed at the site by the branch office was forwarded to the main office, where it was folded into the balance of the estimate before being sent on to management for final review, adjustment, and the addition of the overhead and profit percentage deemed appropriate for the work.

Japanese contractors generally allow seven to fourteen days in which to prepare an estimate for a medium sized project. More complicated projects require more time.

THE ROLE OF THE PROJECT MANAGER IN THE ESTIMATING PROCESS

It is not unusual for a project manager to become involved with a project as soon as the bid proposal is initiated. Based upon his knowledge of construction in general, and his analysis of the project being bid, his opinions and observations are sought out as the project progresses through the estimating department. If the project bid is successful and a construction contract is obtained, the project manager working with the estimating department can easily make the transition from bid to construction activity. Some companies entrust the project manager with the responsibility of seeing to it that at least three bids are obtained for each subcontracted portion of the bid work; however, if he feels that only one or two bids are necessary for certain portions of the work, he can proceed accordingly once he has obtained permission from higher management. There may be a time when a long term subcontractor is temporarily out of work, and the project manager may allow this company to participate in the project and not obtain any other bids for this particular phase of work.

Based upon the project manager's evaluation of the completed estimate, management will look to him to decide whether or not to add a contingency to the final proposal amount.

ESTIMATING DEPARTMENT PERSONNEL

Estimating departments in large contracting offices are generously staffed by American standards, and most estimators are engineers. In one $2 billion plus construction office there were a total of 49 people in the estimating department at the main office—39 engineers and 10 administrative staff. Most of these estimators had prior field experience, having come into the department from previous field positions, or having been assigned to project sites for extended periods of time as part of the company's training program. In this particular construction office there was a separate estimating department for the preparation of conceptual estimates, manned by four estimators.

The following question was posed to a Chief Engineer in the estimating department of a general contractor's office: "When you receive drawings that lack sufficient detail or essential information needed to properly estimate the cost, or when drawings contain obvious omissions, how do you estimate these items? Do you include costs to do what you know must be done, or do you only assign costs to work clearly shown on the drawings?" The Chief Engineer replied, "When there are large sums of money at risk, we look to management for direction." However, this Chief Engineer went on to say that if management decides not to include costs in the estimate for vague details or missing but essential items of work, and if they are the successful bidder, the company must do whatever is necessary or needed to properly build the project. Although the company may have been willing to gamble on including or excluding obvious costs in the estimate, once they have received a contract to construct the project, they cannot claim and will not claim extra costs to include work essential to the proper completion of the job. They are professionals and have a responsibility to assess the bid documents in a professional manner.

AN UNUSUAL ESTIMATING TOOL—THE TOWER CRANE

In Japan tower cranes are commonplace since confined construction sites are the norm, and rarely is there space on the project for mobile cranes working at grade level. These tower cranes are used as sources of estimating data as well as for their ability to lift and distribute materials.

Many of the tower cranes operating in and around Tokyo are guided and controlled by computers, not only for safety reasons, but also for the purpose of extracting and storing information that will later be used in the company's estimating data base. One Japanese contractor astutely observed that there is valuable information being produced each day on every construction site. He said, "Most builders are unaware of the amount and quality of estimating information that is being generated daily at the job site, and very few are taking steps to capture and evaluate the wealth of information that is within their grasp."

Safety First

When a single tower crane is being operated on a confined urban construction site, control over its area of operation must be closely monitored. The arc or swing of the boom must be controlled so that it will remain at a safe distance from adjacent property lines or structures. Boom travel on many of these cranes in Japan is controlled by computer. The configuration of the building to be constructed is programmed into a job-site computer and can be displayed in plan view on a CRT located in the cab of the crane, in the field office, or in both locations. The tower crane will be programmed to travel within designated areas of this building configuration, thereby relieving the operator of this responsibility without relegating the task to some type of mechanical stop or brake on the equipment.

The computer has the ability to display the building elevation on a CRT screen so that vertical travel of a material being lifted can be observed by the crane operator. This ability to know the position of the lifted member is very helpful in developing better communication between the operator and his ground crews. When overload conditions are detected by the computer as the sling or hook engages an item on the ground, hoisting ceases immediately and the crane operator is warned of this condition. These same audio and visual warnings on the CRT screen and in the crane cab warn the operator when he is approaching a point where the crane's arc restrictions are being reached and horizontal travel will stop.

When two tower cranes are in operation on a congested construction site, crane travel must be coordinated to avoid adjoining obstacles, including the other crane. Computers are used to predetermine the areas in which each crane could possibly encroach into the other's arc of travel. A sort of neutral zone will be created, and when this zone is being approached by one crane or the other, each operator will be alerted by alarms, and at a predesignated point of travel, the cranes will slow down and stop as they approach the neutral zone.

Estimating with a Tower Crane

As a crane goes about its task of picking, lifting, and placing materials, it is generating data that can easily be converted into cost information. During the erection phase of a structural steel framing system, for example, a general contractor can capture valuable information concerning the time it takes to erect specific steel members, or the time it takes to complete each floor in the building. All that it takes to obtain such data is to create a system or procedure that will allow the crane operator or an assistant engineer in the field office to record and store certain items of information as the lifting operations take place. Structural steel erection drawings usually indicate where each beam or column will be placed, and each structural member is numbered and coded accordingly. When the code number for each piece is keyed into a computer by the crane operator as the piece is being lifted, and some sort of time and distance measurement is also recorded, an accurate record of the crane's production at the end of each day can easily be ascertained. The computer will have stored data about the length of time it took to place a specific tonnage of steel, and this information can be used to determine erection costs and to create accurate scheduling data.

The Japanese have been using systems of this sort, and Fig. 5-1 is an actual computer printout from a two-crane operation used to erect a five-story steel framed building. The various parts of the chart have been identified with English letters and an explanatory legend added so that the chart is more easily interpreted.

A quick look at the printout shows that columns A, B, C indicate year, month, and day of the activity, i.e., October 8th through 31st plus November 2, 1987. Column D represents the floor to which the steel members were being hoisted, and columns F to J contain the number of pieces placed on a specific day. Columns K through O reflect the actual number of lifts or picks.

As an example, on October 8, 1987, eight pieces of structural steel were lifted onto the third floor of the building with eight picks of Crane No. 1. The total time for all of these lifts is shown in Column E and totals 125 minutes. The lifting time per piece is 15.6 minutes and is listed under Column P. Columns S through AE pertain to the operation and output of Crane No. 2, and by including its production of seven pieces placed on the third floor of the building, an accurate cost and time analysis can be made of the erection operation.

Data such as this can be accumulated for any hoisting operations—

① データ 入力

(LANPLANに入力)

1ピース当たり作業時間

揚重1回当たり作業時間

毎日の
全作業時間
(他材含める)

A	B	C	D	E	F	G	H	I	J	K	L	M	N	O	P	Q	R	S	T	U	V	W	X	Y	Z	AA	AB	AC	AD	AE
年	月	日	階	鉄骨	ピース数					揚重回数					T/P	T/L	PC	ピース数					揚重回数					T/P	T/L	作業時間
				TIME	1	2	3	4	5	1	2	3	4	5	TIME	TIME	TIME	1	2	3	4	5	1	2	3	4	5	TIME	TIME	分/日
87	10	7	0	0											0	0	0											0	0	
87	10	8	3	125	8					8					15.6	15.6	113	7					7					16.1	16.1	
87	10	9	3	186		11					11				16.9	16.9	117		12					12				9.75	9.75	
87	10	12	3	43			2					2			21.5	21.5	64			4					4			16	16	
87	10	13	3	56				4					4		14	14	44				4					4		11	11	
87	10	14	3	59					3					3	19.7	19.7	68					8					8	11	11	
87	10	15	3	0											0	0	0											0	0	
87	10	19	4	123	8					7					15.4	17.6	78	8					6					9.75	13	
87	10	20	4	90		2					2				45	45	115		13					10				8.85	11.5	
87	10	21	4	90			2					2			45	45	84			4					4			21	21	
87	10	22	4	61				6					5		10.2	12.2	58				5					4		11.6	14.5	
87	10	23	4	77					4					4	19.3	19.3	79					8					5	9.88	15.8	
87	10	24	5	94	5					5					18.8	18.8	0											0	0	
87	10	26	5	38	3					2					12.7	19	100	7					6					14.3	16.7	
87	10	27	5	131		2	6				2	6			16.4	16.4	86		3						3			28.7	28.7	
87	10	28	5	0											0	0	75		7					7				10.7	10.7	
87	10	29	5	0											0	0	44			4					4			11	11	
87	10	30	5	40				2					2		20	20	70				7					5		10	14	
87	10	31	5	55					2					2	27.5	27.5	69					7					5	9.86	13.8	
87	11	2	5	0											0	0	0											0	0	

LEGEND: A = Year; B = Month; C= Day; D = Building Floor; F - J = No. of pieces lifted; K - O = No. of Lifts; P = Working time per piece; R = Working time per lift.; S - AE = Data for second crane

Fig. 5-1. Computer printout for tower crane activities during steel erection. Chart is for two-crane operation.

precast plank, architectural precast panels, mechanical and electrical equipment, buckets of ready-mix concrete, and so forth. By creating historical records of hoisting data, not only are distribution costs more accurately determined, but a great deal of scheduling information can be developed.

With precise knowledge of the length of time it takes to place concrete on upper floors of high rise buildings, the scheduling of transit mix trucks can be refined so that mixing time in the barrel of the truck is kept to a minimum, and trucks can be scheduled to arrive at the job site in a more orderly manner and be rapidly unloaded. By having comprehensive time study information available, estimators can make more accurate assessments of exact hoisting requirements for new projects and more accurate unloading and distribution costs can be established. But the potential goes beyond just cost and scheduling functions.

Automated Materials Placement

One executive at the Kajima Institute of Construction Technology recently stated that automated materials placement by computerized cranes is a high priority in Japan. Is it possible that these computer controlled cranes on Japanese construction sites today are just the first step in a carefully planned, fully automated construction process in the future?

Just imagine a total electronically controlled hoisting system where a scanner would pass over a coded steel structural member on the ground in much the same manner as a check out scanner in a grocery store. An unmanned crane scanning a structural beam or column would obtain placement instructions from the code, and according to information carefully preprogrammed would hoist it into place with such precision that some type of magnetically controlled bolt-up robot could deftly insert a fastener into the perfectly aligned moment connection, torque up the nut with one swift motion, and climb up to the next level.

COST INFORMATION MANUALS

Along with cost data generated and stored within the company's database, there are a number of cost data manuals available to Japanese contractors from sources similar to the R.S. Means Estimating manuals so familiar to U.S. contractors. Fig. 5-2 is the cover of *MRC*, a monthly standard building cost index and unit price booklet published by the Management Research Society for Construction Industry.

In addition to this manual, the Construction Price Research Association publishes a monthly periodical called *Price Data for Construction Cost Estimating*. The Ministry of Construction issues their *Standard La-*

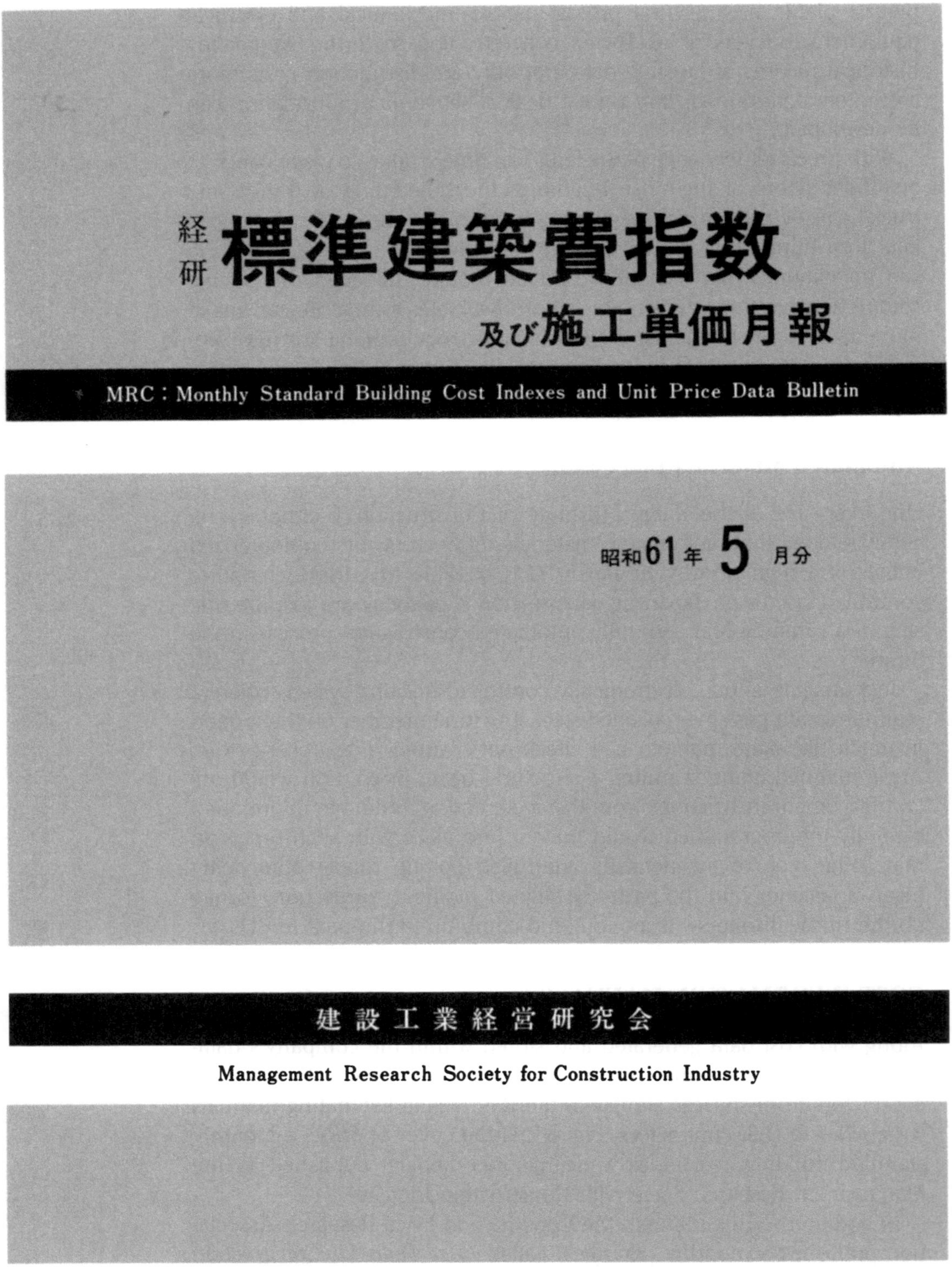
経研 標準建築費指数

及び施工単価月報

MRC: Monthly Standard Building Cost Indexes and Unit Price Data Bulletin

昭和61年5月分

建設工業経営研究会

Management Research Society for Construction Industry

Fig. 5-2. Cover of *MRC*—building cost and unit price data booklet.

bor Index containing labor rates for all types of construction work employed on public projects.

Some representative in-place unit costs are listed below:

Precast concrete exterior wall panels	$21.45 per square foot
Structural steel—fabricated and erected	$1639.74 per ton
Ceramic Tile	
Wall Tile	$4.32 per square foot
Floor tile	$3.97 per square foot
Roofing	
1″ insulation	$1.22 per square foot
Built-up roof	$2.74 per square foot
Concrete work	
Forming walls	$3.60 per square foot
Reinforcing steel—in place	$837.64 per ton
Placing concrete	$79.40 per cubic yard (including cost of concrete)
Fenestration—aluminum window frames, fixed lites	$21.63 per square foot
Gypsum drywall—½″ material	$0.94 per square foot
Carbon Fiber Reinforced Concrete Wall Panels	$25.95 per square foot
	$28.84 per square foot
	$57.67 per square foot

COST REPORTING

Much of the cost reporting activity on Japanese construction sites originates from the subcontractors hired by the general contractor. Most subcontractors are required to submit daily reports containing the amount of work they have put in place and the size and makeup of their work crews. The Project Manager or the Site Engineer checks these daily reports for completeness and a copy is sent back to the general contractor's head office.

As labor, material, equipment, and subcontractor data flow from the field to the main office, cost monitoring will begin. Project cost data will be accumulated on a monthly basis, analyzed, and compared with the construction budget. The project manager is expected to submit monthly project forecasts, which will include cost and scheduling projections. If it appears that problems may be looming on the horizon due to the complexity of the job or the nature of the work being undertaken at the time, the project manager can request support teams from the home office or the company's research facility on a temporary basis.

BIDDING PROJECTS IN JAPAN

The process of preparing and submitting bids for work in the private sector in Japan is similar to that in the United States; it is in the public sector where bidding procedures differ sharply.

The much publicized Kansai Airport project brought the Japanese government's practice of prequalified, selected bidder's lists to the attention of the world, and highlighted yet another pact between government and industry that reinforced the image of Japan, Inc. The process of obtaining a contractor's license in Japan was looked upon by foreign contractors, especially American builders, as an exclusionary measure to prevent them from participating in many of the huge domestic construction projects planned in both the private and public sectors.

If and when it became possible to obtain a contractor's license in Japan, the next hurdle would seem almost as insurmountable—obtaining a spot on the selected bidder's list for a public works project. In order to gain access to the prequalified bidders list a contractor must be able to demonstrate experience in similar projects. Since "similar" projects could rightfully be interpreted as having built jobs of a like nature in Japan, the classic Catch 22 situation will have been created.

In fact it was not until the third quarter of 1988 when U.S. contractors were actually qualified for any meaningful construction work in Japan. In October 1988, it was announced that four U.S. contractors had been "short listed" for consideration for work at Yokohama's massive Minato Mirai 21 port development project, estimated to cost about $16 billion.

Chicago-based Schal Associates, New York's Turner Construction and Tishman Realty and Construction, and the Parsons Group from Pasadena, California were notified that they were considered qualified to bid on various portions of construction work, and it is expected that these four U.S. builders would team up with "short-listed" Japanese contractors to form joint ventures when final bids are submitted.

Eligibility for qualification is based upon many criteria, including Ministry of Construction annual rating system, and contractors that are already top-rated will consistently appear at the top of the yearly lists. Not too many points separate the top contenders, as the list below reveals:

Company	1985 Rating	1984 Rating
Shimizu	466	462
Kumagai Gumi	462	466
Toda Construction	462	462
Maeda Construction	462	462
Kajima Corporation	459	456
Ohbayashi Corporation	459	457

Source: Nomura Research Institute

This selected bidders list system is defended by government and contractors alike, sometimes with impassioned statements that the entire Japanese society will be in peril if the system is changed.

Mr. Kunio Maeda, Director and General Manager of the International Operations division of Tobishima Corporation, an engineering and construction firm, defended the system as being necessary to the very fiber of the Japanese industrial system. Mr. Maeda viewed abolition of a selected or prequalified bidder's list as tantamount to destroying Japanese society by opening the floodgates of cheap imported labor. If foreign contractors were allowed free access to Japan's domestic construction market, many of the newly industrialized nations such as the Philippines, Thailand, South Korea, Singapore, and mainland China would come into Japan with large crews of low wage workers. This would create mass unemployment, it would undermine the stability of small and medium sized contractors, and would create serious social problems throughout the country. No longer would Japan be a homogeneous society; illegal immigrants could be forced to work for low wages by unscrupulous Japanese businessmen. These workers would not pay taxes, would create slum areas, and would disrupt much of the country's harmony. It is interesting to note that while American contractors were busily decrying the practice of selected bidders and prequalification in Japan, one of America's sources of long time construction work was making plans to institute the same type of system in this country.

THE U.S. POSTAL SERVICE AND ITS SELECTED BIDDERS LIST PROGRAM

The United States Postal Service, once a government agency and now a quasi-public corporation, has provided work for many contractors throughout the country ever since Ben Franklin became Postmaster General in 1794. Since becoming a more-or-less private corporation, post office officials have become disenchanted with the open competitive bidding system and the many problems they have encountered with costs, quality of work, and scheduling.

The Postal Service has lifted a page out of the Japanese government manuals and instituted a system of prequalifying bidders, which in their opinion will result in a better end product. With a construction budget of $1.34 billion per year for the period 1988–1993, the U.S.P.S. represents a substantial construction market, and this practice of limited open bidding is viewed by some as a threat to other markets in this country. The Associated General Contractors, who raised the American flag when attacking this practice in Japan, has to use a different approach now. AGC says that the Postal Service plan is "out and out bid shopping" and will nurture scandals. The AGC views this action by the Postal

Service as the beginning of the end for open competitive bidding on U.S. public projects.

In defense of their new approach to awarding construction contracts, Post Office officials claim that the lowest bid at bid time does not necessarily insure the lowest cost after the project has been completed. The assertion by U.S.P.S. executives that some construction firms make a practice of submitting multiple and costly claims during the construction period ultimately drives up the final cost of the project. These Postal Service officials claim that their past experiences with "claims prone" builders has been less than satisfactory. The U.S.P.S. makes no mention, however, of the reasons why these claims seem to proliferate and why their projects seem to become targets for "claims artists," a phrase used by Stanley Smith, Assistant Postmaster General for Facilities.

SIMPLIFICATION OF THE SELECTED BIDDERS LIST PROGRAM

Acceding to mounting pressures brought about to a great degree by the United States government, the Japanese have relaxed, somewhat, their requirements for bidding on public projects. In June of 1988, the Ministry of Construction published a booklet entitled "From Obtaining Construction Business Licenses to Winning Contracts—A Guide to Construction in Japan." Aimed primarily at the U.S. contractor, this booklet contains very clear and concise instructions for obtaining a contractor's license in Japan, the first step in a series of acts required to attain a position on a selected bidder's list. Once a contractor's license is obtained, a prospective bidder must subject his company to scrutiny by the project commissioning agency and be judged qualified to submit a bid proposal. The government booklet states that "Most contracts for public works under the jurisdiction of the Ministry of Construction are awarded through the designated competitive bidding system." As a further explanation, the booklet continues, "In designated competitive bidding, the commissioning entity selects those firms that are thought to be suitable as contractors with regard to funding, credit, operational capacity, and other factors (this selection process is called 'designation'), and the winning contractor will then be chosen through competitive bidding. Since the designated competitive bidding has the advantage of being able to exclude those firms whose operational capacities are insufficient, or who might provide inferior work, or who might commit other breaches of contract, most contracts for public works under the jurisdiction of the Ministry of Construction are awarded through the designated competitive bidding system."

There is another procedure, known as the *discretionary contract system,* that is used by the government when the nature of the project is such that very few firms have the capability of bidding on the work and the commissioning entity may select bidders at their discretion. Where there is a need to award a contract for work when time is of the essence, such as cleanup work after a natural disaster, the commissioning agency has the power to issue contracts without going through any type of competitive bid effort.

In order for a newly licensed contractor to obtain a spot on a prequalified bidders list he needs to submit an application.

THE QUALIFIED BIDDER'S APPLICATION

When requesting consideration for placement on a selected government project bidder's list the applicant must include the following information:

1. A copy of the application for an examination of the company's business conditions.
2. Submission of records showing previous construction experience in similar projects.
3. A list of the applicants business premises, either singular or multiple.
4. A copy of a joint venture agreement, if the applicant is planning to join with a partner in a joint venture proposal.
5. Any other necessary documents deemed necessary by the agency. (This rather vague requirement seems rather obvious, if the government agency wishes to exclude the applicant from the bidder's list, this open-ended requirement will certainly help.)

Once all of these documents have been received by the government agency commissioning the proposed project, they will be examined in order to determine whether the applicant:

1. Is under some type of interdiction, or is in a state of bankruptcy, or is "quasi-incompetent."
2. Has or has not engaged in certain unfair practices.
3. Is not financially unstable.
4. Has not made false statements in the application form.
5. Does, in fact, have a valid contractor's license.

LOW BID IS NOT ALWAYS THE SUCCESSFUL BID

The Japanese federal government, through its own estimating sources, establishes a construction budget for each project being considered for bidding, and a ceiling price is determined for each project. If all bids received exceed this ceiling price, the project may be re-bid with an entirely different group of contractors. Along with this cap or ceiling price, a lower limit for acceptable bids is also established prior to issuing proposals. When a bid is received that is substantially lower than the government's low limit level, the commissioning agency has the right to disqualify the bidder or bidders.

THE TOKYO MUNICIPAL GOVERNMENT

The TMG, as it is known, patterns its bidding practices along the lines of the central government and requires that contractors be prequalified in order to submit bids on municipal work. The Tokyo Municipal Government solicits Requests for Qualifications (RFQ's) from those contractors interested in doing work for the city. Prospective contractors need to submit detailed statements of financial condition, past performance records, experience in various types of projects and information about the company's operations. When a contractor has been adjudged acceptable to the TMG, his application will be placed on file, and when projects are going to be put out for bid, all of these applications are reviewed and the most qualified bidders are sent requests for bid proposal packages. The TMG has a rather extensive engineering staff who prepare construction budgets for all projects going out for bid. These projects, like their federal counterparts, will have an upper and lower acceptable bid range. The bottom limit is usually 80% of the estimated cost, although in some cases it will be lowered to 60% of the engineer's estimate. Bids received below these ranges will be disqualified. No efforts are made to review the bid with the contractor, and although there may be an occasion where one of these low bids is justified, the contractor is not given an opportunity to present his case. The TMG takes the position that the contractor has failed to grasp the scope of the work or has made a serious error in his estimate and that acceptance of any of these unreasonably low bids will only cost the contracting agency more money in the end and will ultimately delay the completion of the project.

Engineers at TMG indicate that change order work does not really represent a major problem because few change orders are issued during the course of a project. The documents issued for bid are thoroughly prepared and reviewed in great detail before they are released to the contractors. Most Japanese general contractors feel that they have a responsibility to fulfill the intent of the contract documents as well as the scope of work actually shown on the plans and in the specifications. When both of these conditions are present, as they are in Tokyo, it is easy to see why TMG has very little concern with change order work.

The TMG takes certain precautions to avoid possible collusion between government workers and private contractors. Every two years, middle managers within the city government are rotated to other jobs, ostensibly to prevent long term relationships from developing between purchasor and vendor, but in keeping with the Japanese tradition of continuing training these shifting of positions might be taking place primarily to fulfill the need to create generalists, and the disruption of relationships might be of secondary importance.

TMG AND DESIGN CONSULTANT SELECTION

When the Tokyo Municipal Government needs to select design consultants for a construction project, their selection process is similar to their contractor qualification procedures. TMG will assemble a committee consisting of engineers and architects to preselect a group of design professionals to be invited to submit proposals. The city engineers will have established a fee structure that is acceptable and fees that exceed or fall below this predetermined fee level will be rejected. The 80% factor for lower limits is generally used, and TMG's rationale is similar to the reasons for rejecting low contractors bids—an unrealistically low design fee may mean that the designer did not understand the scope of the work under consideration, or if he did, will produce work below the quality levels expected by the commissioning agency. On very large and complex design projects, such as the recently awarded $900 million Metropolitan Government Center, which will house the Governor of Tokyo's offices, it was decided that a design competition would be preferable rather than a fee-type proposal. After a winning design has been selected, a fee structure would be negotiated with the successful architect. Renowned Japanese architect Kenzo Tange was ultimately awarded the design contract for this showpiece project consisting of three buildings now under construction on a site formerly occupied by Tokyo's red light district.

QUALITY CONTROL—A DRIVING FORCE IN THE CONSTRUCTION INDUSTRY

Quality control, or TQC (Total Quality Control), as it is referred to in Japan, is a concept developed initially by an American statistician, W. Edwards Deming. Deming, often referred to as the Father of Statistical Process Control, was more or less ignored in the U.S. when he developed his theories about quality control. He felt that management was largely responsible for the quality of a corporation's product and that their influence on the line worker would result in high quality levels on the production line. Deming said that daily worker feedback was critical to the development of a quality control system and that many

U.S. managers were cut off from this important source of information, relying more upon periodic reporting systems as a means of keeping abreast of the company's manufacturing capabilities. Although given short shrift in the U.S., Deming's teachings were wholeheartedly embraced by Japanese industry in the 1950s as it emerged from the war years. The QC concept caught the attention of the Japanese construction industry in the 1970s.

In recent years the Deming concept has been "imported" from Japan by any number of prominent U.S. manufacturers.

In Japan today, the Deming Prize for Outstanding Achievement in Quality Control is the Holy Grail pursued by everyone in the business community, including construction companies. The Deming Medal, engraved with the words, "The Right Quality and Uniformity Are Foundations of Commerce, Prosperity and Peace" is proudly displayed in Kajima Corporation's corporate brochure, having been awarded to that firm in 1982. The quest for this prestigious prize is such that several Japanese builders have included its acquisition as a management goal when formulating their five years plans.

The TQC Concept

Total quality control to most Japanese contractors means more than just checking the flatness of slabs or inspecting the quality of mortar joints. TQC is an all-consuming corporate philosophy that has as its goal the uplifting of every aspect of company life. A total quality control program will strive to:

- Increase yearly sales volume by more aggressive selling and also by analyzing and investigating the ever changing needs of the client.
- Improve job quality and the quality of the finished product.
- Increase gross profit percentages while decreasing corporate costs.
- Continue the pursuit and development of new construction technology in order to achieve and maintain an edge over the competition.

Once management decides to embark on a total quality control program, the first step in the process is the creation of a quality control committee which will organize the program, coordinate all of the QC activities within the company, and monitor the results of the corporate commitment. Departments that are known to have problems will be requested to formulate a plan for self-improvement and forward it to the QC committee for review and approval. In typical Japanese fashion, suggestions and cooperation is elicited from below and as consensus is achieved, the plans that have been formulated are passed along to upper management.

Attacking QC Problems

In the early stages of a TQC program, it is necessary to ferret out the problem areas and recognize the fact that some departments, or even divisions, have shortcomings or deficiencies. As problems surface the process of dissection begins, and each element thought to be in need of special attention is examined closely in order to salvage the strong parts and sort out and discard the weak ones. A period of reevaluation follows where very searching questions are asked. Why did our final costs on that last project exceed the estimate by 30%? Was the estimate in error, did we not manage the project properly, did we have poor performing subcontractors? Why are we hiring more people in the accounting department when workloads seem to be tailing off?

Once the existing problems are uncovered, the quality control circles that will be formed as part of the next phase of the TQC program can be given clearly defined missions.

The QC Circle Strategy

The U.S. automobile industry was one of the first in America to recognize the value of the Japanese concept of QC circles—a formation of small groups of workers who perform work tasks at the most basic level in the manufacturing process. Since workers at the primary level of manufacturing or assembly are the first line of defense against product defects, they ought to act as inspectors, industrial engineers, and designers in order to improve the quality of work directly under their control. Suggestions that originate at the QC level need to be given serious consideration by management and implemented promptly when approved, and manuals have to be created so that all of the quality control procedures can be formalized and followed by succeeding generations of workers.

The QC committee is given the responsibility of collecting all the ideas and suggestions produced by the QC circles and incorporating them into a master plan. On-the-job training and education also have to be initiated so that the quality control measures can be explained to all employees in the office and in the field.

Quality Control and Kajima Corporation

When the current Chairman of Kajima, Mr. Rokuro Ishikawa, was company president in 1978, he introduced the policy of companywide quality control (CWQC), and in his first speech as president he said, "The introduction of CWQC is in line with the spirit of our company's basic policies of scientific rationalism and humanism, which I believe to be of ever increasing importance today. Our resolve to meet the present problems with this firm posture is based upon our strong desire to build a solid spirit and constitution by which we can respond through our own endeavors to any change in environment. I believe that we must attend to its promotion with a strengthened determination."

Companywide quality control was going to concentrate in four main activities

- Personal development and training of all company personnel.
- QC group activities.
- Quality assurance activities.
- Policy management activities.

Education and training, so important to most Japanese companies, was to be expanded to include corporate directors, division and section managers, as well as the other 10,156 Kajima employees. Special courses would be tailored to the individual needs of the group and a multitude of in-house training sessions would begin.

In the office, section chiefs would assume the role of instructor, while at the job site the project manager or superintendent would take on the task of instructor. The subcontractor's foremen would be invited to join in these on-the-job training programs as Kajima would concern itself in every aspect of the construction process—from estimate preparation to after-sale service.

Even today, Kajima has been known to send a team of inspectors to recently completed projects after the client has occupied the building. These Kajima inspectors will randomly tap tile or masonry walls to uncover any loosely adhered materials. They will check concrete surfaces for cracks, chips, flaking, or spalling, and report any defects or pertinent observations back to their QC department.

The intricacy of Kajima's quality assurance program can be seen in their elaborate flow chart shown in Figure 5-3.

In 1981, Kajima took stock of their company to determine whether the CWQC program they had instituted three years earlier had produced any tangible results. They sought to compare several aspects of their operations on a before-and-after basis. Fig. 5-4 graphically displays the results of their program. Cost of repairs of completed work dropped significantly. Productivity, the amount of execution per man per month, increased dramatically, and the labor/cost ratio decreased proportionately. Last, but certainly not least, Kajima's accident rate plummeted. All in all, their companywide quality control program seems to have been a resounding success.

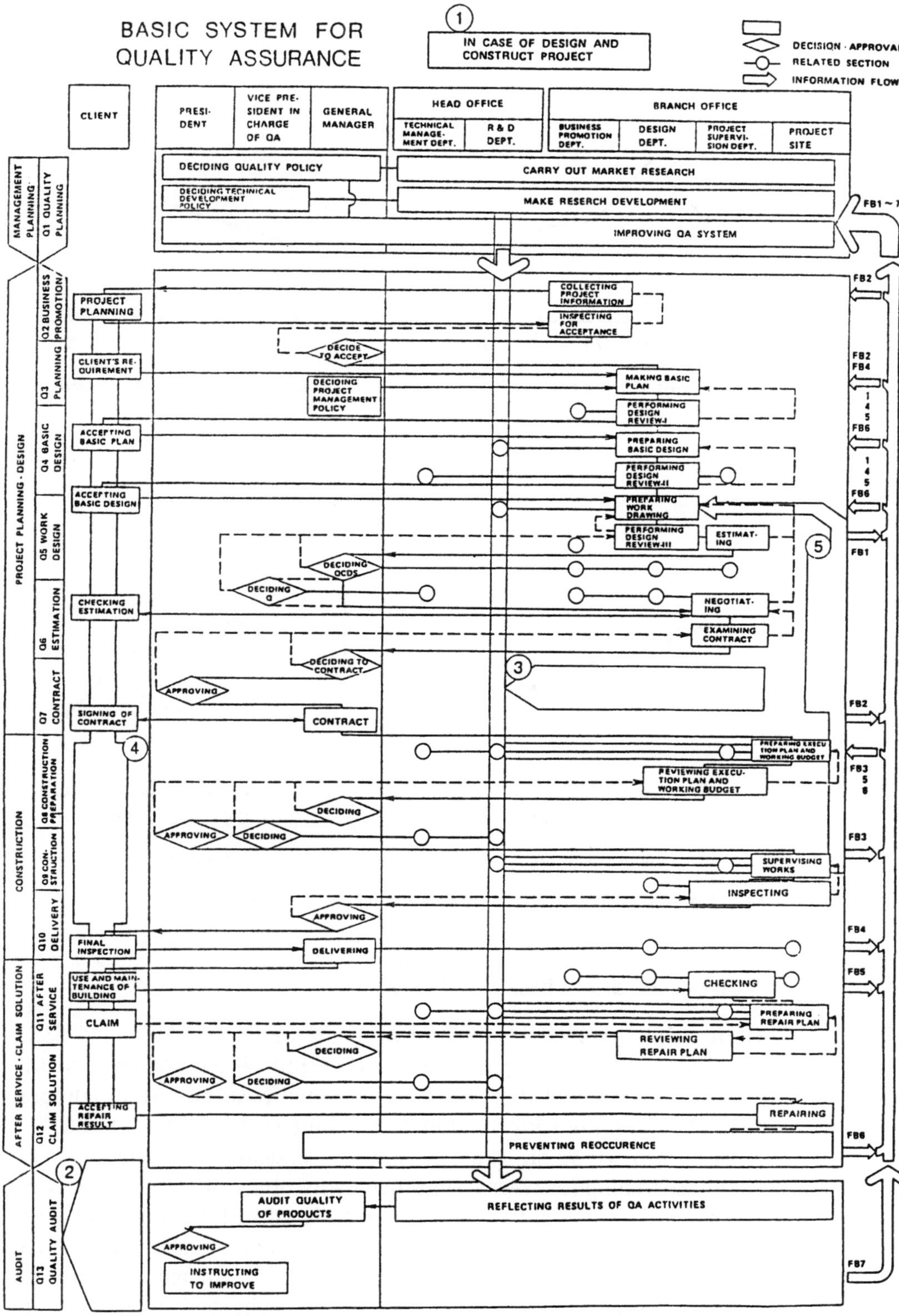

Fig. 5-3. Kajima Corporation basic system for their quality assurance program.

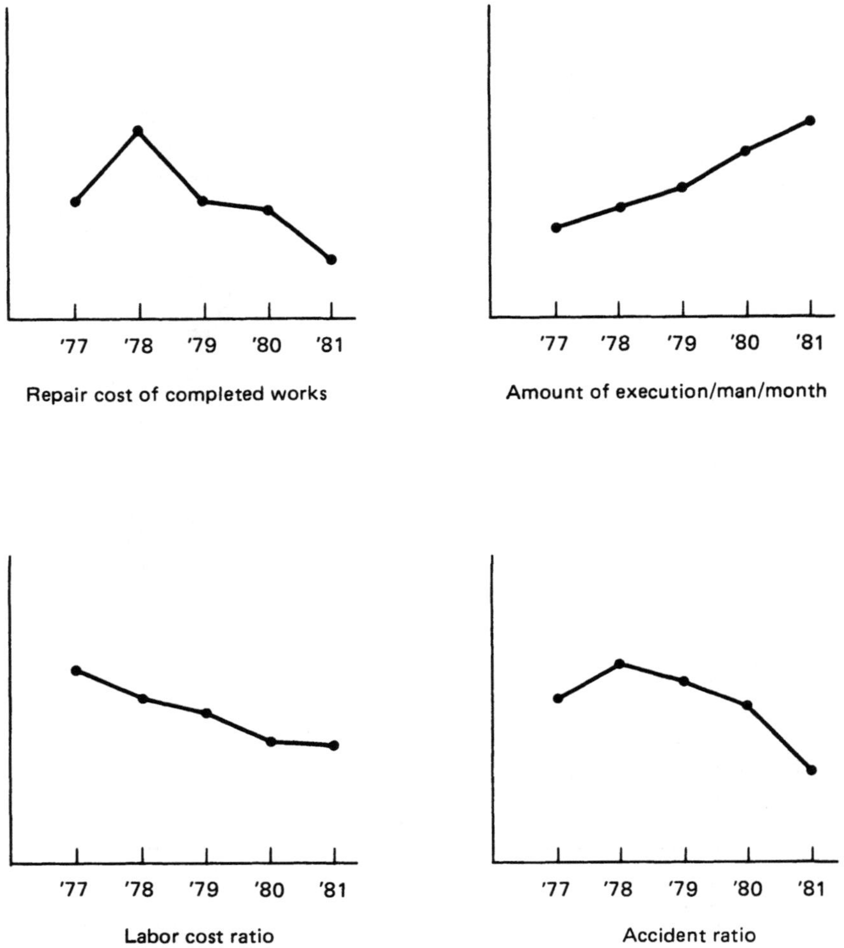

Fig. 5-4. Four-year analysis of tangible results of Kajima's CWQC program.

Ohbayashi and Their SK Program

Ohbayashi's total quality control program was originally called SSK–*sogo* (Overall) *shitsu* (Quality) *kanri* (Control). When someone in the company discovered that SSK was the name of an industrial corporation, Ohbayashi quickly shortened the name of the program to SK.

Ohbayashi views SK as a corporatewide endeavor, and they are constantly searching for ways to improve every phase of company activity. For example, some of their workers were faced with a problem of forming a concrete beam before the suspended concrete slab was formed and poured, and they were looking for a way to increase production while decreasing costs. The workers at the job site devised a method to combine beam and slab forms and produce a quality pour at lowered costs. After the forming system was perfected, the method was formalized and given a special name—the OK form system—and the QC group that devised this new forming method was given companywide recognition. A patent was obtained by Ohbayashi and credit was officially given to the responsible quality control group.

Each year Ohbayashi recognizes the accomplishments of their QC groups during their annual SK Award ceremony, and these SK Award winners are held in high esteem within the company. Ohbayashi sums up their quality control program in three words—question, thought, and action—and they have recently added a new slogan throughout the company—Act SK!

Kumagai Gumi and QC

When concerns for quality are reported to headquarters, they are treated with the importance that they deserve. Kumagai Gumi decided to investigate a better way to produce cast-in-place concrete walls that would not have the voids and airpocket holes usually associated with this type of construction. Although rubbing the walls after stripping could improve the aesthetics of the wall, was there no way to improve appearance and quality and even lower the cost of the end product? Enter the process of quality control, whose purpose would be to improve quality and efficiency and decrease costs of construction.

Kumagai Gumi developed a unique process of concrete wall forming which they named the *textile form method.* Small holes, ranging from 3 mm to 6 mm, would be drilled through the concrete wall panel forming the outside face of the concrete wall that would ultimately be exposed. A specially developed textile mat would be inserted onto the inside face of this form. As concrete was placed in the wall form, excess water and entrapped air, after the form was vibrated, would be allowed to "bleed" through the woven textile form liner and out through the tiny holes in the form. The resulting concrete panel would have a dense, durable surface, and the woven fabric pattern of the textile liner would impart a slight texture to the concrete surface. Kumagai Gumi, through the efforts of their research department, created a classic case of quality control performance.

The Problem:	Increase surface durability and appearance of cast-in-place concrete walls, while maintaining control over costs.
The Solution:	Develop a new system of forming so that excess water and entrapped air can escape from the concrete matrix after it has been placed in a conventional form.

Results:	Costs were decreased. The fabric liner that was developed is reusable up to five times, and any additional costs associated with its use are more than likely offset by the fact that when using this method of forming, the water-cement ratio in the concrete permits much earlier stripping time.
Appearance:	Improved to the point where rubbing is no longer needed in many cases.
Quality of Product:	Surface strength of the concrete wall poured using this method of forming is improved by about 30%.
Durability:	Greatly increased with respect to freeze-thaw resistance: Carbonation (the action of carbon dioxide on the alkalinity levels of concrete) is delayed to a significant degree, and saline permeability is greatly reduced.
Added Benefits:	The fabriclike texture of the finish on the wall surface enhances adhesion if other materials are to be adhered to this surface.

Mitsui's Syncro TT System

Concrete is a primary building material in Japan and all major contractors have devoted considerable amounts of research time and money toward the study of the properties of this material in order to improve its performance at the lowest possible cost—the classic charge of a QC program.

Determining the optimum time to remove forms from recently placed structural concrete has always been of concern to contractors. Strip too soon and the structural integrity of the concrete might be irreversibly damaged; strip too late and valuable production time is lost. When concrete forms are being leased and these costs form the basis of an estimate, delays in maintaining a specific form-pour-strip sequence will increase costs, create scheduling delays, and have a domino effect on all subsequent operations.

Concrete curing time, ambient temperatures and curing methods all have an effect on the compressive strength required before concrete forms can be stripped and move on. The time honored method of determining the proper time for concrete form removal has been to prepare sample test cylinders at the job site and send them to a local testing laboratory for destructive testing at time intervals ranging from two to three days to fifty six days after the samples have been prepared. Mitsui has developed a sophisticated method to determine the compressive strength of newly poured concrete without the use of destructive testing methods. Called Syncro TT, (possibly for time and temperature), this entirely new way of concrete testing is being refined today by Mitsui.

Since the curing of concrete is controlled to a great degree by time and temperature, Mitsui bases their system around monitoring both of these critical factors. The Syncro TT system is set up right at the job site, and consists of a water tank in which the temperature of the water can be regulated and controlled. Thermocouples are provided for placement in the newly poured concrete, and a personal computer containing a special software package designed by Mitsui engineers is brought to the construction site. The water tank is used to cure a test cylinder of concrete at temperatures approximating the ambient temperatures at the job site. The thermocouples placed in the recently poured concrete are used to transmit heat of hydration readings back to the computer.

The test cylinder in the water tank becomes the "standard" and, based upon previous data accumulated from other pours, the software creates a curing curve and compressive strength graph, similar in concept to the sketch shown in Fig. 5-5. As the thermocouple in the concrete relays temperature readings back to the CPU, they are compared with data being received from the test cylinder in the water tank and the computer displays the "standard" rate of curing with the actual rate of curing and compressive strengths being attained. The optimum time to strip forms will be predicted and charted by the computer.

Mitsui claims that the compressive strengths of concrete can be predicted more accurately with their Syncro TT system than by using conventional cylinder break procedures.

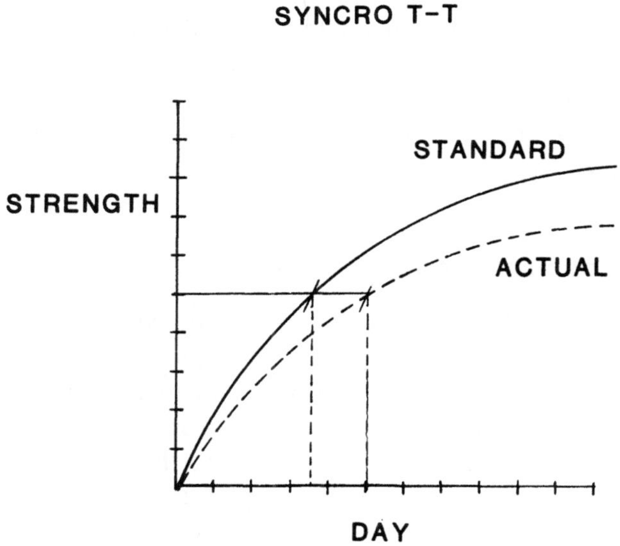

Fig. 5-5. Mitsui's Syncro TT system for optimum concrete form removal sequencing.

Computers and Job Site QC

Computers are being installed on the construction sites in Japan to perform more than accounting and cost reporting functions. Desktop units are used to update CPM schedules, to control material inventories so that just-in-time deliveries can be maintained on the site, and in various other ways to record daily events.

Computers are now being used to instantaneously record the composition of daily work crews on construction jobs. How many times does a superintendent or project manager need to know the exact crew size of each trade working on the job first thing in the morning, to determine whether or not there are enough workers in critical trades so that a tight schedule can be maintained? One Japanese contractor has devised a system whereby he is able to determine the exact composition of each and every work crew the moment they all step through the front gate; just seconds after starting time, the entire daily workforce will be tabulated and displayed on the screen of a CRT or on a printout sheet. This Japanese builder issues every worker an identification card which is electronically coded with the worker's name, subcontractor affiliation, trade designation, and level of experience (apprentice, journeyman, master mechanic, foreman, etc.). These ID cards have all this information encoded on a magnetic strip, and as the tradesman enters through the front gate, he inserts the card in a card reader which is tied into the field office computer. Workers showing up on the job for the first time must go to the field office first and give their information to a clerk who can prepare an ID card on the spot. The general contractor is provided with immediate access to the physical makeup of the day's work crew and by transferring this daily data to memory, a complete history of the project's work force over the entire construction period can be recalled in any number of reports.

TQC and Construction Equipment

Total quality control as it relates to construction equipment, machinery, and tools has been applied with dramatic results in the maintenance shops of several Japanese contractors. When repairs to machines and equipment are made in their shops, complete and accurate record keeping yields many rewards. Records including the exact nature of the repair, the part that failed or needed replacement, the length of time required for the repair, and the length of time that the machine or tool was out of service are carefully maintained.

These records, when accumulated, evaluated and analyzed, can be put to use with many purposes in mind—first of all, for inventory control. Frequently used parts will be stocked in ample quantities, while infrequently used parts will be kept at minimum stock levels. Close control over inventory levels can lower capital investment substantially while assuring field personnel that items sent to the yard will be promptly returned to the field. The life cycle of a tool or piece of equipment can be assessed and the information sent back to the company's purchasing department, where the initial cost of the item can be measured against the length of its useful life. The true cost of one tool over another can be readily discerned and possibly reveal that the low capital cost of an item can be misleading.

Life cycle information is also very valuable in predicting when a piece of equipment or tool might fail, and as the machine or tool approaches the point of potential failure it might be removed from active service in order to avert costly down time at the construction site.

The information contained in the repair shops records may form the nucleus for a realistic preventive maintenance program that pays close attention to the Achilles' heel of certain pieces of equipment with a predeliction for failures. The whole idea behind the extensive record keeping program and TQC in the repair shops of many Japanese contractors is to insure that the company is purchasing the most cost effective machinery and tools, and when tools and equipment are used at the job site, the worker can perform at peak efficiency and with very little loss of productivity due to equipment failure.

The general contractor's watchful eye includes looking after his subcontractor's equipment and also equipment which he has rented or leased. If a subcontractor's equipment breaks down frequently, or his tools are in a constant state of disrepair, this could have a ripple effect on other trades and concurrent, or subsequent operations. Therefore it is in the best interest of the general contractor to assist his subcontractors and offer to bring them into various TQC programs that he has implemented.

One general contractor discussed problems he had been having with some pieces of rental scaffolding. Frequent repairs had to be made to certain sections of the uprights, and although he was being reimbursed for the cost of these repairs, the down time that was created was beginning to affect the progress of the job. The contractor was in the beginning stages of a QC program and he included the scaffolding problems in his on-site quality control activities. By keeping accurate records of the types and frequency of repairs, certain patterns began to emerge. The contractor alerted his scaffolding supplier and together they were able to pinpoint the problem, modify the design, virtually eliminate jobsite repairs, and create a better product.

QC and the Flagpole

Some of the quality control concerns of Japanese contractors might be carried to the point where they will leave U.S. builders scratching their heads. The United States Embassy in Tokyo is located in an area of the city known as Toranomon and is within short walking distance of the beautiful new Ark Hills complex. Almost directly across the street from the embassy is the stately Hotel Okura. The front entrance to the U.S. Embassy is set upon a podium and the rather simple design of the building is enhanced by rose colored reflective glass in the windows. The focal point of the raised entrance podium is a very large and prominent flagpole. It was important to the Japanese contractor installing the glass that the reflection of the flag and the flagpole in the glass would be absolutely true and accurate. In order to do this certain panels of glass would have to be selected from the crates of PPG glass, and precisely placed in wall openings on each floor so that a perfectly straight image of the flagpole would be reflected in the glass.

In order to insure that glass placement would be no less than perfect, the contractor stationed an engineer on the roof of the Hotel Okura, across the street, while the glass was being installed. Equipped with a pair of binoculars and a walkie-talkie, the engineer was able to direct the glass installers to shift a light here or there until a perfect image was reflected in the embassy's windows and a perfect example of Japanese quality control was reflected in the building's construction.

6

The Contract Documents

When was the last time you or your company signed a one page contract for construction that contained slightly more than 120 words and included just the name of the client and contractor, their addresses, the project name and location, the design documents, the start and completion dates, contract sum, and payment schedule—and nothing else?

In Japan, such a simply composed construction contract is commonplace, and although this simple contract format might be accompanied by eighteen pages of general conditions, the emphasis is not so much on the written word as it is on the mutual trust and confidence that has developed between each party to the contract.

THE JAPANESE CONCEPT OF CONTRACTS

Japanese businessmen tend to make verbal commitments rather than written ones, but when a written contract is required, it tends to be vague by American standards, presenting only a broad outline of the agreement between the two parties. The full understanding of their agreement would have been reached during many meetings held previously, in which detailed discussions had taken place. The contract, therefore, is an anticlimax.

The reluctance of the Japanese to create an intricately detailed contract is based upon their historical reliance on good faith to resolve any problems that may occur. The cultural differences between Japan and the United States are amplified by each country's concept of construction contracts. While Japanese businessmen feel comfortable with only a simple contract containing the barest outline of their agreement, this format is very unsettling to the U.S. executive. But by the time the Japanese client and contractor have reached the contract stage, they have each committed themselves to fulfilling all of the obligations that were discussed and the signing of a contract takes on more of a ceremonial meaning. How different these contracts are from American boilerplate documents that really have a life of their own and contain such phrases as "This Agreement represents the entire and complete agreement between the Owner and the Contractor, superseding all previous negotiations or agreements, either written or oral, and this agreement can only be changed by written instrument signed by both Owner and Contractor."

In Japan, the reliance upon good faith and the long relationships that have developed over the years between the parties to the contract takes precedence over the written agreements. This concept has a very long history.

HISTORY OF JAPANESE CONTRACTURAL AGREEMENTS

The written contract in Japan dates back to the Heian Period (794–1191 A.D.), when written agreements were used primarily to buy or sell land. This *shoman* (land purchase agreement) was abandoned as a commercial transaction instrument during the Edo Period (1600–1867 A.D.), when written agreements fell into disfavor and were replaced by the *shinyo torihiki*—an agreement based upon mutual trust. This mutual trust concept evolved from the merchants who conducted business in the 1700s, when trade guilds were being organized in and around the major cities of Japan. The merchant members of these guilds lived by the tradition of good faith, and business transactions between these merchants were based upon verbal commitments. If a member was found to be dishonest in his dealings, or failed to fulfill a commitment, he was expelled from the guild, ostracized, and effectively stripped of his ability to conduct business. In other words, if one went back on his word, one was out of business.

This tradition of establishing business relationships based upon the integrity of the personal relationships between the parties is still very strong today, and forms the basis for 20th Century Japanese contract negotiations.

JAPAN'S LEGAL SYSTEM

The present system in Japan was derived from 19th Century European codes and reflects a system borrowed from one culture and adapted to another. The backbone of the current Japanese legal system is a mixture of codes and statutes accumulated over a period of time from European and American sources and forged into Japanese civil, commercial, and penal codes.

The Civil Code in Japan was established in 1898 and deals with property rights, contract law, obligations, and family law.

The Commercial Code, established one year later, in 1899, and amended in 1981, is concerned with transactions between commercial entities and rules governing corporate functions such as the rights and obligations of company directors, stockholder's rights, recapitalization procedures, law governing monopolies, and fair trade issues. The Commercial Code contains some interesting corporate requirements. There is a provision in the Code that corporations must set aside a reserve fund equal to at least 10% of the company's annual cash dividends until this reserve fund equals 2% of its stated capital. This reserve can not be used for future cash dividends but may be used to reduce the corporation's capital deficit, if necessary.

An Anti-Monopoly Law was enacted and it created the Fair Trade Commission, whose responsibility it is to review agreements between Japanese and non-Japanese businesses regarding the transfer of technology into Japan.

The Foreign Exchange and Foreign Control Law (FECL) of 1949 is strictly a Japanese invention which some say was created solely to baffle foreigners. The FECL is an amalgamation of procedures, ordinances, and regulations that can be used to create an endless stream of reporting requirements for anyone looking to make a direct investment in a Japanese business, or take over a Japanese corporation.

FECL also functions as a conduit to allow a free flow of international transactions and investments into the country. But this organization is sometimes looked upon by foreign corporations as a government creation that can throw up a paper roadblock whenever the need for one arises.

THE ROLE OF PERSONAL RELATIONSHIPS IN AGREEMENTS

More important to the Japanese than a detailed sheaf of papers is the relationship that has been or is being developed as both parties to an agreement get to know one another. It is not uncommon for Japanese businessmen to seem less concerned with getting on with the business at hand than they are about entertaining their business guests. This is not to be interpreted as a lack of interest in business matters, but the Japanese are eager to get to know their new potential business partners, become familiar with the way they think, get to understand their basic values, and start to build that all important feeling of trust.

THE LAWYER IN JAPAN

Unlike Americans, Japanese are not litigious people. They look upon the need to engage a lawyer with some sense of embarassment, a sign that somehow they have failed in their relationships. As one construction executive in Tokyo put it, and he happened to be a lawyer, "If it becomes common knowledge that our company has had to resort to legal action to resolve our disputes with clients, other owners may get the feeling that we cannot manage our affairs properly and they might not care to do business with us any longer."

This lack of the impulse to sue can be graphically illustrated in the number of lawsuits filed against the Tokyo Municipal Government. In 1985, legal claims for negligent road repairs filed against TMG totalled just $19,000. This is in stark contrast to the $20 million in claims of a similar nature filed against the City of New York during the same period of time. Like the Maytag repairman, the construction lawyers in Japan are very lonely people. There are about 13,000 full practice lawyers in the entire country, compared to the 670,000 attorneys practicing in the United States. Japanese lawyers *(bengoshi)* do not spend much time in litigation, since civil lawsuits in Japan are only one-tenth to one-twentieth as frequent as in other countries where common law is the basis of their legal structure.

Rarely are lawyers a part of the construction or contract negotiation team. To introduce one's *bengoshi* at such a time would signal an air of distrust on the part of the party bringing counsel into such a meeting. Until 1868 Japan had no legal profession, per se, and most foreign lawyers were prohibited after 1955, except for a handful left over from practices established during the American Occupation after World War II. As more and more American businesses establish branches in Japan and more and more Western style business transactions take place, more and more American law firms will enter the country. The trend started in 1987, when fifteen U.S. law firms opened offices in Japan, and this trend is most likely to continue.

THE CONSTRUCTION CONTRACT PERCEPTION

The Good Faith concept in construction contracts is exhibited by purposely leaving the details vague so that they can be filled in as construction progresses and the previous understandings between the parties start to unfold. As one Big Six executive told the writer, "The client tells us what he wants, and leaves the details as to how it is accomplished up to us."

The harmony that pervades Japanese daily life is alive and well in construction contracts, and the relative ease with which these contracts are executed is mainly the result of a builder-client relationship that has existed for many years, in some cases over 100 years. During that time contractor and client have grown to know and trust each other, and the relationship in some cases has passed from generation to generation. In many cases, the builder has already completed similar buildings for his client and is very familiar with the owner's requirements, quality levels, and preferred choice of materials and equipment. By making minor changes in a new building program and updating project costs, a new project concept and budget can be prepared with minimal additional owner input. The client, on the other hand, is aware of construction costs and general building design and construction details from previous projects with his contractor, so that a meeting of the minds on the new structure can be simply and quickly attained.

THE CONCEPT OF CHANGE ORDERS

During the course of construction, owners occasionally add or delete work from the project or modify systems or finishes. When changes of a minor nature are requested of Japanese builders, to suggest that a change order ought to be issued might be construed as an act of bad faith; however, if significant changes are requested, it is understood that the contract sum will be adjusted. Change orders are not too common because the vagueness of the contract, and sometimes of the plans and specifications, allows both builder and owner to make certain changes during the course of construction that might deviate from the plans and specifications but not from the intent of the understanding between the contractor and the client.

This nonchalant attitude to changes to the contract does not mean that the Japanese contractor is not cost conscious, but maintaining the quality of the relationship with a client seems to take precedence over concern for loss of profit. The contractor feels that in the long term adequate compensation will be received from the client. When a project has been completed and there have been a series of unusual problems in the construction cycle, or a series of minor changes have added up to a substantial sum, the contractor and client will meet and discuss these matters. During this project review the builder will discuss his cost overruns that occurred not because of his lack of professionalism but due to forces out of his control, as well as changes initiated by the client. In some cases, the owner, recognizing that the builder did not achieve his profit goal, or might have even suffered a loss, may agree to compensate him for his loss at that time or when a future project is to be built.

In the United States this carrot dangling has happened to many contractors, but in Japan the client's promises are kept!

THE CONSTRUCTION CONTRACT

Lump Sum Contract

A typical Japanese lump sum contract form is shown in Fig. 6-1. The terms and conditions of this type of contract are self-explanatory. This contract form is most frequently used when a rather complete set of

The Owner ....

and the Contractor ....

have entered into this Contract for the performance of the work

entitled ....

in conformity with the following clauses, the annexed General Conditions,

......................sheets of the Design Drawings and ....volume(s) of the Specifications.

1. Place of Work ....

2. Term of Work

   Commencement: on the .... day of .... in the year 19 ....,

   or within................days from the date of the Contract

   Completion : on the .... day of .... in the year 19........,

   or within................days from the date of the Commencement

3. Time of Delivery : Within....................... days from the date of Completion.

4. Contract Price : ....Yen

5. Payment of Contract Price:

   Advance Payment : at the time of signing the Contract ....

   Partial Payment(s) : ....

   ....

   The closing date for demand of each partial payment: ....

   At the time of delivery: ....

   ....

6. Other Matters:

Fig. 6-1. Typical Japanese lump sum contract.

この契約の証として本書２通を作り，当事者および保証人が記名押印して，当事者がそれぞれ１通を保有する。 日本語と英語との間に解釈について相違がある場合には日本語の本文による。	In witness whereof, this Contract has been prepared in duplicate and the parties hereto and the guarantors have hereunto put their respective seals and names. Each of the parties hereto shall keep its respective copy of this Contract. In case of any discrepancy of interpretation between the Japanese and English Texts, the Japanese Text shall prevail.
昭和　　年　　月　　日	Date:

発　注　者
The Owner ....

同　保　証　人
The Surety ....

請　負　者
The Contractor ....

同　保　証　人
The Surety ....

監理者としての責任を負うためここに記名押印する。	I have hereunto put my seal and name to assume the responsibility of the Supervisor.

監　理　者
The Supervisor ....

（昭和56年９月改正・四会連合協定用紙）
September 1981 amendment enacted.

Fig. 6-1. *(continued)*

construction plans and specifications have been prepared, either by the client or by the contractor in a design-build situation.

The only condition in the contract that might not be familiar to U.S. contractors is the term Advance Payment listed under Payment of Contract Price. The practice of advancing payments to contractors is discussed more fully in another part of this chapter. Two other documents are usually attached to this lump sum construction contract and they are The General Conditions of Construction Contract, a document very similar in nature to the A.I.A. Document A201 General Conditions to the Contract for Construction, and an Arbitration Agreement. These Japanese General Conditions are reprinted as Fig. 6-2 on the following pages, and a typical Arbitration Agreement is shown in Fig. 6-3

The Joint Venture Contract

The joint venture is a common occurrence on the Japanese construction scene. One major contractor interviewed attributed its popularity to the fact that everyone in the construction community in Japan ought to be able to share in the construction activity, particularly when it comes to public works projects. The joint venture affords the small and medium sized contractor an opportunity to work alongside bigger and more sophisticated builders and gain valuable experience. Joint ventures in the private sector can occur if the client or group of clients have had long term relationships with a number of builders and would like to distribute the work to more than one general contractor.

The actual joint venture agreement is as simple as the standard short-form, one page lump sum contract and will contain just a few additional clauses setting forth the definition of the joint venture, the parties to the joint venture, and the percentage participation of each of the contractors.

Most of the joint venture negotiations will already have been worked out during the many meetings held by the JV participants before the actual contract agreement has been signed. Each party to the potential JV agreement will have spent hours trying to convince the other builders that he, and he alone was the person most responsible for obtaining the contract in the first place, so he should have his choice of the type of participation and, more important, the percentage of participation. During these lengthy and sometimes heated discussions, the roles of each JV participant will be discussed and resolved and the participation percentages will be finalized. There have been times, however, when agreements could not be reached; when this happens, the members of the group may call in another contractor to act as a mediator so that final resolution can be achieved. The need for a simple contract agreement is obvious. If any of the parties to a joint venture were to renege

(text continued on p. 195)

GENERAL CONDITIONS
of
CONSTRUCTION CONTRACT

Original Text:	August 1923
Revisions:	April 1933
	June 1938
	March 1943
Initial Approval:	February 1951
Revisions:	September 1957
	November 1966
	March 1975
	September 1981

In case of any discrepancy of interpretation between the Japanese and English texts, the Japanese text shall prevail.

Approved jointly by

Architectural Institute of Japan,
The Architectural Association of Japan,
Japan Architects Association,
The Associated General Contractors of Japan, Inc.

September 1981 members

Building Contractors Society,
Japan Federation of Architects and Building Engineers Association,
Japan Federation of Architect Office Association,
Japan Federation of Professional Architects Associations,

Fig. 6-2. General conditions of construction contract.

四会連合約款・契約書使用上の留意事項

Remarks Concerning Use of General Conditions of Construction Contract

昭和56年9月

September, 1981

1) 四会連合約款は，発注者の交付する設計図・仕様書などにもとづいて請負代金額が確定した工事についての工事請負契約の一般約款であって，契約書・約款・設計図・仕様書などが一体となって請負契約を構成するものである。この約款は一般的条件を規定するものであるから加除訂正することなく使用することを目的とするものであるが，必要あるときは当事者合意のうえ加除訂正することを妨げない。

1) The General Conditions of Construction Contract are the general conditions of construction contract relating to the work the price for which has been determined on the basis of Design Drawings and Specifications, etc. which are delivered by the Owner, and the construction contract consists of the Written Contract, General Conditions, Design Drawings and Specifications as a whole. The General Conditions stipulate general conditions and are intended to be used without addition, deletion or amendment. If necessary, however, addition, deletion and/or amendment may be made upon agreement by the parties.

2) この契約について特約する場合は，契約書の「6. その他」の欄にたとえばつぎのような事項を記載する。なお，特約事項は別紙とする方法もある。

i 第26条（1）における違約金の額などを特約する場合

ii 第30条（1）における紛争調停者として第三者を選ぶ場合または管轄審査会を特約する場合

2) In the event that a special condition is agreed in connection with this contract, the following matters, for instance, shall be described in the Paragraph "6. Other Matters" of the Contract. The special condition may be stipulated in the attached document.

i. In case special condition is agreed as to liquidated damages in Article 26 (1);

ii. In case a third party is selected or special agreement is made as to the competent Committee, as a mediator of disputes in Article 30 (1).

3) 発注者，請負者，監理者が法人である場合は，契約書に法人名とその代表者名を記載し押印する。

3) In the event that the Owner, Contractor or Supervisor is a corporate body, the name of the corporate body and the name of its representative and their seals shall be put in the Contract.

4) 請負契約の成立を証するために作成された契約書は契約金額に応じ，1通ごとに印紙を貼付しなければならない。

4) Depending on the amount of Contract, the stamp duty shall be put on each Written Contract which is made as evidence of formation of construction contract.

Fig. 6-2. ***(continued)***

INDEX

Fig. 6-2. ***(continued)***

ARTICLE 1. GENERAL PRINCIPLES

(1) The Owner and the Contractor shall perform this Contract sincerely through co-operation, good faith, and equality.

(2) The Contractor shall perform the work in accordance with the Contract, the General Conditions of Construction Contract (hereinafter referred to as the General Conditions) and the Design Drawings and Specifications attached hereto (hereinafter collectively referred to as the Design Documents, which include the on-site orientation records and the written answers to the queries relative thereto) and shall deliver the work to the Owner, and the Owner shall then complete the payment of the contract price.

(3) The Supervisor shall co-operate to ensure the smooth performance of this Contract.

ARTICLE 2. SPACE FOR WORK, ETC.

The Owner shall secure and make available for the use of the Contractor the building site and the spaces which the Design Documents require for the construction (hereinafter referred to as the Construction Area), etc. on or before the date when the construction must be started (or, alternatively, on or before a specific date if such a date is prescribed in the Design Documents).

ARTICLE 3. ITEMIZED STATEMENT OF CONTRACT PRICE; PROGRESS SCHEDULE

The Contractor shall submit for approval an itemized statement of the contract price (hereinafter referred to as the Itemized Statement) and a Progress Schedule to the Supervisor promptly after the signing of the Contract.

ARTICLE 4. PROHIBITION AGAINST ENTRUSTING OR SUBCONTRACTING THE ENTIRE WORK

The Contractor shall not entrust or subcontract the whole of the work or a major part of the work to any third party without the prior written consent of the Owner.

– 1 –

Fig. 6-2. ***(continued)***

ARTICLE 5. ASSIGNMENT, ETC. OF RIGHTS AND OBLIGATIONS

(1) Neither party to the Contract shall assign to any third party nor allow any third party to succeed to his rights or obligations arising from the Contract without the written consent of the other party.

(2) Neither party to the Contract shall assign, lend, mortgage or give as security, the work or approved construction materials thereof (including factory products located at the manufacturing plant, etc.; hereinafter the same) to any third party without the written consent of the other party.

ARTICLE 6. SURETY (Applicable when a surety guarantees the Contract)

(1) In case either party should default, his surety shall be jointly and severally liable with the said party for pecuniary obligations arising from the Contract.

(2) Either party may require the other party to replace his surety, in case it becomes clear that the surety is unable to perform his obligations.

ARTICLE 7. SUPERVISOR

(1) Entrusted by the Owner, the Supervisor shall do the following acts:

a. Make arrangement with the Contractor, and if necessary, prepare Explanatory Drawings etc., and deliver them to the Contractor in order to correctly instruct the Contractor of the Design;

b. Deliver to the Contractor Detailed Drawings and other drawings prepared in accordance with the Design Documents at times necessary for the Contractor to smoothly proceed with the work in accordance with the Progress Schedule;

c. Review and advise on construction plans submitted by the Contractor;

d. Review and approve Full-size Drawings and Shop Drawings (hereinafter referred to as Working Drawings) and models made by the Contractor;

e. Give instructions for the work and attend the work, and inspect and approve construction materials, samples and equipment and fixtures for utilities etc.;

– 2 –

Fig. 6-2. ***(continued)***

f. Confirm that the work is in compliance with Design Drawings, Detailed Drawings and Working Drawings (hereinafter collectively referred to as the Drawings) and Specifications etc.;

g. Check and approve applications for partial or full payments;

h. Check and approve documents regarding changes of contents, time schedules or the contract price of work;

i. Confirm completion of work and attend the delivery of work;

j. Co-ordinate the work with other related works.

(2) When the Owner entrusts the Supervisor with matters different from those stipulated in the preceding paragraph, the Owner shall so inform the Contractor in writing.

(3) The Supervisor shall promptly comply with requests by the Contractor for instructions, examinations, tests, attendances, confirmations, checkings, approvals, opinions or consultations which are based on this Contract.

(4) All notices and negotiations between the parties shall be made through the Supervisor.

(5) When, upon approval of the Owner, the Supervisor appoints his representative for his partial or complete duties or appoints resident assistant Supervisors stationed at the site, the Supervisor shall inform the Contractor in writing, of his or their names and their duties.

(6) The instructions, confirmations or approvals given by the Supervisor to the Contractor shall be in writing.

ARTICLE 8. FIELD REPRESENTATIVE; ENGINEER-IN-CHARGE, ETC.

(1) When the Contractor appoints a field representative, an engineer in charge (superintendent or chief engineer) who is in charge of technical control of the work at the site of the work, and expert engineer (the engineer prescribed in Article 26-2 of Construction Contractors Law, hereinafter the same), the Contractor shall inform the Owner of their names in writing.

(2) The field representative shall be responsible for the management of all matters at the site of the work; he shall inform the Supervisor of important

Fig. 6-2. ***(continued)***

matters pertaining to the operations at the site, such as maintenance of order, safety and sanitation, accident prevention and working hours.

(3) One person may concurrently be appointed as field representative, superintendent or chief engineer, and expert engineer.

ARTICLE 9. OBJECTIONS IN REGARD TO PERSONS CONCERNED WITH WORK

(1) If, among the Contractor's field representative, superintendent, chief engineer, expert engineer or employees, and among the subcontractors or laborers thereof, there is any person whom the Owner, based on the opinion of the Supervisor, deems extremely objectionable in respect to the performance or management of the work, the Owner may require the Contractor to take necessary measures by expressly stating the reason in writing.

(2) If the Contractor deems measures taken by the Supervisor's representative or resident assistant Supervisor at site to be extremely unreasonable, he may require the Supervisor to take necessary measures by expressly stating the reason in writing.

(3) If the Contractor deems a measure taken by the Supervisor to be extremely unreasonable, he may raise an objection against the Owner.

ARTICLE 10. CONSTRUCTION MATERIALS; EQUIPMENT

(1) The Contractor shall use construction materials and equipment and fixtures for utilities which have passed the inspection by the Supervisor, if it is designated that such inspection is required by the Design Documents, and the Contractor shall use only those which have passed the test, if such test is required in the Design Documents. If the Supervisor deems construction equipment to be improper, he may require the Contractor to replace it.

(2) Expenses directly needed for the inspection or test mentioned in the preceding paragraph (1) shall be borne by the Contractor; however, the expenses needed for those inspections or tests which are not specified in the Design Documents and other exceptional expenses shall be borne by the Owner.

Fig. 6-2. *(continued)*

(3) The Contractor shall withdraw rejected materials as instructed by the Supervisor.

(4) When the quality of construction materials is not specified, such materials shall be of medium quality.

(5) Construction materials or equipment delivered to the site of the work shall not be removed from the site without the approval of the Supervisor.

ARTICLE 11. OWNER-FURNISHED MATERIALS; LENT ARTICLES

(1) Materials furnished or articles lent by the Owner shall be those which have passed the inspection or test by the Supervisor.

(2) In case the Contractor has any doubt about results of the inspection or test mentioned in the preceding paragraph (1), he may require re-inspection or re-test thereof.

(3) In case after receipt the Contractor finds that there is any furnished material or lent article which is deemed to be unsuitable for use due to the latent defect which was not found by the inspcction or test mentioned in paragraph (1), the Contractor shall promptly so inform the Supervisor and request his instructions.

(4) The time of delivery of furnished materials or lent articles shall be in accordance with the Progress Schedule, and unless otherwise specified in the Design Documents, delivery shall be made at the site of the work.

(5) The Contractor shall use the care of a good manager in the use and custody of furnished materials or lent articles.

(6) If the manner of use of furnished materials or disposal of surplus furnished materials (except surplus furnished materials paid for by the Contractor) is not specifically indicated in the Design Documents, the Contractor shall follow instructions of the Supervisor.

(7) The place where the Contractor shall return the furnished materials (except furnished materials paid for by the Contractor) which have become unnecessary for the work, or the lent articles which have become of no further use, shall be the site of the work unless otherwise specified in the Design Documents.

Fig. 6-2. ***(continued)***

ARTICLE 12. SUPERVISOR'S ATTENDANCE; PREPARATION OF WORK RECORD

(1) When the Contractor performs the work that requires the Supervisor's attendance by the Design Documents, the Contractor shall inform the Supervisor.

(2) Upon instructions by the Supervisor, the Contractor may perform the work by preparing records, such as photographs of work, in place of the Supervisor's attendance.

ARTICLE 13. DOUBTS AS TO DRAWINGS AND SPECIFICATIONS; CHANGE OF CONDITIONS

(1) The Contractor shall promptly notify the Supervisor in writing in any of the following cases:

a. If the Contractor discovers any indication in the Drawings or Specifications which is not clear and definite, any discrepancy between the Drawings and Specifications, or any error or omission in the Drawings or Specifications;

b. If the Contractor deems it inappropriate to perform the work in accordance with the Drawings, Specifications or instructions of the Supervisor;

c. If the Contractor discovers any physical, geological or underground-water conditions at the site of the work, or any other conditions during performance of the work, or any discrepancy between the conditions of performance indicated in the Design Documents and the actual circumstances;

d. If any unforeseeable condition is found at the site of the work, which constitutes an obstacle to the work.

(2) Upon receipt of the notification mentioned in the preceding paragraph (1), or upon discovery by himself of the facts enumerated in any of the items of the preceding paragraph (1), the Supervisor shall promptly give instructions to the Contractor in writing.

(3) If, in the case mentioned in the preceding paragraph (2), it becomes necessary to make changes in the work, the time for completion, or the contract price, the matter shall be determined by negotiation among the Owner, the

Fig. 6-2. ***(continued)***

Supervisor and the Contractor.

ARTICLE 14. WORK NOT IN COMPLIANCE WITH DRAWINGS OR SPECIFICATIONS

(1) Any part of the work found not to be in compliance with the Drawings or Specifications shall be promptly reconstructed by the Contractor at his own expense in accordance with the instructions of the Supervisor. In such case, the Contractor may not request any extension of the time for completion.

(2) In case the Supervisor has doubts as to the compliance with the Drawings or Specifications of any part of the work, he may, whenever he deems it necessary to do so, uncover such part for examination upon obtaining a written consent of the Owner.

(3) If the part uncovered and examined in accordance with the preceding paragraph (2) is found not to comply with the Drawings or Specificatons, the expenses for uncovering and examination shall be borne by the Contractor; if such part is found to comply with the Drawings or Specifications, the expenses required for the uncovering, examination and restoration shall be borne by the Owner.

(4) The Contractor shall not be responsible for any work not in compliance with the Drawings or Specifications which results from causes falling under any of the following items:

a. When caused by instructions of the Supervisor;

b. When caused by the nature of furnished or specified materials, or lent articles, or by the specified methods of performance;

c. When caused by construction materials which passed the inspection or test by the Supervisor;

d. When caused by any other reason, for which the Owner or the Supervisor is responsible.

(5) Even in the cases mentioned in the preceding paragraph (4), if there has been intent or gross negligence on the part of the Contractor in the performance of the work, or if the Contractor fails to inform the Supervisor of something inappropriate although he knows it, the Contractor shall not be relieved from the

– 7 –

Fig. 6-2. ***(continued)***

responsibility for such work. However, the Contractor shall be relieved from the responsibility if the Supervisor fails to give proper instructions although the Contractor informs the Supervisor of the inappropriateness.

ARTICLE 15. PREVENTION OF DAMAGE

(1) The Contractor shall, at his own expense, take necessary measures, in accordance with the Design Documents and relevant laws, ordinances and regulations in order to prevent damage to the work, construction materials, adjacent structures or a third party, until the final completion and delivery of the work.

(2) If the Owner, the Supervisor and the Contractor agree after negotiation among them, that any measure for or in connection with the protection of structures adjacent to the work is beyond the scope of the measures mentioned in the preceding paragraph (1) and find it unjustifiable to include the expenses thereof in the contract price, such expenses shall be borne by the Owner.

(3) Whenever he considers it necessary for preventing accidents, the Contractor shall take appropriate measures, after asking for the opinion of the Supervisor; provided, however, that in case of urgency, the Contractor shall notify the Supervisor thereof after taking such measures.

(4) Whenever the Owner or the Supervisor considers it necessary to take measures appropriate for the prevention of accidents and has requested the Contractor to do so, the Contractor shall promptly comply therewith.

(5) The expenses needed for the measures mentioned in the preceding two paragraphs (3) and (4) shall be determined by negotiation among the Owner, the Supervisor and the Contractor.

ARTICLE 16. DAMAGE TO THIRD PARTY

(1) If any damage has been caused to a third party by the performance of the work, the Contractor shall compensate the third party for such damage. However, the Owner shall bear the cost of such part of the damages as has arisen from causes for which the Owner is responsible.

– 8 –

Fig. 6-2. *(continued)*

(2) Notwithstanding the provisions of the preceding paragraph (1), the Owner shall compensate for damage done to a third party by causes such as noise, vibration, ground settlement, or interruption of underground water, etc. which are unavoidable even though the Contractor pays due care as a good manager in performing the work.

(3) If and when, any of the cases mentioned in the preceding two paragraphs (1) and (2), has arisen or in the event of any other dispute with a third party arising in connection with the performance of the work, the Contractor shall settle the matter. However, in cases where it is difficult for the Contractor to settle the matter alone, the Owner shall co-operate with the Contractor.

(4) The Owner shall be in charge of settlement and solution of such damage to third parties caused by obstruction of sunshine or radio waves or by wind arising from the object of Contract. If necessary, the Contractor shall co-operate with the Owner. In this case, the Owner shall compensate for damages which such third party sustained.

(5) In any of the cases mentioned in the preceding four paragraphs (1) through (4), the Owner shall, if necessary, extend the time for completion of the work upon request of the Contractor. The number of days of extention shall be decided upon negotiation among the Owner, the Supervisor and the Contractor.

ARTICLE 17. DAMAGES IN GENERAL

(1) If any damage has been caused to the work, construction materials, furnished materials, lent articles, or to the work in general, before the final completion and delivery of the work, the Contractor shall be liable for such damages, and there shall be no extension of the time for completion of the work.

(2) Among the damages mentioned in the preceding paragraph (1), the Owner shall be liable for those arising from any of the following cases, and the contractor may, if necessary, request extension of the time for completion of the work:

a. When the Contractor has been unable to commence the work by

Fig. 6-2. ***(continued)***

the date of commencement due to the Owner's cause, or when the Owner has postponed or suspended the work;

b. When the Contractor has had to wait or suspend the work due to delay of the delivery of furnished materials or lent articles;

c. When the Contractor has not commenced, or has suspended the work due to delay of advance or partial payments;

d. When there exists any other reason for which the Owner or the Supervisor is responsible.

ARTICLE 18. DAMAGES CAUSED BY FORCE MAJEURE

(1) If any damage has been caused to any work, temporary construction facilities for the work, construction materials delivered to the site of the work (including furnished materials paid for by the Contractor) or construction equipment for the work, by natural or human cause for which neither party is responsible, the Contractor shall notify the Owner of the state of such damage promptly after its occurrence.

(2) The Owner shall bear the cost of the damages mentioned in the preceding paragraph (1) which are deemed to be serious after negotiation among the Owner, the Supervisor and the Contractor, and for which the Contractor is deemed to have used the care of a good manager.

(3) If such damage is covered by fire insurance, Contractor's all risk insurance or otherwise, the amount of such coverage shall be deducted from the amount of compensation to be borne by the Owner provided for in the preceding paragraph (2).

ARTICLE 19. LIABILITY INSURANCE

(1) During the performance of the work, the Contractor shall take out and maintain fire insurance or Contractor's all risk insurance upon any work done, construction materials delivered to the site of the work, etc. and shall present the insurance policies to the Owner. The same shall apply to any other insurance prescribed in the Design Documents.

–10–

Fig. 6-2. ***(continued)***

(2) If and when the Contractor takes out and maintains any insurance on the object of Contract and Construction materials, etc. other than those provided for in the preceding paragraph (1), the Contractor shall promptly notify the Owner thereof.

ARTICLE 20. COMPLETION; INSPECTION

(1) Upon completion of the work, the Contractor shall confirm that the work is in accordance with the Design Documents and request final inspection, and the Supervisor shall promptly comply with the request by inspecting the work in the presence of the Contractor.

(2) If the work does not pass inspection, the Contractor shall repair or reconstruct the work before the time for completion or before such time as may be designated by the Supervisor, and shall have the work re-inspected by the Supervisor.

(3) In compliance with the Supervisor's instructions, the Contractor shall remove temporary construction facilities, clean up the site, and perform other tasks before the time for completion or before such time as may be designated by the Supervisor.

(4) If, without justifiable reason the Contractor delays in taking the measures mentioned in the preceding paragraph (3), in spite of the Owner's notices to do so, the Owner may take such measures in lieu of the Contractor and charge the expenses thereof to the Contractor.

ARTICLE 21. PARTIAL USE

(1) The Owner may use parts of the work even during the progress of performance of the work, upon obtaining written consent of the Contractor. In such cases, the Owner shall be responsible for maintenance and custody of such portions as are used by the Owner.

(2) If any damage has been caused to the Contractor due to the partial use mentioned in the preceding paragraph (1), the Owner shall compensate for such damage.

Fig. 6-2. ***(continued)***

ARTICLE 22. APPLICATION FOR PAYMENT, PAYMENT; DELIVERY OF WORK

(1) When the work has passed the inspection mentioned in paragraph (1) of Article 20, the Contractor shall deliver the work to the Owner, and at that time the Owner shall make the final payment of the contract price to the Contractor.

(2) When any part of the work has passed the final inspection, the Owner may take delivery of such part, upon payment of the full amount of the value of such part of the work.

(3) The Contractor may apply for partial payment of the contract price prior to the completion of the work in accordance with the provisions of the Contract. If such partial payments are progress payments, the amount which the Contractor may apply for each progress payment shall be a sum equivalent to nine-tenths (9/10) of the value of any work and of approved construction materials which have passed the inspection of the Supervisor.

(4) If the Contractor applies for the partial payment mentioned in the preceding paragraph (3), he shall have the amount of payment approved by the Supervisor and submit an application to the Owner not later than the final date for the request of payment.

(5) In case advance payments are made to the Contractor, each progress payment due shall be calculated by the following formula:

$$A = \simeq B \times \left(\frac{\text{contract price} - \text{amount of advance payment}}{\text{contract price}} \right)$$

A = amount of payment applied for

B = amount payable under paragraph (3)

ARTICLE 23. GUARANTEE AGAINST DEFECTS

(1) If any defect is found in the work, the Owner may demand repair thereof by the Contractor, specifying a reasonable period therefore, or claim compensation for damages against the Contractor either in lieu of such repair

Fig. 6-2. ***(continued)***

or in addition to such repair. However, if the defect is of minor importance and excessive expense is required for the repair, the Owner may not demand such repair by the Contractor.

(2) The guarantee period for defects provided for in the preceding paragraph (1) shall be one (1) year in the case of wooden buildings, and two (2) years in the case of stone, metal, concrete or similar buildings, or any other permanent structures or ground works from the date of delivery mentioned in two paragraphs (1) and (2) of Article 22, provided, however, that if such defects have been caused by intent or by gross negligence on the part of the Contractor, the one (1) year period shall be five (5) years, and the two (2) years shall be ten (10) years.

(3) The Contractor shall not be liable for defects in equipment and fixtures for utilities, interior decorations, furniture, or other similar items, unless the Supervisor demands the Contractor repair or replace such defects immediately upon inspection at the time of delivery, provided, however, that the Contractor shall guarantee against latent defects for one (1) year from the date of delivery.

(4) Notwithstanding the provisions of paragraph (1), if and when the Owner finds any defect in the work at the time of delivery, the Owner may not demand repair or damages unless a written notice is given to the Contractor without delay; provided, however, that the same shall not apply, if the Contractor was aware of the defect.

(5) If the work has been destroyed or damaged by the defect mentioned in paragraph (1), the Owner may not exercise the rights mentioned in paragraph (1) unless it is within the periods prescribed in paragraph (2) and not later than six (6) months from the date of destruction or damage.

(6) The provisions of the preceding five paragraphs (1) through (5) shall not apply to those defects, destruction, or damage of the work which have been caused under any of the items of paragraph (4) of Article 14; provided, however, that this shall not apply in case the matter falls under paragraph (5) of the same Article.

Fig. 6-2. *(continued)*

ARTICLE 24. CHANGES IN WORK; CHANGES OF TIME FOR COMPLETION

(1) The Owner may, when necessary, order extra works or make changes in the work.

(2) The Owner may, when necessary, cause the Contractor to change the time for completion.

(3) If any damage has been caused to the Contractor by virtue of the preceding two paragraphs (1) and (2), he may demand the Owner to compensate for such damage.

(4) In case of addition or change of work, suspension of work by virtue of paragraph (1) of Article 28, force majeure or other justifiable causes, the Contractor may request the Owner to change the time for completion, clearly stating the reason therefore; the number of days of extension shall be determined by negotiation among the Owner, the Supervisor and the Contractor.

ARTICLE 25. CHANGES IN CONTRACT PRICE

(1) Either party may demand a change in the contract price in any of the following cases:

a. If extra work is ordered, or the work is changed;

b. If the time for completion is changed;

c. If the items, quantities, time of delivery, place of delivery, or place of return of furnished materials or lent articles are changed;

d. If, during the term of the Contract, the contract price is found to have become obviously inappropriate because of enactment, revision or abrogation of the laws, ordinances and regulations, or drastic changes in the economic conditions, etc. arising from unforeseeable causes;

e. If, in a long-term contract, the contract price for uncompleted parts of the work after one (1) year from the date of entering into the Contract is found to have become inappropriate because of enactment, revision or abrogation of the laws, ordinances and regulations, or changes in the commodity prices, wages, etc;

Fig. 6-2. ***(continued)***

f. If, in the case of resuming the work which has been suspended or has suffered from damage by disaster, the contract price is deemed to have become obviously inappropriate.

(2) In changing the contract price, the amount shall be determined by negotiation among the Owner, the Supervisor and the Contractor, based on calculation of decreased portions in accordance with the Itemized Statements, and increased portions in accordance with the current price.

ARTICLE 26. DELAY IN PERFORMANCE; LIQUIDATED DAMAGES

(1) If the Contractor is unable to deliver the work within the term of the Contract due to causes for which the Contractor is responsible, the Owner may, unless otherwise agreed on, for each calendar day of delay, demand of the Contractor liquidated damages at the rate of one-thousandth (1/1000) of the contract price less the contract price proportionate to completed parts of the work and to materials which have passed inspection or test.

(2) If the Owner is in arrears of the payment of the contract price or the proportionate contract price mentioned in paragraph (1) or (2) of Article 22, the Contractor may demand of the Owner liquidated damages at the rate of one-thousandth (1/1000) of the amount of the contract price in arrears for each calendar day of delay.

(3) If the Owner has delayed advance or partial payments, the provisions of the preceding paragraph (2) shall apply.

(4) If the Owner has delayed the payment mentioned in paragraph (2) of this Article, the Contractor may refuse delivery of the work. In such cases, any damage caused to the work, despite the Contractor's having exercised the same care in custody of the work as he would use for his own property, and the expenses specially needed for custody of the work shall be borne by the Owner.

ARTICLE 27. OWNER'S RIGHT TO SUSPEND WORK OR TERMINATE CONTRACT

(1) The Owner may, when necessary, suspend the work or terminate the

Fig. 6-2. ***(continued)***

Contract. The Owner shall compensate the Contractor for any damage the Contractor suffers because of such suspension or termination.

(2) The Owner may suspend the work or terminate the contract in any of the following cases and in such cases the Owner may demand of the Contractor damages:

a. If the Contractor fails, without justifiable reason, to commence the work even after the date of commencement;

b. If the progress of the work is extremely behind schedule and it is deemed unlikely that the Contractor will complete the work within the time for completion or within a reasonable time thereafter;

c. If the Contractor violates the provisions of Article 4 or paragraph (1) of Article 14;

d. If the Contractor fails, without justifiable reason, to comply with the Owner's request for negotiation under paragraph (2) of Article 25 and if it is deemed that he lacks sincerity for settling the matter in spite of the Owner's having given timely notice;

e. If the Contractor violates the Contract in any way other than the provisions of the preceding four items (a) through (d), and if the objective of the contract is deemed unattainable because of such violation;

f. If the Contractor is deprived of or loses the license for construction under the Construction Contractors Law;

g. If the Contractor proposes to terminate the Contract without having any of the reasons mentioned in items of paragraph (2) of Article 28.

ARTICLE 28. CONTRACTOR'S RIGHT TO SUSPEND WORK OR TERMINATE CONTRACT

(1) The Contractor may suspend the work in any of the following cases if it is deemed that the Owner lacks sincerity for settling the matter in spite of the Contractor's having given timely notice:

a. If the Owner has delayed advance or partial payments;

b. If the Owner fails, without justifiable reason, to comply with the Contractor's request for negotiation under paragraph (3) of Article 13, para-

Fig. 6-2. ***(continued)***

graph (4) of Article 24, or paragraph (2) of Article 25;

c. If the performance of the work is impossible because the Owner is unable to make the Construction Area available, etc. for the work mentioned in Article 2, or because of force majeure or other reasons; or

d. If the work is extremely delayed due to reason for which the Owner is responsible other than the preceding three items (a) through (c).

(2) The Contractor may terminate this contract in any of the following cases:

a. If the work is delayed or suspended due to reasons stipulated in the preceding paragraph (1) in excess of one fourth (1/4) of the time for completion, or for more than two (2) months;

b. If the contract price has been reduced by two-thirds (2/3) or more because of the Owner's having remarkably reduced the work;

c. If the Owner breachs the Contract and it is deemed that the fulfillment of the Contract has become impossible because of such breach;

d. If it has become clear that the Owner lacks the ability to pay the contract price.

(3) In each of the cases mentioned in the preceding paragraphs (1) and (2), the Contractor may claim damages against the Owner.

ARTICLE 29. MEASURES TO BE TAKEN UPON TERMINATION OF CONTRACT

(1) In the event of termination of the Contract, accounts shall accordingly be settled by negotiation among the Owner, the Supervisor and the Contractor, on the basis that all works done and the approved construction materials (including furnished materials paid for by the Contractor) are to be delivered to the Owner.

(2) In the event the Owner terminates the Contract under paragraph (2) of Article 27, if there be found any over-payment as the result of the settlement of accounts, the Contractor shall refund to the Owner the sum overpaid together with interest stipulated by laws which has accrued since the date of such payment.

(3) In the event of termination of the Contract, each party shall remove

Fig. 6-2. ***(continued)***

properties belonging to him and take other necessary measures such as cleaning up the site within a time limit as determined by negotiation among the Owner, the Supervisor and the Contractor.

(4) In case of unjustifiable delay on the part of either party in taking the measures mentioned in the preceding paragraph (3) in spite of a notice given by the other party, the latter may take the measures in lieu of the former and demand of him expenses therefor.

ARTICLE 30. SETTLEMENT OF DISPUTES

(1) If any dispute has arisen with regard to the Contract, either or both of the parties shall appoint a third party approved by the other party and entrust to such third party the settlement of the dispute, or the parties shall try to solve the dispute by mediation or conciliation by the Committee for Adjustment of Construction Work Disputes (hereinafter referred to as "Committee") stipulated in the Contract in accordance with the Construction Contractors Law. However, if the competent Committee is not designated in the Contract the Committee stipulated in paragraph (1) or (2) of Article 25-9 of the Construction Contractors Law shall be the competent Committee.

(2) Each party may submit the dispute to arbitration by the Committee on the basis of the Arbitration Agreement in case either or both of the parties deem that there is no likelihood of the dispute being settled in accordance with the preceding paragraph (1), or in case the Committee does not conduct mediation or conciliation or ceases the mediation or conciliation.

ARTICLE 31. SUPPLEMENTARY PROVISIONS

Any and all matters not provided for in the Contract or the present General Conditions shall be determined by negotiation among the Owner, the Supervisor and the Contractor, as needs arise.

Fig. 6-2. *(continued)*

〔裏面参照のうえ建設工事紛争審査会の仲裁に付することに合意する場合に使用する。〕

(This agreement is used in case the dispute is to be submitted to arbitration by the Committee for Adjustment of Construction Work Disputes, referring to the conditions on the reverse side hereof.)

仲　裁　合　意　書

ARBITRATION AGREEMENT

工 事 名 ....
Name of Work

工事場所 ....
Place of Work

昭和　　年　　月　　日締結した上記建設工事の請負契約に関し紛争が生じた場合は，四会連合協定・工事請負契約約款第30条(2)の規定にもとづき，建設業法により定められた下記の建設工事紛争審査会の仲裁に付し，その仲裁判断に服する。

In the event that disputes arise in connection with the Construction Contract performed on ...., for the above work, such disputes shall be submitted to arbitration by the following Committee for Adjustment of Construction Work Disputes stipulated in the Construction Contractors Law, according to Paragraph (2) of Article 30 of the General Conditions of Construction Contract.

管轄審査会名 .... 建設工事紛争審査会
Name of Competent Committee　　Committee for Adjustment of Construction Work Disputes

〔管轄審査会名の記入してない場合は建設業法第25条の9第1項または第2項に定める建設工事紛争審査会を管轄審査会とする。〕

(If the name of Competent Committee is not designated, the Committee for Adjustment of Construction Work Disputes prescribed in Paragraphs 1 and 2 of Article 25-9 of the Construction Contractors Law has jurisdiction.)

昭和　　年　　月　　日

Date:

発 注 者 ....
The Owner

同保証人 ....
The Surety

請 負 者 ....
The Contractor

同保証人 ....
The Surety

Fig. 6-3. Japanese Arbitration agreement.

on their prior verbal agreements, they would be given the same treatment as those 16th Century merchants, and would be ostracized from the group with respect to any future consideration for joint venture work.

The JV Job Site. Once the joint venture is established and the project site is being mobilized by the various contractors the physical signs of a JV abound. Individual company flags with their corporation logos emblazoned in bright colors will be flown side by side on the construction site. Project signs on the construction fencing around the site will announce "JV" in big, bold letters. The project engineers from each of the joint venture companies will swap their own distinctively colored hard hats with corporate emblems for ones on which the letters JV appear. The JV logo might also be embroidered on the site engineer's uniforms as if to announce to everyone that this is truly a joint venture effort, an entirely new entity created by the assemblage of the three, four or five construction companies whose corporate flags fly proudly side by side in a highly visible area on the job site.

Dango. There is a darker side to these joint ventures that remains somewhat vague and elusive, and that is the so-called practice of *dango*—a system whereby joint consultations are allegedly held by contractors before the submission of a bid on public works projects to determine whose "turn" it is to win the contract.

In years gone by, it was rumored that contracts on public works projects were awarded not so much on the basis of the tender system as by the *dango* system. The "successful" bidder, having been determined before the bid opening, would show his appreciation to his fellow contractors in one form or another, and in turn, if money was exchanged, a portion of the funds channeled back to the other contractors would somehow find their way to an appropriate political individual or political entity.

The practice of *dango* has been declared illegal by the federal government in Japan for some time now. The Anti-Monopoly law makes this kind of collusive activity a criminal offense and assigns the responsibility for policing to the Fair Trade Commission. The Japan Civil Engineering Industry Contractor's Association, known as *Dokokyo,* takes the position that *dango* is merely a leveling device to keep the market forces in check. It helps in keeping Korean and Chinese contractors, and their cheap labor force, out of the domestic Japanese construction market, thereby stabilizing wages.

Although illegal, *dango* is said to be an acknowledged practice on the Japanese construction scene today, parceling out work to many contractors and acting as a conduit for a gigantic political action committee where funds are channeled back into the political system to insure that the flow of public works projects continues when and where needed. The name *dango* might not be too familiar to American contractors, but the term "pork barrel" is one that is certainly familiar to quite a few politicians and to many U.S. contractors. Sometimes East does meet West.

The Bill of Quantities Contract

When contracting with government agencies outside of Japan, a type of contract known as the Bill of Quantities is often used. Sometimes private clients will also request that Japanese contractors use this form of contract when working on projects in other Southeast Asian countries.

The Bill of Quantities contract (Fig. 6-4) owes its origins to the British and the influence they exerted in East Asia during the era of colonialism in the 18th and 19th Centuries. Essentially this type of contract is nothing more than a lump sum contract which contains a detailed list of all items included in the estimate along with corresponding unit prices for each of the estimated items.

This detailed list of labor, material, and equipment elements from the estimate becomes an integral part of the contract and actually defines the scope of the work represented by the lump sum agreement, since it represents the entire sum of items included in the lump sum price.

The Bill of Quantities Listing will almost be a mirror image of the contractor's detailed estimate, but with certain refinements. Whereas a general contractor may use a subcontractor's lump sum price in the calculation and assembling of a lump sum price, in the Bill of Quantities Listing, the detailed breakdown of each subcontracted item of work must be clearly shown, along with a unit price alongside each line item.

The Bill of Quantities Concept. The concept behind this type of contract makes a great deal of sense. The contractor will have based his lump sum price upon a detailed takeoff of all the necessary parts contained in the plans and specifications which were given to him by the owner so that an estimate could be made. The client is then entitled to receive all the parts that are contained in the Bill of Quantities, and if the actual quantities put into place are less than those spelled out in the Bill of Quantities, the owner is due a credit for the deleted items at the unit cost contained in the listing.

Conversely, if more items of a particular trade or trades are added to the work, the contractor is then entitled to additional compensation based upon the number of these items in excess of the number in the listing multiplied by their unit prices. This concept is not too dissimilar from a standard contract in the U.S. with unit prices included. But

(text continued on p. 204)

Standard Contract Form for Overseas Works by International Tender

GOVERNMENT OF ....

CONSTRUCTION & COMPLETION OF ....

PUBLIC WORKS DEPARTMENT CONTRACT NO. ....

ARTICLES OF AGREEMENT made and entered into this .... day of .... BETWEEN the Government of .... (hereinafter called "the Government") by .... of the one part and .... (hereinafter called "the Contractor") a Company incorporated in accordance with the Laws of .... and having its registered or principal office of business at .... of the other part.

WHEREAS:—

A. The Government is desirous that such Works as are herein mentioned for the construction of .... and other works in connection therewith shall be executed constructed completed and maintained in the manner and during the period hereinafter provided.

B. The Contractor has submitted a Tender dated the .... day of .... to execute construct complete and maintain the Works for the sum of .... or such other sum as may become payable under the terms hereof upon the terms during the period and in the manner hereinafter mentioned which Tender has been accepted by the Government.

C. It has been agreed between the parties hereto that these presents shall be entered into by way of Contract and that the Contractor shall secure the due performance of the Contract by the joint and several Bond of himself and in tho sum of $ ....

NOW THIS DEED WITNESSETH AND IT IS HEREBY AGREED AND DECLARED AS FOLLOWS:—

1. IN these Articles of Agreement words and expressions shall have the same meanings as are respectively assigned to them in the Conditions of Contract.

2. THE followinn documents shall be deemed to form and bo road and construed as part of these Articles of Agreement viz.

(a) The Tender above recited together with tho Letter of Acceptance thereof.

(b) The Conditions of Contract hereunto annexed.

(c) The Specification hereunto annexed.

Fig. 6-4. Typical bill of quantities contract.

(*d*) The priced Bill of Quantities and Schedules thereto.

(e) The Drawings.

3. IN consideration of the payments to be made by the Government to the Contractor as hereinafter mentioned the Contractor hereby covenants with the Government to execute construct complete and maintain the Works in conformity in all respects with the provisions of the Contract.

4. IN consideration of the execution construction completion and maintenance of the Works the Government hereby covenants to pay to the Contractor the said sum of .... or such other sum as shall become payable under the terms hereof by such instalments at such times and in such manner as is prescribed in the Contract.

5. THE Engineer's order to commence work was received by the Contractor on the .... day of ....

IN WITNESS whereof ....

.... on behalf of the Government of .... has hereunto set his hand and seal and the Common Seal of the Contractor has been hereunto affixed the day and the year first above written.

SIGNED SEALED and DELIVERED
by the said
in the presence of:—

For and on behalf of the Government

The COMMON SEAL of the Contractor was hereunto affixed in the presence of:—

Fig. 6-4. *(continued)*

GOVERNMENT OF ____________

(Ministry of ____________

Tender for CONSTRUCTION AND COMPLETION OF ____________

FORM OF AGREEMENT

THIS AGREEMENT made the .... day of .... 19 .... between THE MINISTER OF .... Government of ...., (hereinafter called the "Employer" or "the Administration") of the one part and .... (hereinafter called "the Contractor") of the other part

WITNESSETH THAT:

WHEREAS, the Administration is desirous that certain Works should be constructed, viz. the CONSTRUCTION AND COMPLETION OF .... and has accepted a tender by the Contractor for the construction, completion and maintenance of such Works.

NOW, THEREFORE it is agreed as follows:

1) In this Agreement words and expressions shall have the same meanings as are respectively assigned to them in the Legal Clauses and Conditions hereinafter referred to.

2) The following documents shall be deemed to form and be read and construed as part of this Agreement, viz.:

DOCUMENT	I	:	TENDERING PROCEDURE:
	I–1)	:	Invitation to Tender
	I–2)	:	Instructions to Tenderers
	I–3)	:	Form of Tender and Appendix
	I–4)	:	Plant Register
	I–5)	:	Contractor's Staff
	I–6)	:	Statement of Works
	I–7)	:	Initial Bond
	I–8)	:	Final Bond
	I–9)	:	Form of Agreement
DOCUMENT	II	:	GENERAL CONDITIONS:
	II–1)	:	Legal Clauses and Conditions
	II–2)	:	Tenders Law (Law No.......)

Fig. 6-4. ***(continued)***

DOCUMENT III : TECHNICAL SPECIFICATIONS AND APPENDICES:

III-1) : Technical Specifications:-

III-1-2/1 : Technical Specifications for Civil and Building Works.

III-1-2/2 : Technical Specifications for Electrical Installations.

III-1-2/3 : Technical Specifications for Plumbing, Gas and Oil Tank Installations.

III-1-2/4 : Technical Specifications for Heating, Ventilating and Air-Conditioning.

III-1-2/5 : Technical Sprecifications for Power Supply Equipment and Power Generating Facilities.

III-2) : The Appendices which Include:-

III-2-1 : The Drawings.

III-2-2 : The Schedule of Rates, priced Bill of Quantities and priced Schedules:

III-2-2/1 : The Schedule of Rates and priced Bill of Quantities for Civil and Building Works.

III-2-2/2 : The Schedule of Rates and priced Bill. of Quantities for Electrical Installations.

III-2-2/3 : The Schedule of Rates and priced Bill of Quantities for Plumbing, Gas and Oil Tank Installations.

III-2-2/4 : Priced Schedules for Heating, Ventilating and Air-Conditioning.

III-2-2/5 : Priced Schedules for Power Supply Equipment and Power Generating Facilities.

III-2-3 : The Analysis of Prices which is submitted by the Tenderer.

Supplement I Dayworks

3) In consideration of the payments to be made by the Administration to the Contractor as hereinafter mentioned the Contractor hereby covenants with the Administration to construct complete and maintain the Works in conformity in all respects with the provisions of the Contract.

4) The Administration hereby covenants to pay the Contractor in consideration of the construction completion and maintenance of the Works the contract sums at the time and in the manner prescribed by the Contract.

Fig. 6-4. ***(continued)***

IN WITNESS whereof the parties hereto have hereunder set their respective hands and seals the day and year first above written.

Signed by the Administration: ....

(MINISTER OF ....)

In the presence of: ....

Address: MINISTRY OF..

Designation: UNDER-SECRETARY

LOCAL PARTNER OR AGENT:	THE TENDERER:
Name:	Name:
Signature:	Signature:
Address:	Capacity:
	Address:

Both duly authorised to sign Tenders for and on behalf of Messrs. ....

....

Dated this .... day of .... 19.................

In the presence of: ....

Address : ....

Designation : ....

Date : ....

Fig. 6-4. ***(continued)***

PRELIMINARIES

PRICING

The Contractor shall satisfy himself as to the meaning of every item in the Bill of Quantities and the rates and prices inserted by him are deemed to cover all his obligations under the Contract and all matters and things necessary for the proper construction, complection and maintenance of the works. No claim for additional payment will be allowed for any error or misunderstanding by the Contractor in these respects.

Any item not priced in the Bill of Quantities shall be deemed to be covered by other prices or rates in the Bill of Quantities.

In the event of any discrepancies between the details on the drawings and the description in the Bill of Quantities then the items are deemed to be priced in accordance with the descriptions in the Bill of Quantities.

All materials or items specified as being "as manufactured by" (or otherwise similarly described) with reference to a particular company, manufacturer or trade name shall be priced as being obtained therefrom.

Should the Tenderer wish to offer an alternative material or item he shall supply details of such alternative at the time of tendering in a seperate form which shall have no effect on the tender.

The Bill of Quantities shall be read with conjunction with the drawings listed in the Technical Specifications.

On the request of the Engineer or any person nominated by him, the Contractor is to produce a detailed analysis of all or any of the unit rates contained in his Bill of Quantities within fourteen days of such request.

SITE OF WORK

The site of the work is as shown on the drawing of site plan. The Contractor shall be deemed to have visited the site, to have taken into consideration all local and existing conditions and to have made himself thoroughly acquainted with the position and accessibility of the propsed work and the conditions under which the work will have to be carried out. No claim for want of knowledge in this respect will be entertained.

DESCRIPTION OF WORKS

The work comprises the construction and completion of following buildings with electrical installations, plumbing works, heating ventilating and air-conditioning, lifts, power supply and power generating facilities, paving and ancillary works.

Fig. 6-4. ***(continued)***

PREAMBLE

PRICING

The Contractor shall satisfy himself as to the meaning of every item in the Bill of Quantities and respective items in the Technical Specifications.

The rates and prices inserted by him are deemed to cover all his obligations under the Contract and all matters and things necessary for proper construction, completion and maintenance of the work.

They shall also be deemed to include the cost as follows:

a) Labour and all costs in connection therewith, including supervision.

b) Materials, goods and all costs in connection therewith (e.g. conveyance, delivery, handling and storing).

c) Fitting and fixing materials and goods into position.

d) Use of all plant, scaffolding, tools and equipment and all temporary work.

e) Cutting and waste of materials; work at any height or depth; all work in narrow widths, confined spaces or small quantities.

f) Establishment charges, overhead charges and profit.

g) All other costs, risks, expenses and obligations relating to the particular item, set forth or implied in the contract documents.

NOTE: 1) U. R. L. columns shall be filled by tenderer with unit rates in letters.

Fig. 6-4. ***(continued)***

instead of the usual three to four pages of unit prices common to many U.S. contracts, particularly on speculative office buildings where tenant fit-up unit prices are commonplace, the Bill of Quantities listing contains almost every single nut and bolt needed to build the project—for every trade.

The Electrical portion of the Listing will include lineal meters of each size of conduit to be used, with unit prices for each size. The number and types of wiring devices will be displayed, along with unit prices for same. Lighting fixtures will be listed by type and price, panel boxes will be shown by type size and price, along with other distribution equipment, transformers, and so on—each with their unit price. When sprinkler work is included in the project, the Listing will include a complete quantity listing of each size main, branch, and riser, with unit price, the number, type, and cost of hangers and supports, as well as an itemized, priced list of valves by size, type, and material of construction. Pumps, motors, hose cabinets, and sprinkler heads are also included in much the same manner.

Costs to be Included in the Bill of Quantities Listing. The unit price for materials and equipment contained in the Bill of Quantities Listing is to include the following:

1. Labor rates incorporating all other costs related to the cost of labor, such as taxes, cost of benefits, and other items generally considered labor "burden" costs in the U.S.
2. Material prices are to include delivery costs, handling, storage, and distribution costs.
3. The cost to fit and place all materials and equipment into the project, including the waste factor generally factored into various materials.
4. Scaffolding costs, small tools, and equipment required for the installation of all materials and equipment.
5. The cost of labor should take into account any inefficiencies caused by working in confined spaces or at abnormally high elevations.
6. Corporate overhead and profit.
7. Any and all other costs associated with the work must be included such as service costs anticipated for maintenance of equipment during the warranty period.

There are clauses in the Bill of Quantity Contract that prohibit a contractor from claiming any additional payment for items that were not included in the Listing due to his misinterpretation of the scope of the work and therefore inadvertently omitted. If there is a discrepancy between the details shown on the drawings and the description of the work contained in the Listing, the client will accept, as the contract obligation, the items and their prices in the Bill of Quantities.

If a contractor wishes to submit alternate materials and/or equipment, he is free to submit a list of these substitutions when he tenders his bid, and if any are accepted they are listed in the Bill of Quantities on a separate form for Alternates.

Format of a Bill of Quantities Contract. A typical Bill of Quantities Contract was inspected by the writer at the headquarters of Ohbayashi Corporation in Tokyo. Ohbayashi chooses to keep their construction contracts in the form of bound volumes much like a book. Each contract has its own blue leatherlike binding, imprinted with the name of the client and the project identification. This particular job was a bank building in a Southeast Asian country outside of Japan.

Mr. Takashi Akaogi, Manager of Public Communications for Ohbayashi, explained that prior to bidding this job a pre-construction conference was held at the job site with all bidders and questions were solicited from all attendees. The client then responded to all of these questions in writing, and when the contract was finally awarded, the complete list of questions and answers became a part of the contract documents. Mr. Akaogi explained that this procedure is not adhered to when working in Japan, but it is frequently employed when Japanese contractors work elsewhere in Southeast Asia.

The Ohbayashi Lump Sum–Bill of Quantities bound contract consisted of the following elements:

1. The list of pre-bid questions and answers.
2. The simple one page construction contract with signatory page.
3. A copy of the standard general conditions to the construction contract (Japanese version).
4. A brief one page description of the project.
5. A completed, reduced set of contract drawings along with the specifications.
6. The Bill of Quantities list with a summary heading for each major component of construction, such as General Requirements, Concrete, Carpentry and millwork, Mechanical and Electrical, etc., under which the detailed breakdown and unit prices were listed.

Mr. Akaogi said that during the progress of this particular project, if the client added or deleted any of the items included in the Bill of Quantities, it would be done based upon the established price for the item. Certain ornamental door pulls were included at a unit cost of 184,000 yen (about $1415) and if the owner had wished to add three more to the job, the contract sum would have been increased by 552,000 yen.

The Bill of Quantities for this project showed the actual estimated square meters of 15 mm sheetrock and other interior finish work such as the area of vinyl flooring and the thickness of the flooring material. A very extensive unit price breakdown for mechanical, electrical, and sprinkler work was also in this listing. The quantities and unit prices in these listings are not only used by the client to add or delete items from the scope of the project; the Bill of Quantities also acts as a guide when the monthly requisitions are being reviewed by the owner. This particular project was not a very large one, so the Bill of Quantities was not that complex or extensive, but Mr. Akaogi said that there are times when a Bill of Quantities Listing will contain several hundred 8½″ × 11″ pages of detailed breakdowns.

The Contract Between General Contractor and Subcontractor

The contract between general contractor and subcontractor, as illustrated in Fig. 6-5, is a rather simple form, but it does contain at least two unique clauses, one dealing with advance payments and the other with method of payment. Long standing relationships notwithstanding, the subject of payment from general contractor to subcontractor is of concern in Japan, and in fact the procedures to be followed have been stated in a federal law governing construction. The Construction Contractors Law enacted by the federal government in Japan in 1949 and revised several times since devotes several articles to the subject of subcontractor payments. The text of portions of Articles 24-3, 24-4, and 24-5 of the Construction Contractors Law are reproduced below in English translation.

Construction Contractor Law As It Relates to Subcontractor Payments

Article 24-3—(Payment of Subcontract Cost)

1. When the main contractor has received the payment for contract cost for the piece work completed on or after completion of all the work, he shall pay the subcontract cost equivalent to the rate of the amount paid to him against the piece work completed and equivalent to the piece work completed by the subcontractor concerned to the subcontractor who executed the construction work for the payment concerned within one month, or as short period as possible, after he received such payment.
2. When the main contractor has received advance payment, he shall take thoughtful consideration for payment of expenses as an advance so that the subcontractor can purchase materials, recruit workers, and start the construction work.

Standard Form of Sub-contract Agreement between Prime Contractor and Prime Sub-contractor

1. Name of the Project:

2. Construction Site:

3. Term of the Works: Commencement Date:

   Completion Date:

4. Contract Sum:

5. Time and Method of Payment of Contract Sum

   (1) Advance payment: Within ( ) days after signing of the Contract.

   Amount: Cash or Draft: (Type of draft, ratio, usance)

   (2) Partial payment: Month: Last Date:

   Cash or Draft:

   Month: Due Date:

   (3) Payment at the time of the delivery of the Works:

   Within ( ) days after receipt of a demand for payment. Cash or Draft:

   * The amount proportionate to labour cost shall be, as a general rule, in cash.

6. Others

   Of the ( ) Works ordered by the Employer ( ), the above mentioned Works shall be duly executed by the undersigned under the annexed provisions of the Contract hereon concluded on equal status by both parties concerned.

   As evidence of the Contract, this Contract shall be made in duplicate each one of which shall be kept by each party concerned after being signed by both parties (and sureties).

   Date:

   Primary Contractor: (Surety for pecuniary liability:)

   Address:

   Name:

   Subcontrator: (Surety for pecuniary liability:)

   (Surety for completion of the Works:)

   Address:

   Name:

Fig. 6-5. Typical contract between general contractor and subcontractor.

Article 24-4—(Inspection and Delivery)

1. When the main contractor is notified from the subcontractor of completion of the work, he shall complete the inspection for confirming completion of the work within twenty days, or as short period possible, after he receives such notice.

Article 24-5—(Date of Payment, etc. for Subcontract Cost by Specific Contractor)

1. The date of payment for subcontract cost in the subcontract for which a specific contractor is the person placing an order (except where the contractor in a subcontract is a specific constructor or corporation of which the capital exceeds the value specified by the government ordinance, the same shall be applicable to this Article) shall be fixed so that it is within fifty days, or as short period as possible after the request specified in Paragraph 2 of the preceeding Article (in the case of the proviso of the same Paragraph, the said certain day, the same shall be applicable to this Article).
2. In the subcontract in which the person placing an order is a specific constructor, the date of the request specified in Paragraph 2 of the preceeding Article when the date of payment for the subcontract cost is not fixed or the fiftieth day counting from the date of the request specified in Paragraph 2 of the preceeding Article when the date of payment is fixed in violation of the preceeding Paragraph shall be deemed as the fixed date of payment for the subcontract cost.
3. For the payment of the subcontract cost with regard to the subcontract for which the person placing an order is a specific constructor, the said constructor shall not give a note which is deemed to be difficult to receive a discount from general banking facilities (organization which operates receiving deposits or savings and financing) until the date of payment for the subcontract cost concerned.

[Author's note: This practice of paying subs via promissory note is discussed more fully in this chapter.]

4. The specific constructor shall pay the subcontract cost with regard to the subcontract for the person placing an order is the specific constructor concerned until the date of payment specified in Paragraph 1 or Paragraph 2. When the specific constructor concerned does not make payment, he shall pay to the subcontractor the amount, as a respite interest, equivalent to the unpaid amount multiplied by the rate specified by the Ordinance of the Ministry of Construction for the term from the fiftieth day after the request specified in Paragraph 2 of the preceeding Article until the date of actual payment for the subcontract cost concerned.

Although the existence of a federal law mandates prompt and proper payment from contractor to subcontractor and even dictates that interest must be paid on all overdue payments, there seems to be an incongruity in the entire payment process.

Subcontractor Payment and the Promissory Note. Just as in the United States, it is fairly typical in Japan for a subcontractor to submit a requistion for payment at the end of each month which includes the cost of his labor, materials, work in place, and equipment stored on the jobsite for the preceeding thirty days. Payment is expected within a 30 day period following the submission of the requisition. But here is where American and Japanese practices differ. In many cases it is standard practice in Japan for a subcontractor to be paid an amount equal to 30% of his monthly requisition in cash or check and the balance of 70% in the form of promissory note due in 120–150 days. When requisitions are paid in this fashion, it is, however, customary for the subcontractor to receive cash or check for 100% of the cost of labor in the requisition.

The subcontractor now has a promissory note which will not become due and payable for four to five months, and if he cannot wait the 120 to 150 days for the note to come due, he can take the note to a bank and discount it. There is a provision in the Construction Contractors Law dealing specifically with the practice of issuing promissory notes to subcontractors and it states that the "constructor shall not give a note which is deemed to be difficult to receive a discount from general banking facilities." Since interest rates in Japan are relatively low, a subcontractor could use one of these promissory notes as collateral for a short term loan if cash were needed.

This somewhat odd system of paying for goods and services and materials and equipment via the use of promissory notes does not stop at the general contractor–subcontractor level, but extends beyond this point. When a subcontractor needs to purchase materials or equipment from his customary source of supply he has the option of paying for his purchases by giving his vendor a promissory note in the amount of 100% of the total order placed. So it seems as though the entire construction industry is floating on a sea of short term promissory notes.

Although the percentage of cash to note is a negotiable issue between contractor and subcontractor at the time of final negotiations and contract signing, this practice seems to be incongruous with the nurturing and guiding principles that the Japanese government and general contractors proport to pursue in their relationships with subcontractors.

The Government's Influence on the Relationship Between Contractor and Subcontractor. The Ministry of Construction has no authority to dictate how relationships and contractural arrangements are conducted between contractor and subcontractor, but they seem to have gone to great lengths to make suggestions. The Ministry of Construction recommends that all general contractors issue formal written contracts to their subcontractors (Fig. 6-5) and not rely on just verbal commitments. The Ministry also suggests ways in which general contractor and subcontractor should relate to each other after the contract is signed, and they have formalized these suggestions in a series of guidelines which were first printed in 1978 and are still being recommended today. The Ministry published these guidelines in a document which has been translated by Professor Yasuyuki Hippo, Faculty of Economics, Rissho University, Tokyo. It is reproduced below with Professor Hippo's permission.

Standard Form of Sub-contract Agreement between Prime Contractor and Prime Sub-contractor

1. Name of the Project:
2. Construction Site:
3. Term of the Works: Commencement Date:
   Completion Date:
4. Contract Sum:
5. Time and Method of Payment of Contract Sum
   (1) Advance payment: Within ( ) days after signing of the Contract.
   Amount: Cash or Draft: (Type of draft, * ratio, usance)
   (2) Partial payment: Month: Last Date:
   Cash or Draft:
   Month: Due Date:
   (3) Payment at the time of the delivery of the Works:
   Within ( ) days after receipt of a demand for payment. Cash or Draft:

*The amount proportionate to labour cost shall be, as a general rule, in cash.

6. Others

Of the ( ) Works ordered by the Employer ( ), the above mentioned Works shall be duly executed by the undersigned under the annexed provisions of the Contract hereon concluded on equal status by both parties concerned.

As evidence of the Contract, this Contract shall be made in duplicate each one of which shall be kept by each party concerned after being signed by both parties (and sureties).

Date:

Primary Contractor: (Surety for pecuniary liability:)

Address:

Name:

Subcontractor: (Surety for pecuniary liability:)
(Surety for completion of the Works:)

Address:

Name:

GUIDELINES FOR RATIONALIZATION OF THE RELATIONSHIP BETWEEN CONTRACTOR AND SUBCONTRACTOR

1. *Basic Principles*

Construction works are conducted by the combination of various types of works and therefore it is unavoidable to involve subcontractors depending upon the nature and scale of works.

In case of part of the work to be done by subcontractor, the contractor must secure proper execution of work by subcontractor so that the contractor may carry out the works properly in accordance with the contract with the owner.

For this purpose, the contractor needs to select the competant subcontractor and to establish the reasonable relationship between both parties.

In the meantime, the subcontractor is expected to improve execution capability, management of workers, work safety and control, etc., in order to ensure the proper execution of work according to the contract with the contractor.

Further, the contractor is requested to provide guidance and advice in respect of the above-mentioned measures to be taken by subcontractor.

Aforementioned measures are necessary not only to secure the proper implementation of work but also to facilitate further development of competent subcontractors, which in turn would lead to the sound development of the subcontractors as well as to the development of the construction industry as a whole including contractor and subcontractor.

In view of the above principles, this guideline spells out the necessary items concerning measures to be taken by both contractor and subcontractor.

2. *Prohibition of Blanket Subcontracting*

The Construction Contractors Law prohibits in principle general subcontracting because it would create a number of problems such as unreasonable exploitation of profit in the process of subcontracting operation resulting in low quality of works and aggravated labor conditions of workers under subcontractors, unclear location of responsibility in the actual execution of work, act against the owner's trust, etc.

All efforts should be made to avoid general subcontracting even in the case of work approved by the owner.

Further, multi-layered subcontracting which is not called for should also be avoided since it would also bring about similar problems.

3. *Selection of Subcontractor*

When the contractor (who is the orderer in the subcontracting contract including not only the major contractor who directly receives the work from the owner, but also other contractors who indirectly receive the work under the subcontracting contract) selects the subcontractor (who receives the order under subcontracting contract including not only one who receives part of the work from the person directly contracted by the owner, but also other subcontractors who indirectly receive work under the subcontracting contract), a subcontractor who has not received an official permit under the Construction Contractors Law with reference to the execution of the work concerned or who is prohibited or suspended to carry out the operation should not be qualified for selection. Furthermore, the contractor should make efforts to select competent subcontractors taking into account:

(1) Capability of execution of work.
(2) Position of management of workers and work safety and control.
(3) Position of worker's welfare.
(4) Position of business transaction with subcontractors.

In case of selection, care should be taken to fully meet the following aspects:

(1) Satisfactory work record.
(2) Technical capability to execute the work.
(3) Security of work force adequate to execute the work.
(4) Security of machinery and tools adequate to execute the work.
(5) Security of legally qualified staff adequate to execute the work.
(6) Satisfactory financial record.
(7) Appointment of a personnel manager at each workplace.
(8) Preparation of work rules and submission of the same to the Labor Standards Inspection Office in case of a subcontractor who regularly employs more than 10 workers at one workplace.
(9) Low rate of work accidents.
(10) Low possibility of failure to pay wages.
(11) Preparation of lodging rules and submission of same to the Labor Standards Inspection Office in the case of a subcontractor who lodges workers in a dormitory.
(12) Low possibility of failure to pay the subcontracting amount in case part of the work has to be re-subcontracted due to its nature.

4. *Agreement of Reasonable Subcontracting*
Prior to the commencement of the construction works, both contractor and subcontractor shall enter into a written subcontracting agreement in accordance with the Provisions of the Standard Subcontracting of Construction Works recommended by the Central Council on Construction Contracting Business in 1977.

5. *Payment by the Contractor*
The contractor is requested to properly put into effect the items covered by the concerned subcontracting agreement and also to observe the items relating to the subcontracting agreement prescribed by the Construction Contractors Law as well as the items stipulated in the following:

(1) Upon completion of the subcontracting agreement, the contractor should not reduce the amount to be paid to the subcontractor without due reason.
(2) In case of the necessary materials for the work to be purchased from the contractor, he should not request the subcontractor to make the payment for them prior to the due date of the payment without justifiable reason.
(3) Payment shall be made most preferably in the form of cash. In case payment shall be made in both cash and promissory note, payment equivalent of wages should at least be made in cash.

(4) Due date of the promissory note shall be as short as possible.

6. *Personnel Management in Subcontracting*

The subcontractor is requested to properly put into effect the items stipulated in the concerned subcontracting agreement and further to observe the items indicated in the following.

Also the contractor who directly receives an order from the owner is requested to implement proper management of work progress, to observe the law concerning the improvement of employment status of construction workers and the Industrial Safety and Health Law, to take action as to the payment of insurance premiums for labor accidents, as well as to provide guidance, assistance, and other help to the subcontractor in the following items.

Further, the contractor who indirectly receives an order from the owner is also requested to extend his cooperation to the major contractor in providing guidance, assistance, and other help.

(1) A personnel manager shall be appointed, for whom efforts shall be made to gain and improve his knowledge of personnel management.
(2) Recruitment of workers shall be organized in a proper manner.
(3) In the recruitment of workers, proper working conditions shall be prepared which shall be presented at the time of recruitment, and an official document relating to employment shall be issued.
(4) Work rules shall be prepared which shall be submitted to the Labor Standards Inspection Office in the case of a subcontractor who employs more than 10 workers at one workplace.
(5) Contractors to whom the previous item (4) does not apply shall also be encouraged to prepare similar working rules.
(6) *Full* wages shall be paid *directly* to the workers in *cash more than once a month* at a *prescribed date.*
(7) A list of workers and a wage register shall be prepared in order.
(8) Management of work hours shall be properly maintained.
(9) Technical training shall be provided to workers.
(10) Safety and health education shall be provided to workers who are newly recruited, who receive revised a job assignment, who carry out dangerous or toxic operations, or who are newly appointed as supervisor.
(11) In addition to the aforementioned item (10), safe operation of work shall be maintained in accordance with the Industrial Safety and Health Law.
(12) Accidents should be reported to a contractor who receives an order from the owner either directly or indirectly.
(13) Payment of insurance premiums relating to employment insurance, health insurance (including insurance applicable to daily-wage workers), and welfare annuity insurance shall be made in a proper manner. Guidance shall be provided to encourage those workers to whom the health insurance does not apply to enroll in national health insurance such as the national civil engineering and construction health insurance society, and those workers to whom the welfare annuity insurance does not apply, to join the national annuity system.

(14) The subcontractor is encouraged to join the worker's compensation insurance system so that compensation for accidents is fully provided for.
(15) The subcontractor is encouraged to join the construction industry pension cooperative so that pension systems may be established.
(16) A medical check-up of regular workers shall be conducted at the time of employment and at regular intervals, and efforts shall be made to carry out similar check-ups on those workers who are not employed on a regular basis.
(17) When workers stay at lodging facilities annexed to the workplace, lodging rules shall be prepared which shall be submitted to the Labor Standards Inspection Office. Facilities have to be properly maintained.
(18) In case of the above (17), the standards relating to facilities and safety and health prescribed under the regulations concerning construction industry lodging facilities shall be observed.
(19) In addition to the items prescribed above from (1) to (18), item 3 of Article 7 in the Construction Contractors Law shall also be observed.

7. *Guidance Given to Other Contractors by the Major Contractor who Directly Receives an Order from the Owner*
The contractor who directly receives an order from the owner shall provide guidance to other contractors involved in the process of the work so that they may also observe the items mentioned in this guideline from Articles 2 to 5.

SETTLING DISPUTES IN JAPAN

Where there are contracts there is the potential for contract disputes, and the Japanese are no exception to the rule. In a country where lawyers are not used with frequency to settle disputes, grievances involving construction matters need to be resolved without resorting to legal action and the courts. When disputes arise and cannot be settled in the normal course of business, reconciliation by mediation, conciliation, and/or arbitration provides effective ways to obtain a resolution, and the Japanese government stands ready to assist. The Ministry of Construction, recognizing the need to create an agency where complaints can be filed and the need to create a forum in which the complaintants can air their complaints, stepped into the dispute picture in an advisory capacity.

The Committee for Adjustment of Construction Work Disputes

In 1956, the Construction Contractors Law established the Committee for Adjustment of Construction Work Disputes, more commonly known as the Examination Council. Recognizing that the number and types of construction disputes can range from very minor, local situations to complicated disputes, the law established procedures for handling both types of complaints. Table 6-1 shows the range of disputes handled at all levels in Japan in 1987.

Since contractor's licenses are issued by the Ministry of Construction when a construction company has branch offices in more than one prefecture, disputes involving these licensees would tend to be complicated. Licenses issued by the prefectures would generally indicate smaller, local contractors and possibly less complicated disputes. This distinction, based upon the licensing authority, was the basis for the types of mechanisms established by the Ministry to deal with complaints.

The Central Examination Council was set up to deal with disputes involving one or more parties that had been issued a contractor's license by the Ministry. A Prefecture Examination Council was established in each prefecture in the country to handle disputes registered by one or more parties with a contractor's license issued by the governor of that prefecture. Where a contractor's license was not required by either the Ministry of Construction or the prefecture because the size of the construction company and the dollar value of the work did not require a license, the case would be heard at the prefecture level.

The Councils at both the federal and prefecture level were established to assist individuals, contractors, subcontractors, and corporations in settling disputes by offering them an opportunity to meet with a Council and state the nature of their complaint. The Council would then suggest one of three ways in which the disagreements should be heard, through proceedings involving conciliation, mediation, or arbitration. If both parties to a dispute agreed to discuss their complaints in the forum suggested by the Council, an application for a hearing would be filed along with an application fee. If the Council felt that there might be extraordinary expenses incurred during the hearing process, they would request that the parties prepay an estimated sum to cover these expenses.

At that time, a three member committee would be formed by the Council consisting of a lawyer, an engineer, and a contractor so that the parties to the dispute could commence with the hearing. Depending upon which type of hearing was selected, the Council might ask both parties to submit written documentation to support their claim, or the committe members might decide that they should visit the site where the disagreement occured to inspect the problem first hand. At the Ministry level it may be decided that the best way to handle a particular dispute is to appoint a single Contractor Dispute Officer, rather than appoint a committee to listen to both sides of the complaint. The nature of the hearing, whether it be conciliation, mediation, or arbitration, is informal, and the recomendations of the committee or Dispute Officer are not binding in any way.

Two members of the Ministry of Construction explained the purpose

Table 6-1. Construction Disputes Registered at Federal and Prefecture Levels in 1987.

PARTIES FILING DISPUTES

Party	Central Council No. of Cases	Central Council %	Prefecture Council No. of Cases	Prefecture Council %	Total No. of Cases	Total %
Individual client/contractor	19	40%	79	43%	98	42%
Corporation/contractor	6	13%	11	6%	17	7%
Contractor/client	11	23%	41	22%	52	22%
Contractor/corporation	2	4%	19	10%	21	9%
Subcontractor/prime	9	19%	29	16%	38	6%
Prime/subcontractor	0	0%	3	2%	3	1%
Other	1	2%	2	1%	3	1%
Total	48	100%	184	100%	232	100%

DISPUTES BY TYPE OF CONSTRUCTION PROJECT

Project	Central Council No. of Cases	Central Council %	Prefecture Council No. of Cases	Prefecture Council %	Total No. of Cases	Total %
General construction	39	81%	164	89%	203	88%
Civil engineering	6	13%	5	3%	11	5%
Other	3	6%	15	8%	18	7%
Total	48	100%	184	100%	232	100%

CASES IN WHICH GUIDANCE WAS REQUESTED

Party	Central Council No. of Cases	Central Council %	Prefecture Council No. of Cases	Prefecture Council %	Total No. of Cases	Total %
Individual client/contractor	214	68%	3348	76%	3562	75%
Corporation/contractor	20	6%	99	2%	119	39%
Contractor/client	5	2%	165	4%	170	4%
Contractor/corporation	6	2%	83	2%	89	2%
Subcontractor/prime	48	15%	502	11%	550	12%
Prime/subcontractor	4	1%	25	1%	29	1%
Others	17	5%	199	5%	216	5%
Total	314	100%	4421	100%	4735	100%

GUIDANCE REQUESTS BY TYPE OF CONSTRUCTION PROJECT

Project	Central Council No. of Cases	Central Council %	Prefecture Council No. of Cases	Prefecture Council %	Total No. of Cases	Total %
General construction	264	84%	4052	92%	4316	91%
Civil engineering	30	10%	199	5%	229	5%
Other	20	6%	170	4%	190	4%
Total	314	100%	4421	100%	4735	100%

Source: Ministry of Construction, 1988

of these hearings in the following way: "We try to give advice to contractors and their clients on ways to settle their disputes among themselves. We give advice in the same manner as one family member would give to another member of the family, but we try not to get too involved in the details of the disagreement. Sometimes it takes only a call from the Ministry of Construction to get an owner and a contractor to agree to talk to each other and resolve their differences. At other times we have to separate each party and hear each side of the complaint in different rooms because emotions run high. We listen, and try to advise them on the best course of action. Many times we are effective just by acting as a calming influence. In all cases we do not have the power to enforce decisions. They must be resolved by the parties bringing the dispute to us. Just as good will is required when a contract is formed, so is good will required to resolve contract disputes. Our aim is to provide an atmosphere where good will can be fostered."

7

The Construction Process

There are several things that strike the eye of an American builder visiting Tokyo for the first time. The area surrounding a typical fenced in urban construction site is immaculate. Walk by very early in the morning and one can see a construction worker dressed in coveralls, wearing all of his safety equipment, including hard-hat with chin strap secured, picking up small pieces of paper and debris on the adjacent sidewalks, or near the roll-up entrance door to the construction site. Walk by the construction site late in the afternoon when activity is beginning to wind down and the large doors in the metal fence enclosure around the site are wide open. Truck drivers can be seen backing their vehicles into perfect alignment with each other off to the side of the main entrance aisle. Rows of construction materials and supplies are neatly stacked on pallets. More than likely, two or three laborers will be observed manning fire hoses and brooms as the concrete slab on grade acting as the staging area is being swept and washed down. A glimpse inside the time keeper's shack reveals a row of fire extinguishers lined up as perfectly as West Point cadets.

Perhaps neatness and orderliness were forced upon the Japanese because of their crowded living conditions; or maybe this trait is just inbred. The writer was traveling to a meeting with the Overseas Contractors Association of Japan and his taxi stopped at a traffic light across the street from a construction site in midtown Tokyo. A construction worker crossed in front of the cab and lifted a manhole cover in the sidewalk to either turn on or turn off some underground utility. The distinctive logo on his hardhat identified him as an Ohbayashi Corporation employee. When the manhole had been replaced and the worker started to recross the street to his work site, he spotted a cigarette butt on the pavement, reached down and picked it up, placed it in a nearby trash container, and proceeded on his way.

The second thing that strikes the eye of an American observer is the seemingly strong work ethic of the Japanese worker. It is a Sunday morning, about 7:30 A.M. The weather is clear and warm in late July in Tokyo, and it is about 10 days before the summer holiday season known as *Obon.* Most companies close their doors during a period ranging from 5 to 10 days so that their employees can celebrate this annual holiday by returning to their ancestral home to visit with family and friends. On this balmy summer morning, activity can be seen on the tenth floor of a massive new high rise office building under construction on Sakurada-Dori Avenue in the Toronomon section of the city. Iron workers are welding brackets and clips onto steel beams and columns to prepare for Monday's precast concrete wall panel installation. There is a schedule to be met, and the interruption caused by the *Obon* holidays will require extra hours of work now to compensate for lost time when the project is shut down in a week or so. On another building site in Tokyo, construction workers can be seen toiling away as late as 10:00 P.M. during the week and also on the weekend so that they can get ahead of schedule before the upcoming holiday season.

Walk though a building under construction. There are no radios blaring rock music. There are no informal tete-a-tetes between workers, and there seem to be no long winded discussions about the state of the Mets or who will be pitching for the Angels tonight. There is a strange silence by American construction site standards. Screw guns can be heard, the occasional whine of a high speed drill, a power saw ripping through plywood, the high pitched whir of a masonry saw in the background—and not much else.

The average Japanese worker does not seem to work at a frenetic pace, but they do seem to work at a steady, measured rate throughout the day. And it appears that construction workers are not the only part of the workforce with a sense of commitment. Stand on a street corner near a subway station exit at 7:30 A.M. on a weekday morning in a part of town where office buildings abound and watch the people coming out of these exits. Many of them can be seen running across the street toward their offices, and other office workers who don't see the need to run are walking at more than a leisurely pace so that they can be at their work stations before the normal starting time. It appears that Japan, Inc. is composed not only of industry and government, but includes the hard working Japanese people who play a crucial role in this alliance.

THE JAPANESE CONSTRUCTION EMPLOYEE

Sometimes statistical data can present a clear picture of certain segments of an industry, and because the Japanese society is fond of information, there is a plethora of construction related statistics available pertaining to the makeup of the industry's workforce. In 1988 there were 5.5 million people employed in the construction industry in Japan. About 1.2 million were either self employed or worked in small, family owned companies. There were about 3.6 million other permanent employees in the industry, and approximately 680,000 temporary and unskilled workers in the workforce. (Table 7-1).

The construction industry in Japan plays a much greater role in the nation's economy than it does in four other major free world economies. Construction in Japan accounts for just about 9% of the country's total workforce, exceeding West Germany's 7%+ and the U.S's 6.6%, as shown in Table 7-2.

The construction industry in Japan has been characterized, in recent years, as a low paying, accident riddled business, and as a result it has been difficult to attract both white and blue collar workers. Other, more attractive and glamorous high-tech industries are luring university graduates, and semi-skilled workers find that construction work is often hazardous and physically demanding, and both groups are looking to other careers. Table 7-3 reveals that since 1984 college graduates have found other areas of employment more to their liking, and that 20–29 year olds, the nucleus of future management pools, are not considering construction work as a career.

Table 7-1. The Construction Workforce Breakdown in Japan.

YEAR	Total Workers, millions	Self Employed	Small Family Business	Permanent Workers, millions	Temporary Workers	Day Labor (Unskilled)
1980	5.48	920,000	290,000	3.46	260,000	550,000
1981	5.44	910,000	290,000	3.45	260,000	530,000
1982	5.41	880,000	310,000	3.46	250,000	510,000
1983	5.41	880,000	300,000	3.47	260,000	480,000
1984	5.27	860,000	300,000	3.42	240,000	450,000
1985	5.30	880,000	290,000	3.45	230,000	450,000
1986	5.34	890,000	300,000	3.50	210,000	430,000
1987	5.33	900,000	310,000	3.49	220,000	400,000
1988[a]	5.54	890,000	324,000	3.64	260,000	420,000

[a]First five months only

Source: Ministry of Construction.

Table 7-2. Percentage of Employed Persons by Industry in Five Industrialized nations (as of 1984).

Industry	Japan	United States	U.K.	West Germany	France
Agriculture, hunting, forestry	8.6	3.3	2.5	5.1	7.0
Mining and quarrying	0.2	0.9	1.3	1.2	0.6
Manufacturing	24.2	19.6	26.3	31.0	21.1
Electricity, gas, and water	0.6	1.4	1.3	0.8	1.0
Construction	8.9	6.6	6.4	7.2	6.7
Wholesale/retail, food service	22.3	20.6	15.9	14.9	14.6
Transportation/communication	5.7	5.3	6.0	5.7	5.8
Finance, insurance, real estate business services	6.5	9.5	6.0	6.1	7.0
Community, social/personal service	19.9	30.3	26.8	25.9	25.0
Not adequately defined	0.4	—	—	1.1	—
Unemployed	2.7	1.0[a]	6.3	0.9	9.9
Armed forces	—	1.5	1.2	—	1.3
Total	100.0	100.0	100.0	100.0	100.0

[a]Number of unemployed not previously employed.
Source; ILO Year Book for Labour Statistics—1985.

According to the information in Tables 7-4 and 7-5, there are other reasons why the construction industry in Japan is less attractive than other jobs; the average employee in construction works longer hours and receives a smaller paycheck than his counterpart in the manufacturing sector. The hourly pay rate of the average Japanese construction worker in 1986 did not compare favorably with either union or open shop rates in the United States during that same period of time:

Trade Classification	U.S. (Union)	U.S. (Open Shop)	Japan
Laborers	$13.50*	$6.00–$7.55	$7.31
Carpenters	17.00*	9.20–12.20	8.22
Masons	17.50*	9.67–13.13	8.70
Plumbers	—	9.70–12.23	7.74
Electricians	—	9.50–12.50	8.70
Ironworkers	—	8.63–12.45	7.91
Equipment operators	—	9.50–12.75	7.83

*Union fringe benefits not included.

Wage levels have increased significantly for some trades in 1988, partially due to the fluctuation of the dollar. (An hourly rate of 2470 yen at the 1986 conversion rate of 155 equates to $15.00 per hour, but this same hourly rate becomes $19.00 per hour at the 1988 rate of 130 yen to the U.S. dollar.) The current construction boom in Tokyo has also

Table 7-3. New Construction Industry Entrants by Educational Level (numbers in 1,000s).

	1980	1981	1982	1983	1984	1985	1986	1987
New High School Graduates	29 (2.7%)	28 (2.5%)	27 (2.4%)	29 (2.7%)	25 (2.4%)	20 (2.0%)	23 (2.1%)	25 (2.3%
New University Graduates	23 (2.2%)	22 (1.95%)	24 '2.16%)	22 (2.0%)	20 (1.9%)	19 (1.9%)	18 (1.64%)	19 (1.7%)
Entrants with less than high school diploma	1007	1075	1056	1014	988	961	1055	1016
Total	1059	1125	1107	1065	1033	1000	1096	1060

New Construction Industry Entrants by Age Level (percentage of workforce).

Age Group	1980	1981	1982	1983	1984	1985	1986	1987
19 years old	2.2	2.0	1.8	2.0	2.1	1.7	1.9	1.9
20–29 years	19.9	18.9	17.9	17.2	16.3	15.3	14.8	14.9
30–39 years	28.6	28.7	28.5	28.3	27.7	28.1	29.0	28.0
40–49 years	25.9	25.9	26.2	25.7	26.2	26.1	25.5	25.7
50 and over	23.4	24.4	25.5	26.8	27.6	28.7	28.8	30.0

Source: Ministry of Construction

Table 7-4. Average Number of Hours Worked per 6 Day Week by Construction Workers.

Year	Construction	Manufacturing
1981	46.58	44.37
1982	46.52	
1983	46.32	
1984	47.17	45.14
1985	46.85	
1986	47.35	
1987	47.67	44.90

Average Monthly Salary of Construction Worker Compared with Manufacturing (in yen)

Year	Construction	Manufacturing
1981	204,628	226,190
1982	206,506	233,632
1983	207,056	241,982
1984	219,967	253,313
1985	226,661	258,986
1986	234,088	263,762
1987	242,331	270,354

It would be misleading to convert each year's monthly salary into U.S. dollars because of the significant change in the conversion rates from 1981 to 1987. By using a 1987 average rate of exchange at 145 yen, the average monthly salary of a construction worker would be $1671.25, or $417.81 per week.

been responsible for some increases in worker's hourly rates; some 1988 rates are shown below:

Carpenters	$19.00 per hour
Steelworkers	17.00 per hour
Tile setters	$24.00 per hour
Laborers	11.50–12.50 per hour

A listing of other typical construction trades hourly rates, as of August 1986, are set forth in Table 7-5.

MANAGERIAL AND PROFESSIONAL STAFF SALARIES

According to information received from several Big Six contractors, managerial and professional salary levels in Japan in 1988 are significantly lower than they are in the U.S., except possibly for project management positions:

Salary Ranges for Construction Company White Collar Workers, Degreed personnel, and Project Managers (based on Y130 per U.S.$).

High school graduates		
Starting salary:	120,000 yen/mon.	$923.00
Salary in four years:	150–160,000 yen/mon.	$1153–1230
University graduates		
Starting salary:	150–160,000 yen/mon.	$1153–1230
Salary in two years:	170–180,000 yen/mon.	$1308–1385
Master's degree		
Starting salary:	170–180,000 yen/mon.	$1308–1385
Salary in three years:	200,000 yen/mon.	$1538
Doctor of philosophy		
Starting salary:	200,000 yen/mon.	$1538
Project managers,[a] yearly		
10 years experience:	4.5 million yen	$34,615
20 years experience:	7–8 million yen	$53,846–$61,538
30 years experience:	10 million yen	$76,923

[a]Including twice a year bonus.

Added to the above weekly, monthly and yearly wage levels are the Japanese construction company fringe benefit package, which can contribute to real annual income in a significant manner.

An engineer with 6 or 7 years' seniority might be earning $2000–$2500 per month or $24,000–$30,000 per year. Based upon a typical company annual bonus plan amounting to five times monthly salary, the engineer's compensation is increased to $34,000–$42,500 per year. Now add in the effect of subsidized housing and subsidized family vacations each year and the annual remuneration of the Japanese engineer comes much closer to American compensation packages.

SAFETY AND CONSTRUCTION

Safety on the construction site assumes a very high priority among Japanese builders. Their concern for safety is apparent as a construction site is being approached. In urban areas, solid metal fences at least 12–15 feet high completely enclose the site at ground level. Cartoonlike characters are displayed on safety posters prominently posted on these fences as a good-natured warning that anyone entering the construction site must wear a hard hat and abide by strict safety rules and regulations (Fig. 7-1). A glance upward will reveal a building completely wrapped, cocoonlike, with either brightly colored extruded plastic netting or with solid, reinforced plastic sheeting. Just inside the front gate to the project,

Table 7-5. Japanese Construction Worker's Wage Levels and Hours Worked as of August 1986.

Trade Classification	Average Daily Earnings, yen	Average Daily[a] Earnings, U.S.$	Average Hours Worked Daily
Laborer, Heavy	9,360	$58.50 ($7.31/hr.)	8.5
Light	7,330	45.81 ($5.72/hr.)	8.2
Carpenter	10,520	65.75 ($8.22/hr.)	8.3
Scaffold maker	10,680	66.75 ($8.34/hr.)	8.3
Mason	11,130	69.56 ($8.70/hr.)	8.1
Plasterer	10,150	63.44 ($7.93/hr.)	8.1
Electrician	11,140	69.63 ($8.70/hr.)	8.7
Plumber	9,910	61.94 ($7.74/hr.)	8.5
Painter	10,240	64.00 ($8.00/hr.)	8.3
Truck driver	9,080	56.75 ($7.09/hr.)	8.4
Sheet metal worker	9,870	61.69 ($7.71/hr.)	8.6
Welder	10,700	66.88 ($8.36/hr.)	8.7
Machine operator	10,028	62.68 ($7.83/hr.)	8.5
Steel worker	10,120	63.25 ($7.91/hr.)	8.8
Drilling/blasting worker	15,450	96.56 ($12.07/hr.)	9.1
Tile setter	11,310	70.69 ($8.84/hr.)	8.4
Concrete finisher	10,570	66.06 ($8.26/hr.)	8.0
Joiner	9,750	60.94 ($7.62/hr.)	8.5
Roofer	10,550	65.94 ($8.24/hr.)	8.1
Caisson worker	14,880	93.00 ($11.63/hr.)	8.3
Boring machine operator	9,420	58.88 ($7.36/hr.)	8.6

[a]Average daily wage in U.S. dollars based upon 160 yen average conversion rate in 1986. Wages expressed in both yen and U.S. dollars based upon gross wages before deductions for income tax and any other benefits including social insurance contributions.

Source: Ministry of Labor, Tokyo, Japan

the staging area created on the ground level slab-on-grade will be found stocked with various construction materials and supplies. Bags of mortar or spray fireproofing materials will be placed on pallets in an orderly manner and covered with tarpaulins secured with rope. The center aisle of these staging areas is kept open so that delivery trucks can back completely into the site to unload or load materials, keeping the street clear for normal traffic flow.

Walk up the stairwell onto upper floors where various operations are taking place and you will see the same orderly stocking of materials. Plastic bags filled with construction debris are securely tied and stacked in neat rows against a wall, awaiting removal. It is easy to see why Japanese safety records have improved so dramatically over the years. The clean and orderly construction site is a major contributor. Tripping

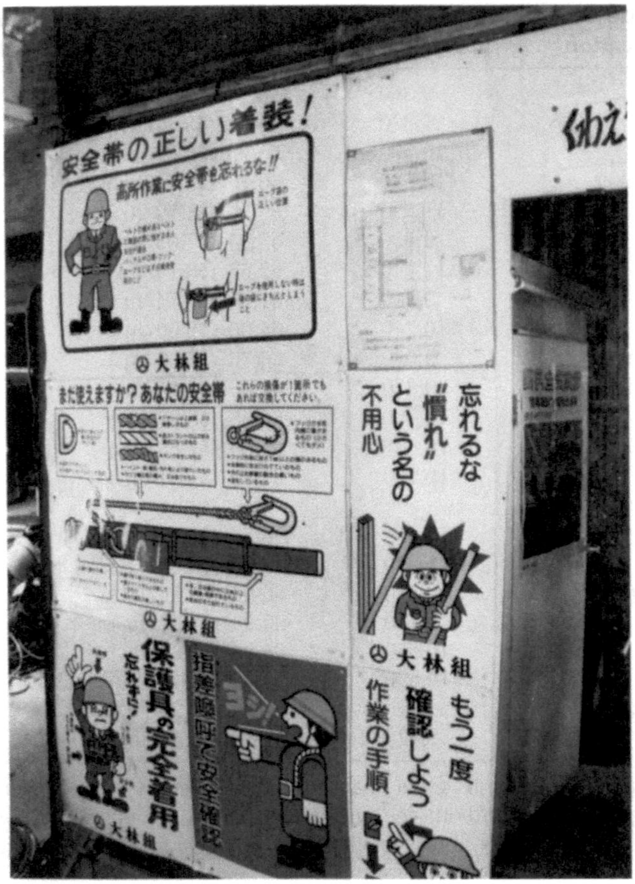

Fig. 7-1. Posters with safety slogans on construction fence display obvious messages and warnings.

hazards are almost nonexistent, dust is at a bare minimum, and work areas are uncluttered as workers stock materials and equipment on the floors.

It was not always this way.

THE UNSAFE WORKPLACE

During the period from 1951 to 1960, workplace fatalities in Japan reached a level of about 6,000 annually as industry raced to increase production capacity. By 1961 the government became alarmed at the high incidence of job related injuries and deaths and began to take an active role in industrial safety. But, primarily because of the large influx of relatively unskilled workers into the workforce, industrial fatalities continued to hover around the 6,000 per year mark. The construction industry was responsible for about 40% of these fatal accidents.

It was not until 1971 that the Japanese government dramatically restructured their system of safety education and enacted several laws requiring construction equipment operators to be licensed and to require employers to institute safety programs on their own.

Contractors were faced with several serious labor related problems as the 1970s faded into the 1980s. Japan, like many other industrialized nations, began to realize that they had a "graying" labor market because people were living longer, were more active, and birth rates had diminished, which would ultimately reduce future labor availability. Many younger workers entering the workforce were seeking jobs in the electronics industry, service industries, and other occupations suppposedly less physically demanding than construction. Construction work was looked upon as unglamorous and hazardous, and unless the industry could improve its image and its work practices, a severe labor shortage loomed on the horizon. Contractors finally realized the seriousness of the problem and embarked on a concerted effect to make the construction work site a safer place. The need to attract younger workers was one of the reasons for the rush to increase R&D efforts to develop robots that could perform many of the unsafe, unpleasant, and dangerous jobs previously relegated to humans.

The combined government-private sector emphasis on safety in the work place paid off, and by 1984 construction related deaths had dropped below the 1,000 level, where it has remained. Injuries, while still hovering around the 40% mark of all industrial injuries had also dropped dramatically in sheer numbers, if not in percentages. From 1965 to 1987 the number of injuries in the construction industry in Japan dropped by 45,089 (see Table 7-6 and Fig. 7-2), a decrease of 57%, and the number of fatalities went from a high of 2430 in 1975 to 983 in 1987, a resounding 60% decrease.

Table 7-6. Record of Industrial Injuries and Fatalities for the Period 1965–1987.

Year	Total Industry		Total Construction Industry			
	Injuries	Fatalities	Injuries	Percent of Total	Fatalities	Percent of Total
1965	408,331	6,046	113,444	27.8	2251	37.2
1970	354,444	6,048	102,340	28.8	2430	40.2
1975	322,322	3,725	99,406	30.8	1582	42.5
1976	333,311	3,345	103,683	31.1	1451	43.4
1977	345,293	3,302	111,026	32.2	1464	44.3
1978	348,826	3,326	118,568	34.0	1583	47.6
1979	340,731	3,077	116,487	34.1	1404	45.6
1980	335,706	3,009	112,786	33.6	1374	45.7
1981	312,844	2,912	100,281	32.1	1173	40.3
1982	294,319	2,674	91,170	31.0	1113	41.6
1983	278,623	2,588	85,175	30.6	1106	42.7
1984	271,840	2,635	79,781	29.3	1083	41.1
1985	257,240	2,572	73,595	28.6	960	37.3
1986	246,891	2,318	71,602	29.0	927	40.0
1987	232,953	2,342	68,355	29.3	983	42.0

Source: Ministry of Construction

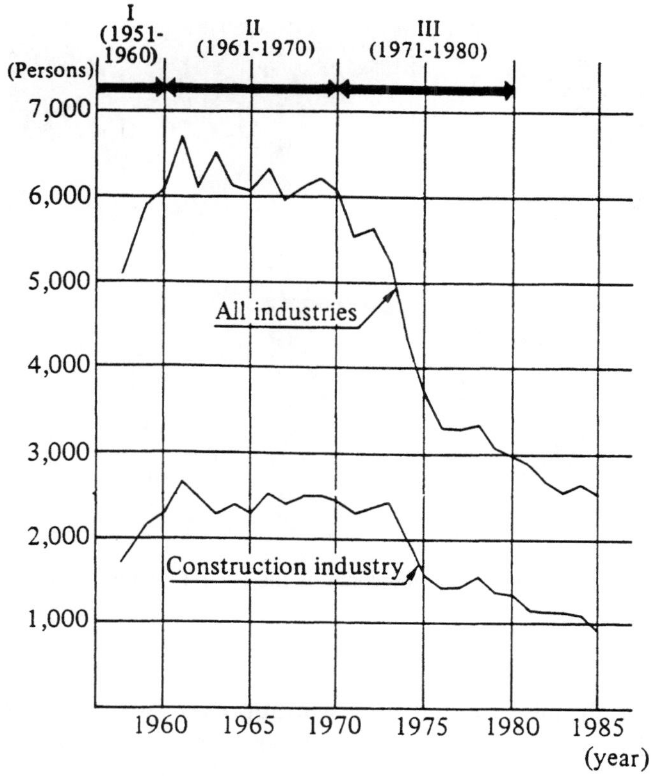

Fig. 7-2. Trends in the number of construction and other industrial fatalities. *(Source: The Japan Society of Industrial Machinery Manufactures.)*

By way of contrast, the U.S. National Institute of Occupational Safety and Health reported that the construction industry in America was responsible for an average of 1428 fatalities per year during the period between 1980 and 1984.

SAFETY ON THE JOB SITE

The success of some safety programs is directly proportional to the redundancy of their warnings. Constantly reminding workers to wear hard hats, to avoid tripping hazards, to use eye and ear protection, and to lift loads properly, might ultimately get safety messages through.

In Japan these messages are augmented by other means. If a worker incurs an accident and has not been wearing his hard hat, his insurance coverage might be void. If for no other reason, the wearing of hard hats on most Japanese construction jobs is at the 100% compliance level. Fig. 7-3 shows a hard hat with the wearer's blood type, an intelligent idea.

At most Japanese construction sites, safety meetings are held at various times through the daily work day, and specific meetings covering specific topics are scheduled on a weekly and monthly basis. The first safety meeting on most construction projects begins first thing in the

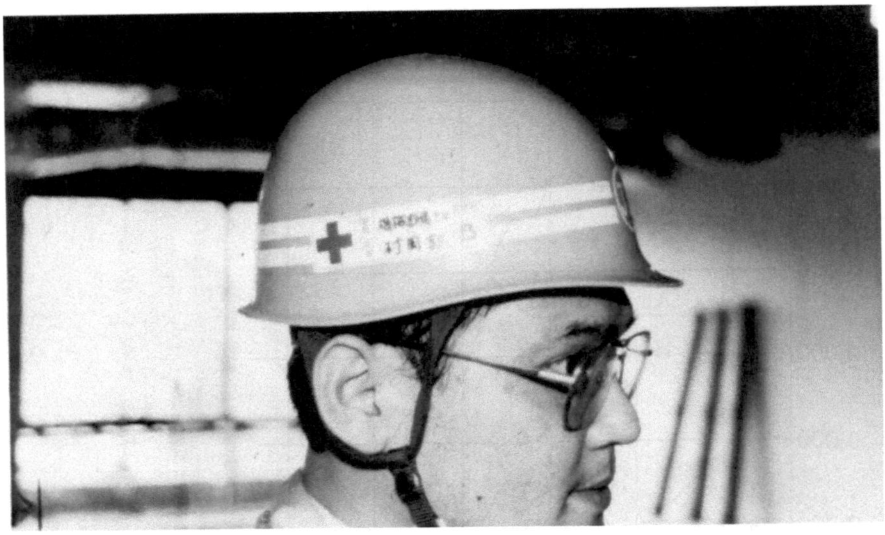

Fig. 7-3. A hard hat with a message. This one contains the wearer's name, address, and blood type.

morning, generally right after the morning warmup exercises, and this safety meeting will only last about 10–15 minutes. Each company has its own tailor made safety program, but a typical one is outlined below.

Safety Meeting Schedule

Daily Meetings	8:10–8:20 A.M. General meeting attended by all workers and staff to briefly run through safety procedures. 8:20–8:30 A.M. Tool box meeting held with selected subcontractors. 10:30 A.M. Tour of building conducted by Project Manager or Chief Engineer to inspect safety practices and check out the site environment. 2:00 P.M. Review of specific safety concerns and procedures with specific groups of workers and/or subcontractors. 4:55 P.M. Five minutes clean-up, straighten-up period before workers end their shift. Project Manager and/or Chief Engineer conduct coordination meeting with all job foremen.
Weekly Meetings	1. Every Monday morning, a safety meeting is held with all foremen to review the coming week's activities and discuss potential safety concerns for the week. 2. Once each week conduct an inspection of all heavy equipment and vehicles stationed on the site.

3. Conduct an inspection of all temporary structures and of all materials storage areas.
4. Devote one day each week to a thorough cleaning of the entire site (usually performed on Friday or Saturday)

Monthly Meetings

1. On the first Monday of each month, review the proposed activity for that month and discuss safety concerns and potential safety hazards that may occur.
2. Devote one day each month to a complete inspection of all temporary wiring systems, to include GFI's, circuit breakers, temporary wires, plugs, and switches.
3. Devote one day each month to an explanation of company safety concerns and company safety goals to all foremen on the job.
4. Devote one day each month to a safety and health discussion with the safety directors from each subcontracting firm.
5. Devote one day each month to a meeting in which all recent accidents and injuries are reviewed with all personnel.

In addition to these regularly scheduled safety meetings, special safety programs are implemented during the month to provide training sessions for newly hired workers, to provide medical check-ups for these newly hired employees, and to conduct a fire drill and mock building evacutation drill for all personnel on the site.

OHBAYASHI'S SAFETY SCHEDULE

Ohbayashi Corporation, one of Japan's Big Six, maintains a similar safety schedule, but they start off the day with an individual worker safety equipment check. Along with the required hard hat and chin strap, Ohbayashi requires that all workers on high rise buildings wear a safety belt. These belts are made of heavy duty nylon webbing, about 2 inches wide, and are secured around the waist with a formidable looking steel buckle. A nine inch, spring loaded steel hook is attached to the belt via a five foot long, 5/8″ braided nylon rope, and when working near the building perimeter on upper floors, a worker is required to attach this hook onto perimeter steel safety cables. Whether working on the ground floor or the twentieth floor, all workers must keep this belt on during the day.

At the Ohbayashi job site morning safety meeting, each worker will have had his safety belt securely fastened around his waist, his hard hat firmly placed on his head and chin strap pulled up snugly. Workers will line up in two rows facing each other and check out one another's safety equipment. "Hard hat on—Check! Chin Strap Secure—Check! Safety Belt On and Secure—Check!, Tool Belt in place and Secure—Check! Work Boots in Good Condition—Check!" When a worker facing his partner has run down the list, and has, in turn been checked out, the crew will break ranks and go to their assigned work stations.

Ohbayashi conducts a series of other safety meetings throughout the day as shown in Fig. 7-4, a copy of their daily safety schedule:

9:30–10:00 A.M.	The safety group, consisting of two Ohbayashi staff members, one subcontractor and foreman, tour the building checking on safety compliance.
1:00–1:15 P.M.	The first daily cleanup session takes place.
1:15–1:45 P.M.	The safety group conducts their second walkthrough.
5:00–5:10 P.M.	The final cleanup period of the day takes place before workers leave the job site.
5:15 P.M.	Ohbayashi staff members conduct their final walk-through safety check of the day.

The last Thursday of every month is when the project manager conducts an all-encompassing safety meeting with each subcontractor's safety director, and the last Friday of every month Ohbayashi holds a safety meeting and a scheduling meeting with all the subcontractor's formen. The constant reminder and periodic inspection forms the core of this safety program.

CONSTRUCTION SAFETY OFF THE SITE

A great deal of construction work involving utilities replacements and repairs and assorted road work is performed by construction workers during the evening hours in Japan's major cities. In several areas in Tokyo, water, gas, and telephone lines are buried under asphalt paved sidewalks. When new installations or repairs are required, night crews go about their work outside of the flow of traffic. They also can resurface the sidewalk as soon as work is complete, so the next morning it is often difficult to see any evidence of work that has been performed. This practice of placing certain utilities under sidewalks substantially reduces the need to rework trenches in the street that have settled, and reduces the potential for potholes.

When it is necessary to work in the street during daylight hours, the areas under construction are clearly marked far in advance of the actual work area (see Fig. 7-5) and there is almost always a policeman or construction worker nearby directing traffic. And when the nature of the work is such that it may take days or weeks to complete this road

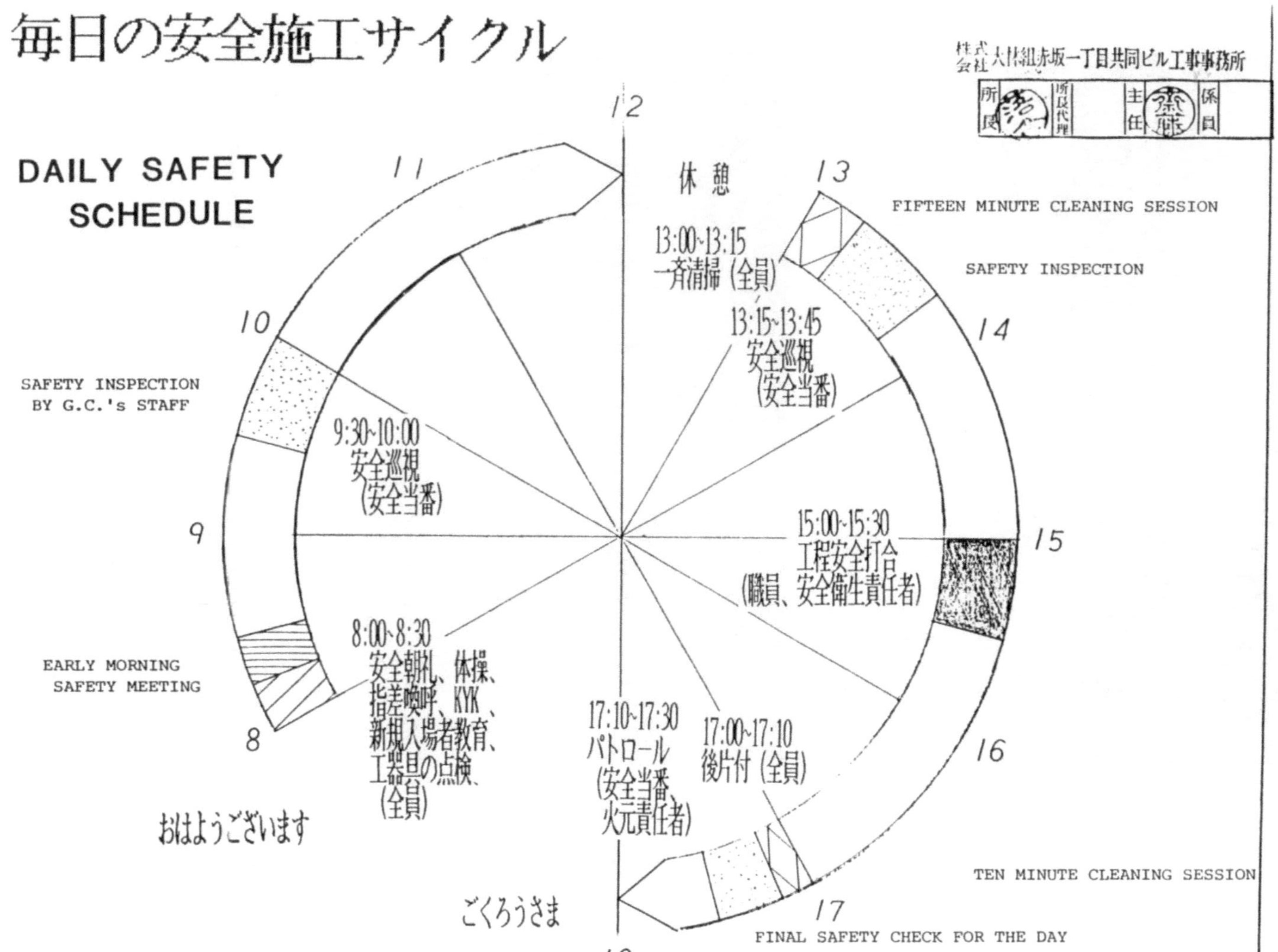

Fig. 7-4. Ohbayashi's daily safety schedule.

Fig. 7-5. When road repairs cannot be completed overnight, barricades warn drivers well in advance of the construction zone.

work, areas under construction are well marked during daylight hours and after dark they light up like the stage at the Roxy.

NIGHT TIME SAFETY PRACTICES

At times it becomes necessary for emergency crews to work in busy city streets during evening hours and traffic flow needs to be directed around these construction zones. The Japanese employ a system of internally lighted traffic cones, day-glow traffic tape, and brightly lit signs, and traffic directors waving fluorescent wands, as shown in Fig. 7-6 and 7-7. Some utility company workers were seen on a side street where these lighted traffic cones were used. In addition, these workers carried light wands and wore internally lighted plastic vests as they worked. Both stationary and moving objects could be plainly seen from at least 100 yards away.

CONSTRUCTION EQUIPMENT SAFETY

Japan went through an accelerated modernization program in the construction field during the past twenty years, and it was not until the early 1970s that construction equipment such as bulldozers, excavators, cranes, and concrete placement machinery began to supplant hand labor in any significant numbers. But like so many things that the Japanese do, once they embark on a program, they do it with a vengeance, and local contractor purchases of mechanized construction equipment skyrocketed. Annual purchases of construction equipment increased from 50 billion yen in 1960 to about 400 billion yen in 1980, as graphically displayed in Fig. 7-8.

As the rush to modernize increased, a severe shortage of properly trained operators was created, and field related injuries and fatalities shot up because of the low skill levels of many of these equipment operators. The Ministry of Construction estimated that at least 30% of all injuries and deaths occurring in the construction industry in 1970 and 1971 were caused by the improper use of powered equipment. In 1972, the government established a technical safety program which was to be conducted in conjunction with a federal licensing system for all construction equipment and machinery operators, and in 1978 the government expanded the training program to include hoisting procedures, scaffolding installation, and small power machine operations.

The effect of their program appeared to be instantaneous and meaningful. Fatal accidents began dropping rapidly, and from 1965 to 1975 fell by almost 30%. And by 1987, total construction injuries were 42% lower than the highwater mark year of 1978.

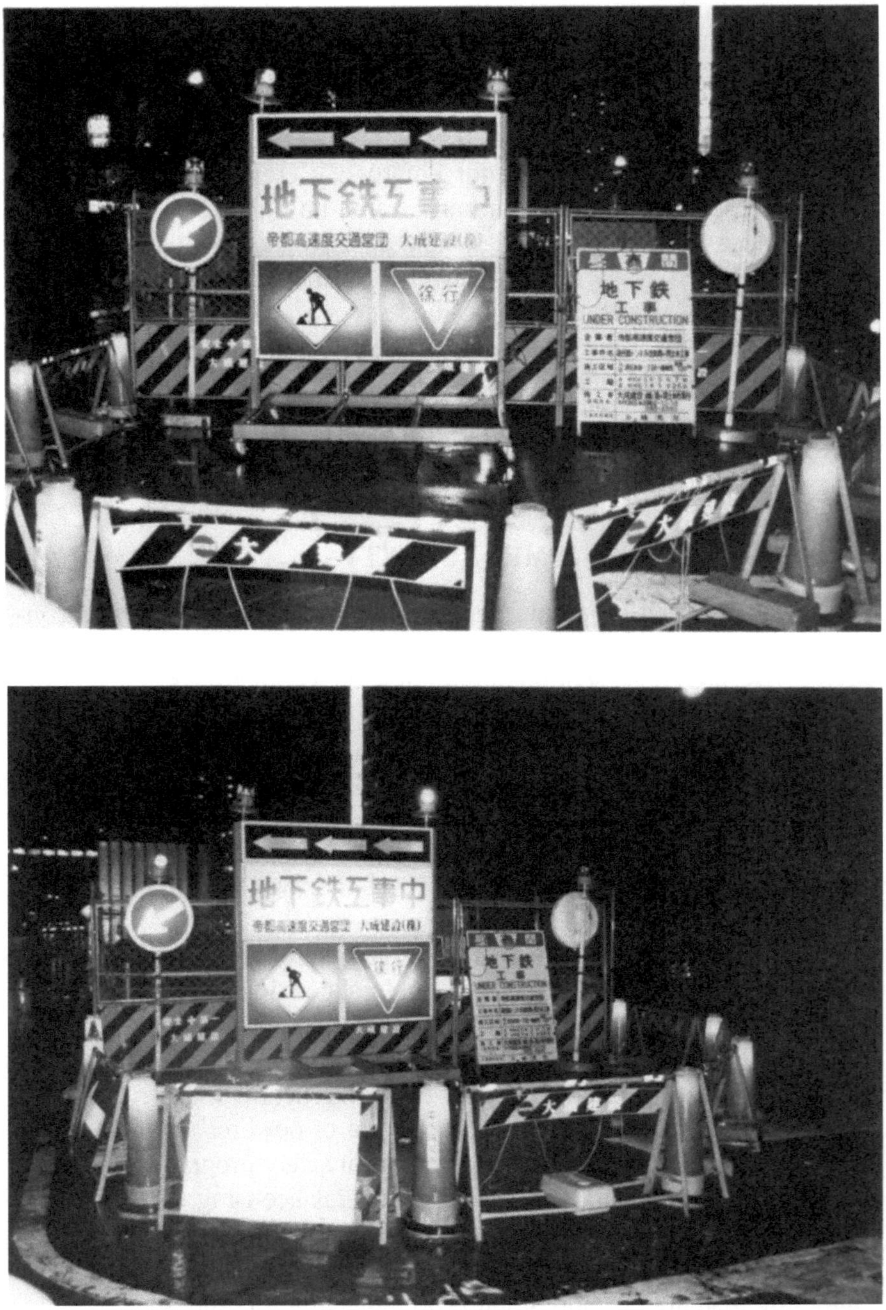

Fig. 7-6. Well lit and delineated work areas at night allow motorists ample warning.

(a)

(b)

Fig. 7-7. (a) Lighted pylons, reflective clothing, and illuminated wands protect workers when evening road repairs are made. (b) On darker side streets, utility workers wear illuminated vests as added protection.

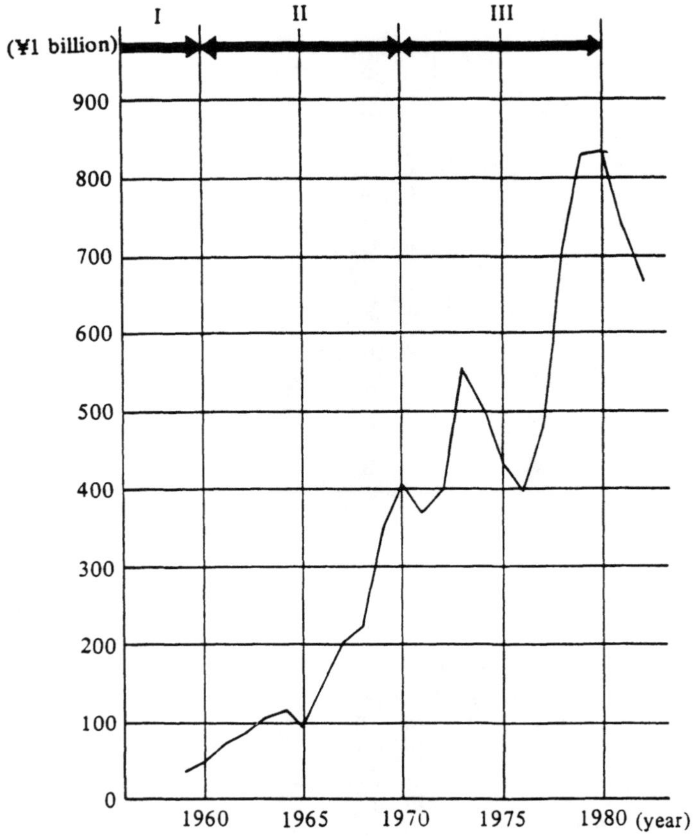

Fig. 7-8. Trends in annual protection of construction machinery in Japan for domestic use. *(Source: The Japan Society of Industrial Machinery Manufactures.)*

The Safety Education Program for Operators

The Ministry of Construction mandated a safety education program, but wisely relegated the implementation of the program to the private sector. Private training schools were to be established, but their curriculum, instructor qualifications, and class room and field training facilities would have to be approved by the government. Strict requirements would be placed upon the school's instructors, who according to government policies "must be experts in their field." The curriculum for each course of instruction would be aimed at a specific piece of construction equipment, and many of the school's staff would come from the membership rolls of the Japan Society of Industrial Machinery Manufacturers. These training schools began to proliferate, and by 1987 there were 180 government approved training schools throughout Japan.

The technical safety education program in these institutions focuses attention upon both classroom and field operating instruction. The schools provide their students with theory and practice relating to the following areas of construction equipment:

1. General construction equipment operation—bulldozers, loaders, excavators.
2. Foundation equipment of a specialized nature—pile drivers, caissons.
3. Compaction equipment—rollers, vibratory compactors, etc.
4. Mobile crane instruction and proper sling techniques.

Program Eligibility

Anyone, male or female, over the age of eighteen can enroll in these training school programs. The complete course of study consists of about 13 hours of theoretical instruction in the classroom and about 38 hours of practical instruction in the field. The school's fees are either paid by an individual seeking employment as an equipment operator in the construction field, or by an employer wishing to upgrade an employee's job skills. The Japanese government will in some cases provide a subsidy to lower the cost of the program to certain individuals.

A Typical Curriculum

The thirteen hours of classroom instruction will usually be spread over two days and will include about four hours devoted to the structure and handling of moving equipment, with a like amount of instruction time spent on stationary working equipment. General knowledge of the bulldozer will occupy four hours, and about one hour of instruction will be devoted to laws and regulations governing the operation of equipment.

Moving out to the field, students will have about 25 hours of actual equipment operation instruction spread over four days, and they will spend no less than 20 hours practicing on moving construction equipment. About five hours will be devoted to the operation of stationary equipment.

Upon successful completion of the course, the student will be issued a license as a qualified equipment operator.

The Program Effectiveness

Since the Ministry of Construction started this program in 1972, several million Japanese workers have completed the course of instruction. Its effectiveness can be measured in a number of ways, and a comparison between U.S. and Japanese accident records is inevitable.

The frequency rates for labor accidents is calculated by dividing the number of casualties by the total hours worked, and then multiplying by 1,000,000. In 1970, both the U.S. and Japan experienced a frequency rate of 10, but by 1984, Japan's rate had dropped to 2.77 while the rate in the U.S.. crept up to 10.75. As accident rates have dropped productivity has increased, so it appears that this training and qualifying program has been a success.

Continuing Education

After the basic training session has been completed, education and equipment training cannot just stop. New equipment is being introduced into the field on a regular basis, and standard pieces of equipment are constantly being upgraded and improved. As mechatronics comes of age, more sophisticated training programs will be needed since electrical theory and electronics will play a vital role in the complete understanding of this advanced equipment. The new breed of equipment operators will also have to become familiar with computer programming and basic computer knowledge. The Japanese recognize the need to maintain and upgrade their work skills, and the technical training program takes this need into account. Equipment licensees are required to enroll in refresher courses five years after they have received their license so that they can remain alert and proficient.

THE FIELD ORGANIZATION ON THE CONSTRUCTION SITE

The size of a typical Japanese field office is an indication of how their builders organize and staff a project in the field (see Fig. 7-9). These field offices are very generous by American standards. They are usually two story, steel framed buildings with metal siding, containing ample working areas for secretarial pools, expediters, engineering and project management staff, computer operator stations with numerous terminals, and a small kitchen, lounge, and lunch room area.

The sturdiness of these field offices was demonstrated in a rather unusual way. While visiting a job site in the Tokyo Bay area, the writer was on the second floor of the project's field office when the building started to move, almost imperceptibly. In a matter of seconds the shaking increased rapidly and windows started rattling, doors slammed shut, papers fluttered and started falling to the floor. Loose objects on desks rolled to the floor and a few chairs started sliding on the vinyl floor. An earthquake measuring 5.3 on the Richter scale had just occurred, and although the metal siding on the outside walls was flexing as the building moved to and fro, it had only received a good shaking. It seemed that the same kind of seismic considerations for large buildings had been incorporated into these temporary structures.

The staffing of Japanese projects seems to be as generous as the size of their field offices. A $25–35 million office building project might be staffed by a project manager, an assistant project manager, an expediter and coordinator, an engineer to coordinate all work performed by the trades, a personnel clerk, and two or three secretaries. Another project of similar size will be staffed by a project manager, a job superintendent, a foreman, a chief engineer, a first class and a second class engineer, and a clerical crew consisting of an accountant and one to three secretaries.

A DAY ON THE JOBSITE WITH OHBAYASHI

Just a few blocks way from Komatsu's Tokyo headquarters building with its distinctive giant yellow bulldozer on the roof, Ohbayashi Corporation is building the Akasaka 1-chome Kyodo Office Building.

The Kyodo building is a thirteen story structural steel framed building with a precast concrete exterior wall panel skin faced with an aluminum finish (Fig. 7-10 and 7-11). The total floor area of the building is about 180,000 square feet, not including the three levels of underground parking.

The structural steel frame has been completed for some time and floor slabs and stairwells have been installed. Windows are being installed on the upper floors and mechanical and electrical rough-ins are being put into place, and sheetrock is being screwed to steel stud framing on these floors. Ceiling grid systems are going up and lay-in lighting fixtures are being installed.

The mechanical equipment has been almost completely installed on the roof and is partially hidden behind some decorative truncated pyramid stair tower enclosures.

Although marble panels are being installed on the ground level near the entrance, not much finish-type work is taking place in the building.

The Project Manager assigned to this building is Mr. Junichi Ochiai, and his right hand man is Mr. Saitoh, Chief Engineer. The field office is located within the building on one of the lower floors, and before entering this rather spacious room, shoes must be removed and slippers placed on the visitor's feet from a rack near the doorway. Behind a knee wall that serves as a leaning post while changing from shoes to slippers, there is a coffee table, leather covered black sofa, and two black leather chairs. A full height room divider separates this visitor's area from Mr. Ochiai's office.

In the middle of this large room there are several rows of filing cabinets and behind the filing cabinets there is an area that serves as a kitchen, complete with sink, stove, and refrigerator. A plain gray clerk's desk separates this cooking area from the file cabinets, while off to the side is a table and chairs set aside for staff meals. Prominently displayed

(a)

(b)

Fig. 7-9. Typical Japanese construction field offices. (a) Interior view. (b) Field office at Yokohama project. (c) Field office at Tokyo Bay Bridge job. (d) Field office in downtown Tokyo, elevated above street level.

(c)

(d)

Fig. 7-9. ***(continued)***

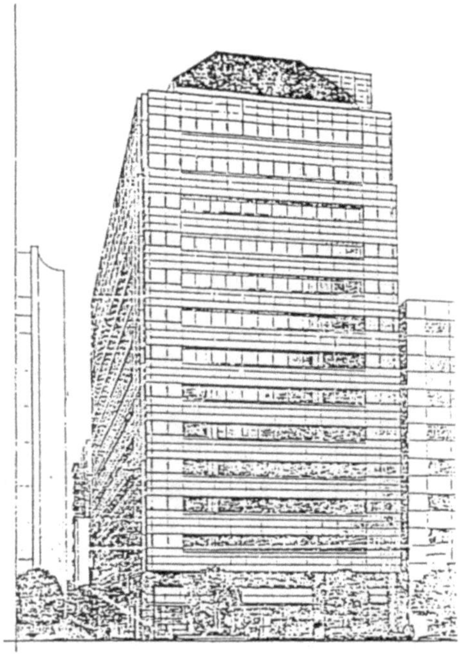

Fig. 7-10. Sketch of Ohbayashi's Kyodo Building project.

on the wall near the project manager's office is the project schedule and a framed safety directive signed by Mr. Ohbayashi.

The staffing for this project is as follows:

Project Manager
Deputy Project Manager
Chief Engineer for Construction
Chief Engineer for Coordination
Chief Administrator who will act as Personnel manager, accountant, and material scheduler.
Two secretaries and two other staff members whose sole function it is to coordinate all drawings.

Before beginning a tour of the building, Mr. Saitoh reviews the project's weekly and monthly schedule of meetings (Fig. 7-12). Along with the daily and weekly safety meetings, there are several significant meeting scheduled throughout the month.

Every Monday, at 1:00 P.M., the project manager conducts an important staff meeting to review the schedule, safety matters, and items relating to construction quality. (It is interesting to note that this meeting is identified with a star—that is how they treat the project manager!)

Daily at 3:00 P.M., a 1/2 hour meeting is held between the Ohbayashi staff and the foremen for each subcontractor. Daily progress versus scheduled progress is reviewed and any mid-course corrections needed will be discussed and acted upon if necessary.

Every Wednesday between the hours of 5:30 and 7:00 P.M., a meeting takes place to review the construction drawings and any revisions that might have taken place during the week. Quality and QC matters are also a part of this agenda.

Every Friday, from 5:30 to 7:00 P.M., Ohbayashi staff members meet and review the past week's events and the coming week's planned activities.

On the last Thursday of every month, the Project Manager holds a two hour and forty-five minute safety meeting which begins at 1:15 P.M. This meeting is attended by each subcontractor's Safety Director.

The last Friday of every month a special safety meeting is held by the Project Manager for all workers on the site and all company staff members. This meeting lasts just 1/2 hour.

On the last Friday of every month, a one hour foreman's meeting is conducted from 12:00 noon until 1:00 P.M. and safety and scheduling is reviewed and discussed in detail.

Every Tuesday, from 1:30 to 2:30 P.M., a meeting takes place with the owner and his design consultants at the site. Questions relating to drawings clarification are raised, shop drawings are reviewed, and the construction schedule is scrutinized.

The Project Manager's Role

As the "star" on the job, the project manager has a great deal of authority and responsibility. It is he who decides how various phases of work will be done, by having a subcontractor furnish labor, equipment, and materials or only labor, with Ohbayashi supplying everything else. On this project, Mr. Ochiai thought it best to purchase the fabricated steel structure from Nippon Steel and make arrangements to bring their own crane to the site and hire ironworkers from a local subcontractor to erect the steel.

The project manager is held accountable for schedule compliance and it is he who will discuss subcontractor work crew make up and working hours with all of his subs. If, in his opinion, it becomes necessary to work extra hours in order to maintain the schedule, the subcontractors are obliged to do his bidding. Rarely is any mention made of premium pay for these overtime efforts. If a subcontractor is uncooperative, the project manager has it in his power to recommend that the recalcitrant sub not be given any more work in the foreseeable future.

The chief engineer is another critical cog in the staffing machinery. Mr. Saitoh said that he is responsible for insuring that all equipment on the site is operating properly and safely. Before any worker can start up a piece of equipment, the operator must first run down a checklist

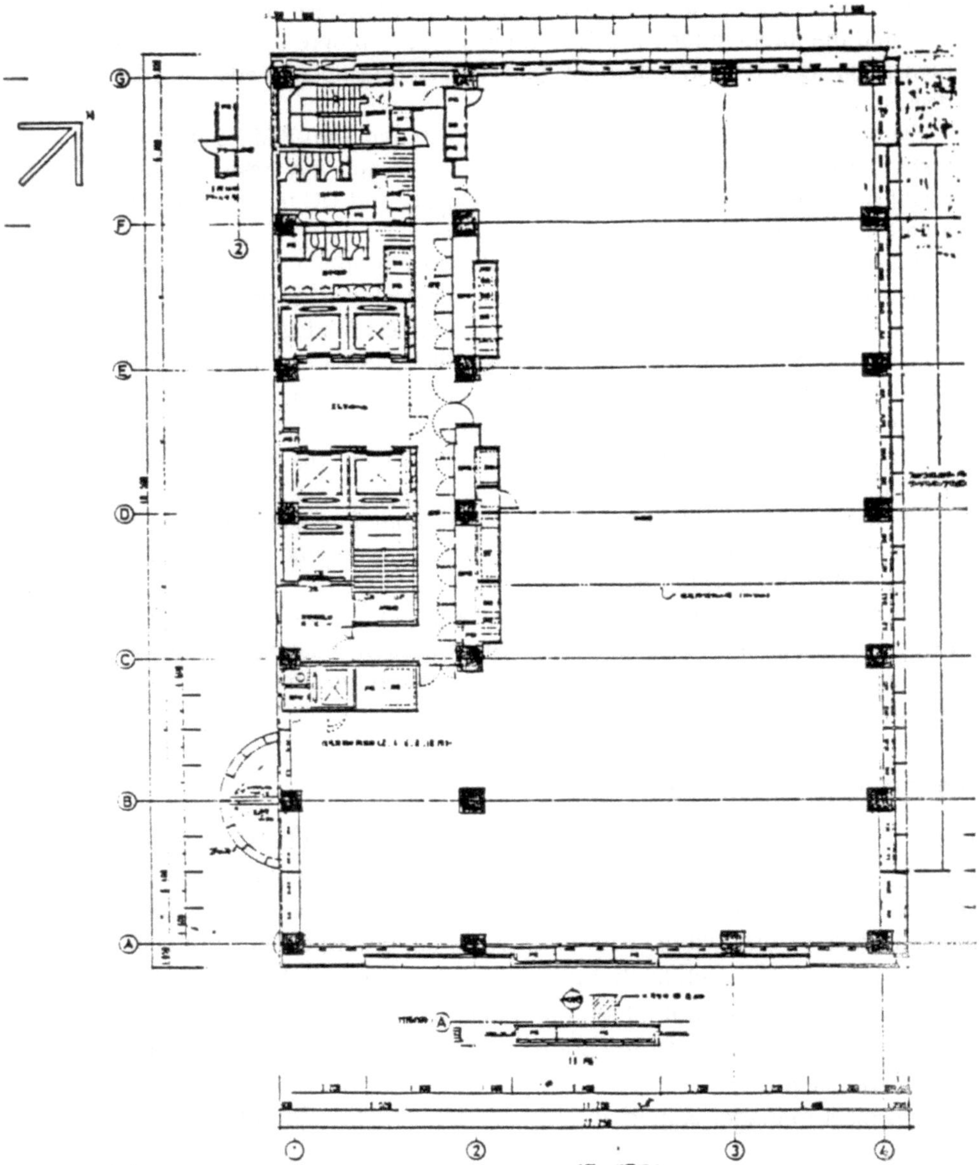

(a)

Fig. 7-11. Kyodo Building. (a) Typical floor plan. (b) Elevation, including underground parking.

to insure that the machine is safe to operate. At the end of each day these reports filter back to Mr. Saitoh, and if a machine is in need of repair, he must see to it that it is either repaired or replaced.

He also oversees the proper manning of various subcontractor work crews and if he cannot convince a particular foreman to increase his

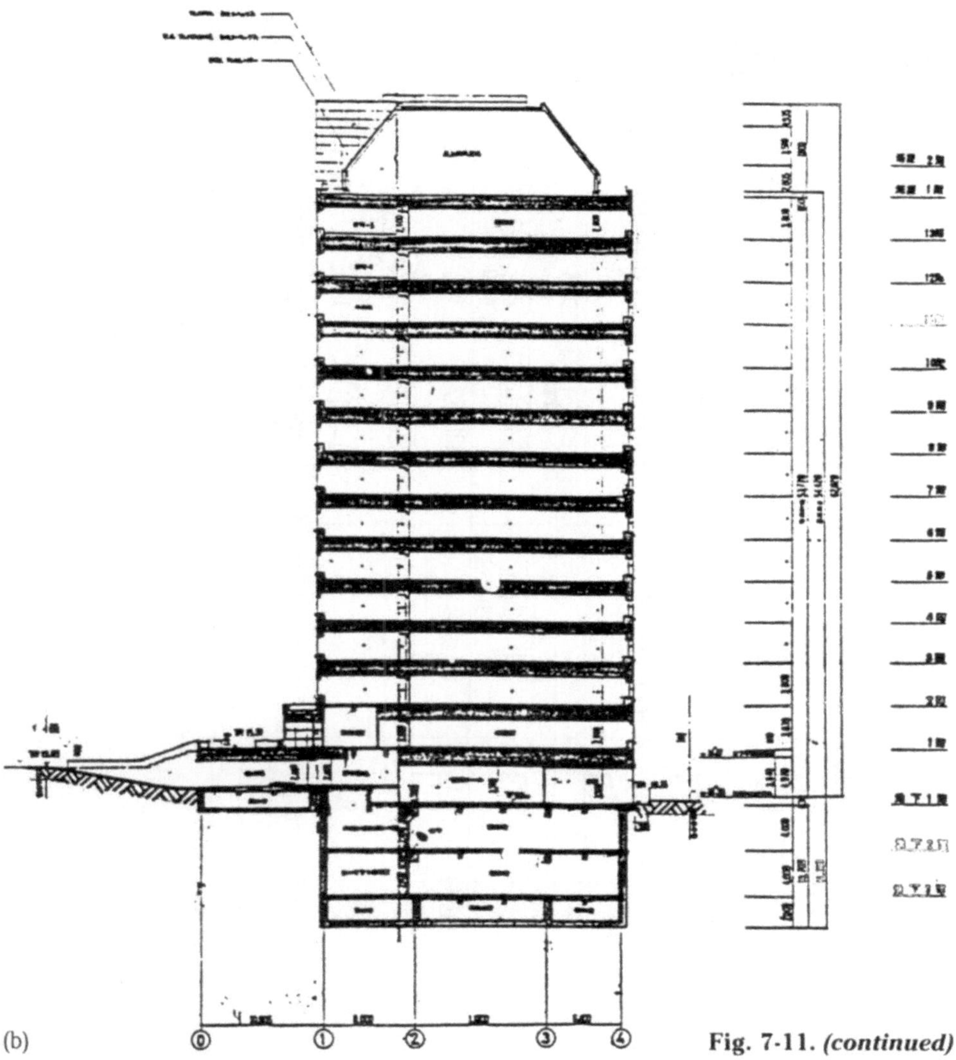

(b) **Fig. 7-11.** ***(continued)***

work crew or extend his working hours, he will place a call directly to the subcontracting firm's president, advising him that his foreman needs some assistance. The threat of no more work from Ohbayashi, a threat that *will* be carried out, usually brings results.

A Tour Through the Building

Walking onto a floor where sheetrock operations were taking place, it appeared that only full sheets of gypsum board were on the floor; the usual half and quarter sheets and miscellaneous scrap pieces normally in abundance on most American construction sites were missing. There was virtually no clutter on the floor, but plastic bags filled with drywall

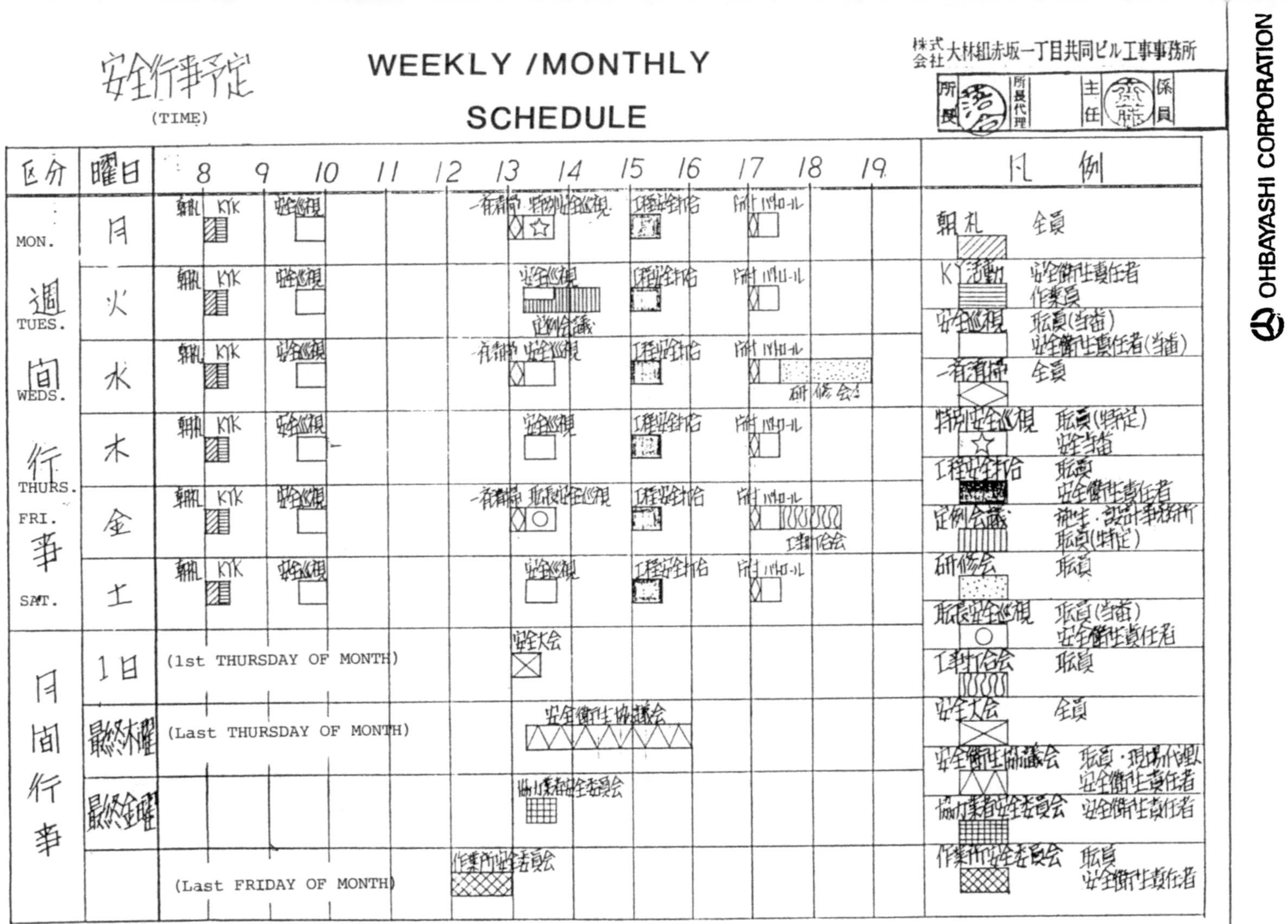

Fig. 7-12. Ohbayashi's weekly/monthly project meeting schedule for the Kyodo Building.

scraps, securely tied, were placed in neat rows against a wall awaiting pickup and disposal.

Two carpenters were operating screw guns and methodically fastening sheets of drywall to steel stud framework. They worked at a steady pace, without talking, and appeared to be oblivious to everyone else working on the floor.

On another floor two sections of alumimum framed windows, already glazed, had been joined together to form a section about eight feet long. These windows were resting on a pump dolly with wheels that could be jacked up hydraulically so that the prefabbed two-window section could be raised to sill height and eased forward onto the precast panelled opening.

Windows were also being installed in the stairwell on this floor, but these were special operable windows. They had very large, hopper-type frames, hinged at the bottom and opening outward when manually cranked open. It was explained that these were special fire-rated windows especially for stairwell installation. When a fire or resulting smoke enters a stairwell, all of these special windows will open automatically and allow the smoke to vent outside. Fig. 7-13 contains photographs taken at the Kyodo job site.

At 12:00 noon, all work stopped and the workers on the upper floors went down to the first floor level of underground parking where there was an area set aside as their lunchroom. A long table and a dozen or more chairs were placed around the table and everyone except the Ohbayashi staff were expected to eat their meals in this area. The surface of the concrete walls in these below grade parking levels was as smooth as glass, and Mr. Saitoh explained that they routinely use plastic coated plywood wall forms or steel forms in garage areas where the finish of the concrete will be exposed.

After lunch, a visit to the tenth floor where ceiling grid and lay-in light fixtures were being installed clearly showed that mechanics had taken some extra time to line up the runs of round ductwork, electrical conduit, and various plumbing pipes above the grid system. Control wiring installed above the ceiling, but not yet connected to the VAV boxes was neatly coiled and suspended from nearby ductwork.

And everywhere there was an absence of trash and construction debris. Materials stored on the floor were stacked in orderly piles out of the normal flow of worker traffic patterns.

Passing this same site at about 10:00 P.M. that evening, the front entrance gate was open and workers could be seen stacking bags of mortar as they were being unloaded from a delivery truck deep inside the building. Construction days do not end early in Japan.

(a)

(b)

(c)

Fig. 7-13. Scenes from the Kyor's Building project. (a) Preparing to lift aluminum clad precast panels in place. (b) Installing special hopper fire windows in stairwell. (c) Orderliness of construction area on upper floor.

A DAY ON A MITSUI BRIDGE PROJECT

Mitsui Construction Company has entered into a joint venture with two other companies, Goyo and Shiraishi Company, Ltd. The joint venture, called MGS JV is under contract with the Tokyo Municipal Government to build a bridge across the deepest end of Tokyo Bay, from Daiba Island to Shibaura (Fig. 7-14). The concrete and substructure work is worth about 20 billion yen ($148 million at 135 yen). Mitsui is a 40%

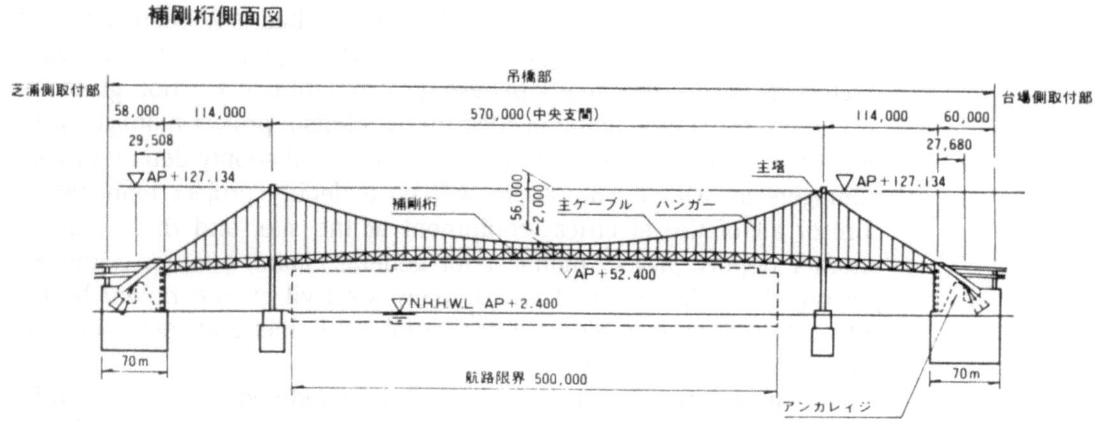

(a)

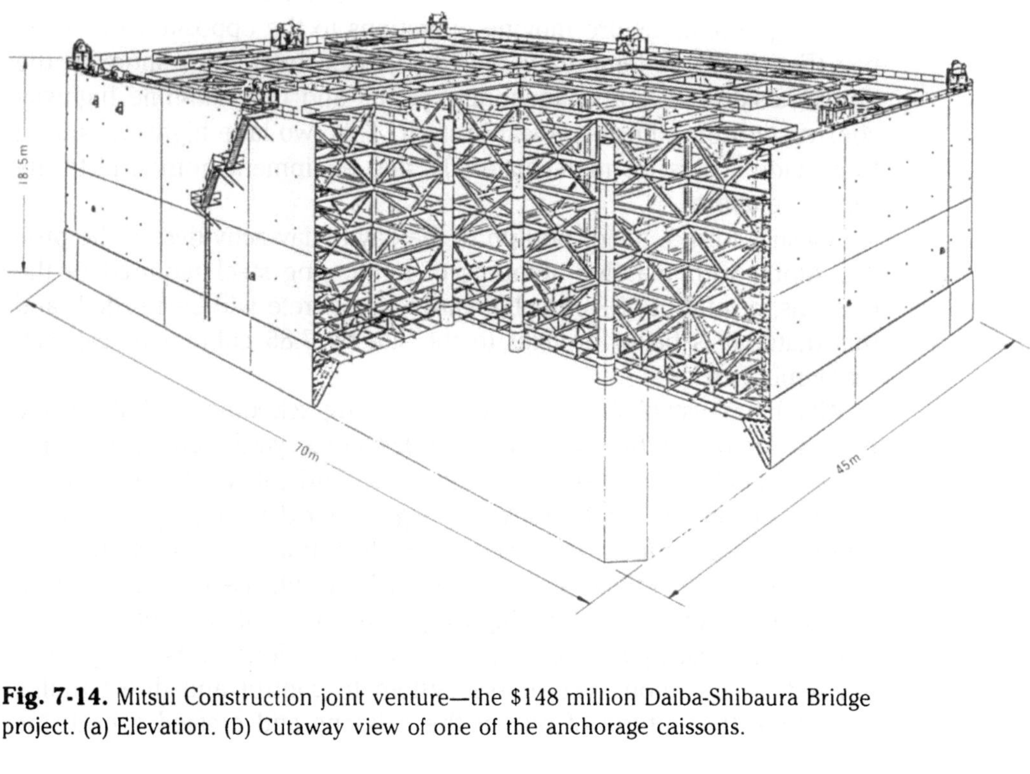

Fig. 7-14. Mitsui Construction joint venture—the $148 million Daiba-Shibaura Bridge project. (a) Elevation. (b) Cutaway view of one of the anchorage caissons.

particpant, Goyo is a 30% partner, and Shiraishi is the remaining 30% partner. In keeping with Japanese custom, the prime or major participant in a JV contributes the first letter of the joint venture's name, hence MGS JV, which is embroidered on staff shirts and lettered on hard hats. One of the islands included in the span, Daiba, dates back to the 1860s when it was created by Japanese labor during the Edo Period as a defense fortress to be used against America's Black Ships. Almost 130 years later, Americans may be welcomed back to the same area where they had previously been unwelcome. According to one Mitsui engineer, the steel superstructure for this project had not been awarded since this bid proposal might be one of those opened to foreign contractors as part of the new Japanese government construction glasnost.

The project staff is being headed up by a Mitsui project manager with twenty years' experience. Shiraishi and Goyo will supply deputy project managers as will Mitsui. There will be a deputy project manager to control all of the electrical equipment at the site, and there will be another deputy project manager for general affairs, another name for personnel administrator. Approximately six civil engineers will be assigned to this project, along with a number of chief engineers and first and second class engineers.

The substructure for the bridge work is scheduled for completion in 1991 and the superstructure is planned to be completed by 1993. Work is currently in progress to create one of the main tower caissons and anchorage points before moving operations to the opposite side of the bay (Fig. 7-15). A temporary road had to be built from shore to the Shibaura-side tower and anchorage points and this half-mile highway construction on pilings is a feat in itself. This two lane highway is used to provide access for men, materials, and equipment from land to the caisson platform.

Logistics form a major portion of the day-to-day activities on the project. Along with massive amounts of reinforcing steel required in the caissons, a total of 65,400 cubic yards of concrete will be needed, and this material is being brought to the site by 7.85 cubic yard capacity ready-mix trucks.

Mitsui has established a concrete pouring schedule of 3,000 cubic meters per day, which equates to 3,924 cubic yards, and since each truck is capable of transporting 7.85 cubic yards, it will become necessary to unload about 500 trucks per eight hour day, slightly more than one truck per minute. The need to create a traffic control system became obvious to Mitsui since these trucks would be traveling over a narrow two lane temporary highway and any mishap could cause a traffic jam of staggering size and costly proportions. Mitsui engineers devised an ingenious yet simple system to control and direct traffic. They took the outline of one of the caissons and created a grid line

(a)

(b)

(c)

Fig. 7-15. Work being done on the massive bridge anchoring caissons. (a) Temporary roadway leading to the project. (b) Tower cranes on the caisson. (c) The maze of reinforcing steel at the caisson.

(a)

(b)

(c)

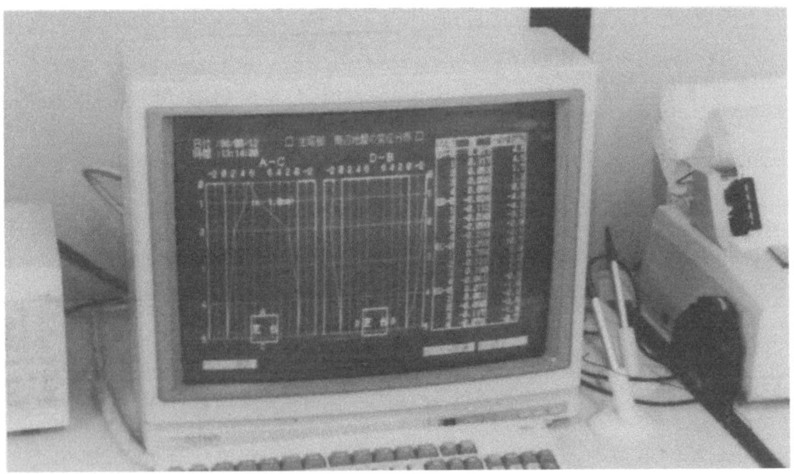

Fig. 7-16. TV cameras and personal computers play decisive roles in Mitsui's daily construction activities. (a) The orderly construction site—typically Japanese. (b) Remote cameras monitor the action at the caisson. (c) A visual display of the reaction force monitoring of the caisson.

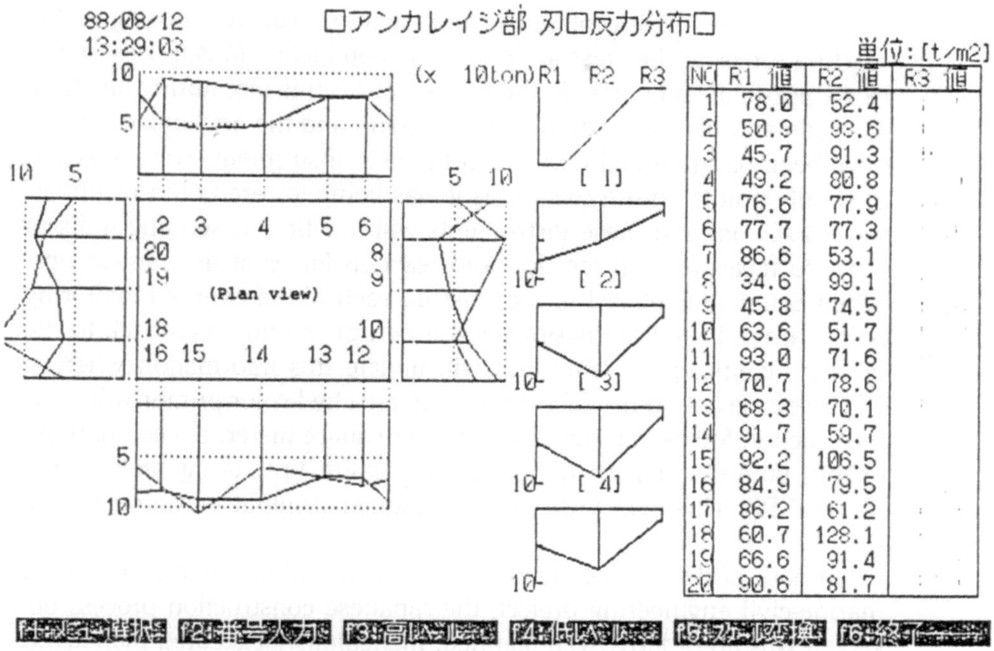

NO	R1 値	R2 値	R3 値
1	78.0	52.4	
2	50.9	93.6	
3	45.7	91.3	
4	49.2	80.8	
5	76.6	77.9	
6	77.7	77.3	
7	86.6	53.1	
8	34.6	99.1	
9	45.8	74.5	
10	63.6	51.7	
11	93.0	71.6	
12	70.7	78.6	
13	68.3	70.1	
14	91.7	59.7	
15	92.2	106.5	
16	84.9	79.5	
17	86.2	61.2	
18	60.7	128.1	
19	66.6	91.4	
20	90.6	81.7	

Fig. 7-17. Caisson monitoring computer printout. Plan view of caisson showing location of measuring points (1–20). R1, R2, R3 diagram at top shows location of these points on outline of caisson tip. Table at right shows reaction forces in tons/m² on points R1, R2, R3 at locations 1–20.

system across a north-south and an east-west axis line. One axis line would be assigned letters and the other numbers. A copy of this grid pattern was given to each ready mix company who in turn gave a copy of each truck driver. Daily delivery schedules would be set up by telling each concrete supplier which area would be worked on that day. Drivers would be given this information and as they drove along the temporary road to the actual discharge area, they would stop at a station on the roadbed about halfway on their travel to the dumping point. At this station there would be a visual display of the grid system and a microphone in case any last minutes changes were to be made.

Mitsui would have their traffic engineer at a computer console back in the field office to direct the flow of trucks coming and going. As a concrete truck arrives at the site and travels toward its destination, it stops at the station on the bridge and the traffic engineer in the field offices keys into his computer the discharge location, i.e., A-3.5, the exact location for this load. This information is transmitted visually to the truck driver, who then proceeds to that point. A television camera picks up the activity at the caisson and displays the scene back to the Mitsui engineer, who continues to orchestrate the day's concrete deliveries (see Fig. 7-16).

As concrete is being placed in the caisson shell, monitoring of the reaction forces at the caisson tip is being conducted. Mitsui uses one of its six personal computers in the field office for this monitoring function. A profile and plan view of the tower caisson and the anchorage caisson have been programmed into one of the PCs. Instruments were attached to twenty points on the caisson as it was being lowered onto its underwater location, and these instruments will function as sending units in order to measure both location and reaction forces at the caisson tips. When concrete is placed in the caisson, each of the twenty monitoring points around the base perimeter will be transmitting data back to the PC. The central processing unit will update this information where it will be displayed on the CRT screen. An attached color printer will chart the reaction forces, registered in tons per square meter, at each of these twenty points and particularly at the caisson tips, designated as R1, R2, and R3. This plan view of the computer printout is reproduced in Fig. 7-17.

On both of these construction sites, one an office building and one a marine-civil engineering project, the Japanese construction process observed was an orderly, well planned, meticulously executed maneuver. Workers seemed joined together in a common cause and went about their daily routine with a fair amount of intensity. Problems that cropped up seemed to be viewed as challenges rather than annoyances—and these were among other observations that struck the eye of an American builder visiting Japan for the first time.

8

Unionism in Japan and the Construction Unions

The thought of unions and the union movement in America conjures up visions of Walter Reuther being pummeled by Henry Ford's goon squads outside of the automaker's Dearborn factory in 1937, or of giant steel mills lying cold and quiet as angry steelworkers march in picket lines around the plant. Labor and management seem to have been at odds with each other in the United States forever, and the union movement was born and took strength from the discord that existed between worker and boss.

How does a union movement gain momentum in a country where workers work an average of 40.6 hours a week, 52 weeks of the year—willingly? Or where absenteeism is a scant 4 days per year, and where workers take only 50% of their allotted vacation time for fear of letting the company down?

The Japanese miracle occurred primarily because of the incredible bond that developed between employee and employer, and the unions that began forming in Japan after the Second World War seemed to enhance that bond. Although Japanese workers worked 2111 hours in 1987, a rate 11% higher than those in the U.S. and a rate 31% higher than West Germany's labor force, a poll conducted in Japan showed that long hours were not of major concern to the average worker. Japan has an absenteeism rate 50% lower than the United States, 73% lower than France, and 81% less than workers in West Germany. This indicates diligence unmatched in the industrial world. Employees in Japan were allotted, on average, 14.9 vacation days in 1987, but refused to take more than 7.5 days—"Because my workload doesn't permit it," responded a large number when being surveyed. Then there is the phenomenon of unpaid overtime reported by 42% of 5419 workers polled in 1987, simply because many of the employees did not feel that all their overtime was requested by the company, but instead was self-imposed.

People in Japan want to take pride in their work and want fair treatment in return—the same as workers everywhere in the world. It seems as though Japan has been the only industrialized nation in the 20th Century to have attained both goals.

The union movement in Japan is one of those unique developments which would be next to impossible to duplicate anywhere else in the world.

THE HISTORY OF UNIONS IN JAPAN

Labor unions were first formed in Japan in 1897. However, a law known as the Law For Maintenance of Public Peace was enacted in 1900 which virtually declared unionism illegal, and the union movement went underground.

In the early days of industrialization in Japan, many employers provided their workers with free housing near the factories. Meals were also provided for workers at little or no cost, and in some cases a rudimentary pension plan was also devised for the workers. These "fringe benefits" were offered by factory owners not as an act of benevolence, but so that workers could put in longer hours, since they would not have to travel great distances to their homes and hence could stay at their weaving machines much later in the day. Employers offered low daily wages, allaying workers concerns by telling them that a portion of their daily wage was being set aside for their retirement years in what may be the earliest recorded use of what is now called "golden handcuffs."

After World War I, as industrialization began to accelerate in Japan, the Japanese Federation of Labor was formed. Known as Sodomei, this labor organization was greatly influenced by the Russian revolution and assumed socialistic and communistic ideologies. The movement, with its leftist leanings, did not endear itself to the center of power being formed in the country at that time. During the 1930s Japan's emerging industrial complex began to focus its energies on a war economy and the concept of a Greater East Asia Co-Prosperity Sphere. At this time the labor movement was pushed far into the background, and it was not until the end of the Second World War that it emerged from the shadows and began to develop with any rapidity.

Unions did not become legal again until December 1945, when the Trade Union Law was enacted, and free and open labor union activities were mandated by law and encouraged by the Allied Occupational Forces provisional government. In November 1946, Japan's new constitution was created. Article 28 of that document guaranteed workers three basic rights: to organize, to bargain collectively, and to strike. Table 8-1 contains statistics relating to the number of unions in Japan as of 1986 and the total union membership as of that period of time. Table 8-2 reflects the percentage of unionism by type of industry as of 1986.

Table 8-1. The Number of Labor Unions in Japan and Their Memberships.

Year	No. of Unions	Union Membership	No. of Persons in Each Union
1980	72,693	12,369,262	170
1981	73,694	12,471,270	169
1982	74,091	12,525,619	169
1983	74,486	12,519,530	168
1984	74,579	12,463,755	167
1985	74,499	12,417,527	167
1986	74,183	12,342,853	166

Source: Ministry of Labor.

Table 8-2. Percentage of Unionism in Japan by Industry.

Industry	With Unions	Without Unions
Mining	31.4%	68.6%
Construction	13.6%	86.4%
Manufacturing	6.4%	93.6%
Wholesale/retail	6.0%	94.0%
Finance, insurance, real estate	24.6%	75.4%
Transportation/communication	14.9%	85.1%
Utilities	26.3%	73.7%
Service	13.5%	86.5%

At the end of June 1987, according to the Ministry of Labor, the rate of unionization for the entire nation was 27.6%

As a means of comparison with the United States, union membership reached its peak of 35.5% during the period following World War II. It declined to 21% in 1979 and dropped to 18% in 1985 and slightly lower, to 17.5%, in 1986.

Union membership in the U.S. construction industry has also been in decline in recent years, falling from 23.4% in 1986 to 22.2% in 1987.

THE ENTERPRISE UNION

The enterprise union, the dominant form of union organization in Japan, is a labor organization formulated on a company-wide (enterprise) basis rather than on a craft or industry-wide guild basis, the prevalent form of unionism in Western industrial nations.

The union formed in this manner is created solely for the benefit of the employees of a particular company, and includes both blue and white collar workers regardless of their job responsibility or trade classification.

The enterprise union has only one basic membership requirement: permanent employment in the company in which the union is being formed. The only members of the enterprise who are not permitted to join the union are the top echelon of management. Craft unions or labor organizations formulated along trade or craft lines are, in the main, nonexistent in Japan. The exception to this rule is the All Japan Seaman's Union, known as Kaiin, whose membership is granted directly to seamen without regard for their place of employment.

THE ROLE OF UNIONS

The two major concerns of unions in Western industrial nations—wage and salary adjustments and job security—do not have to be addressed with the same intensity in Japan. Wages, historically, are adjusted annually by a process known as the Spring Offensive, and the pact between employers and employees implies that everyone's job will be secure until they reach the age of mandatory retirement.

Seniority provides an employee with the security of knowing that barring a catastrophe in the company's financial condition, yearly wage increases will be automatically forthcoming. As an employee's seniority increases so will his skill level, as systematic company training courses and educational programs are provided and experience is absorbed from co-workers and supervisors. New technologies are more readily accepted in Japan, particularly when they make an employee's job safer or easier, since technological advances do not threaten the worker's job security.

The bond of solidarity that holds the enterprise union together is, in effect, the sense of community that exists within the Japanese corporation.

Although enterprise unions within the same industrial category have formed federations, the enterprise union remains an autonomous organization owing its allegiance solely to the workers within the company that formed the union. Groups of enterprise unions that have common bonds or similar occupational interests have joined together in Japan to form industrial federations. A carpenter employed by a carpentry subcontractor will belong to that company's enterprise union. That carpentry enterprise union might, in turn, join a federation of carpentry unions. A federation known as Zenkensoren, the Japanese Wooden House Building Workers Union, is one such organization; it is the largest federation of construction related workers enterprise unions in Japan, counting almost 300,000 members within the group. Zennikken is another federation comprised of enterprise unions in the heavy construction industry. Zenkoku Ippan is a federation consisting of small and medium size construction firms and has a membership of more than 130,000.

One of the main functions of federations such as Zenkensoren is to act as a central clearing house for the gathering and dissemination of information relating to wages and working conditions within the industry that they serve. A secondary function of the federations is to assist member unions in dealing with problems that are common to the industry as a whole. These federations aid member enterprise unions in developing coordinated and unified courses of action as they attempt to resolve problems on the local scene. They provide members with advice on how best to handle labor disputes and, through the many surveys and studies they conduct, they can disseminate information about industry wage trends and working conditions that exist in other parts of the country.

Union federations such as Zenkensoren, Zennikken, and Zenkoku Ippan are part of the construction union pyramid that is topped by Kensetsu-Roren, a national center for all construction related union activity.

KENSETSU-ROREN

This central council is composed of 19 unions representing construction workers in both the private and public sector. Kensetsu-Roren is the central clearing house for all construction related union activity. It conducts cooperative conferences throughout the year to develop philosophical approaches to the plight of workers, their social rights, and directions that member unions need to pursue to fulfill the obligations placed upon them.

The composition of Kensetsu-Roren is listed below, along with the approximate membership of each of its affiliates:

Wooden House Builders Workers Union—Zenkensoren	300,000
Heavy construction workers unions	
Zennikken	16,000
Zenkokuiro	8,200
Construction related industry workers (materials, tools, equipment)	
Zenkoku Cement	14,300
Kensetsukanren Rokyo	8,100
Kensetsu Ippan	100,000
Zenkoku Ippan	130,000
Staff workers unions of public construction related corporations	
Zenkenro (staff at Ministry of Construction)	6,500
Zenkoren (Port Department, Ministry of Transportation)	2,800

Kokuro Kenkyo (Japanese National Railways)	4,500
Zensuido (National Water Works)	37,000
Zenkaihatsu (All Hokkaido Development Bureau)	13,000
Shukoro (Metropolitan Highway Public Corporation)	11,000
Nohkaikyo (Agriculture and Development Public Corpora tion)	495
Suishiro (Water Resource Development Public Corporation)	1,000
Nichijuro (Japanese Housing Development Public Corp.)	38,000
Other unaffiliated organizations	
Dentsukyotoh (observers of Kensetsu-Roren)	334,000
Nikkenkyo (staff workers of big construction firms)	70,000
Kensetsu Domei (outsiders of Kensetsu-Roren)	7,000

Although these various industry federations and councils appear to be functioning solely as central clearing houses for the gathering of industry data and serving as conduits for the distribution of information to their members, they have made one lasting contribution to Japanese labor-management relations—the Spring Offensive.

THE SPRING OFFENSIVE

Through the combined efforts of the union federations, a once-a-year synchronization of individual enterprise union wage negotiations throughout the country was created. Begun in 1955 and firmly entrenched in Japan by 1965, wage negotiations are conducted nationwide each year at about the same time. Nearly 80% of the entire country's salary adjustments take place annually between the months of March and May, hence the term Spring Offensive, which has become a part of the Japanese vernacular. The wage negotiations conducted as part of the Spring Offensive greatly influence those of the remaining 20% of Japan's corporations that do not participate in this yearly ritual.

The iron and steel workers start the ball rolling and usually comprise the first industrial group in the nation to finalize wage negotiations. The wage adjustments in this group establish trends for the industries that follow, typically the large enterprise unions. These large companies in turn serve as a pattern for smaller enterprise unions, which create the parameters for many public employees, and finally, the adjustment to the statutory minimum wage decided each year by the prefecture governments.

April is the month when corporations hire new employees. High school graduates, college and university degreed people, and high school "leavers" (persons who do not finish high school) will enter their new place of employment as a new class. This time of the year is regarded as the start of another year's employment for established workers within the company, and wage adjustments are made at that time as part of the company's seniority policy.

During the Spring Offensive, the Japanese Diet is usually deliberating the national budget for the forthcoming year, and the knowledge of how wage levels are being adjusted is helpful to the legislators as they prepare various budgets for the next fiscal year.

Since the percentage increase of most industrial wage adjustments is affected somewhat by the prosperity of the individual company, and possibly the entire industry as well, the government has an unusual opportunity to observe the state of the industry by analysing how each industry has adjusted its wage structure.

NATIONAL FEDERATIONS OF CONSTRUCTION UNIONS—ZENKENSOREN

The role of the union federation was explained in detail by Mr. Tadayoshi Katoh, General Secretary of Zenkensoren, during an interview held in his office in the Shinjuku section of Tokyo.

Mr. Katoh, in addition to his position as head of Zenkensoren, also holds the title of Vice President of the Liaison Council of Independent Unions of Japan (Churen). According to Mr. Katoh, Zenkensoren fulfills five important functions in their role as advisor of enterprise carpenter unions:

1. They assist in the collective bargaining process that occurs within the individual company by supplying the enterprise union leaders with wage information and wage settlements as they occur in other parts of the prefecture and other parts of the country.
2. The federation provides a source for low cost insurance to all of its members.
3. The federation lobbies in the federal and prefecture governments on behalf of the construction industry and the carpentry trade in particular.
4. Educational services are provided to individual members and enterprise unions where the practice of safety and a safe working environment is stressed. The federation maintains liaison with carpenter subcontracting firms and various government agencies to make them aware of the need to increase their efforts to provide carpenters with safe working conditions.
5. Twice a year, Zenkensoren conducts a national conference in which representatives from all of the prefectures attend, to discuss working conditions and safety practices in the industry. Training programs are introduced at this time and both slide presentations and video

tapes are made available for member unions to take back to their prefectures.

Membership Fees

There is no initiation fee for those unions or individual members desirous of joining Zenkensoren, but monthly dues are collected from members of enterprise unions and other persons belonging to the federation. These dues are collected on the local level and passed along to the national organization.

Individual enterprise union dues vary from city to city and from prefecture to prefecture, and Tokyo appears to have the highest dues rate in the country. On average, a carpenter belonging to an enterprise union in Tokyo will pay 3,000 yen in dues each month. This is equivalent to $23.00 based upon an exchange rate of 130 yen to the U.S. dollar. Zenkensoren will receive 90 yen from the monthly dues paid by each member of the enterprise union belonging to the Federation.

Table 8-3 lists the average monthly unions dues for all types of unions in Japan for the period 1975–1985.

Assistance in the Collective Bargaining Process.

Mr. Katoh said that wage levels for carpenters vary considerably from area to area and from small towns to the major cities such as Tokyo, Osaka, Nagoya, and Yokohama. Zenkensoren attempts to establish more

Table 8-3. Average Monthly Union Dues per Member in Japan.

Year	Average Monthly Dues (yen)	Percentage of Monthly Wage
1975	1,884	1.5%
1977	2,437	1.6%
1979	3,014	1.83%
1981	3,181	1.73%
1983	3,285	1.67%
1985	3,746	1.62%

Source: Institute for Social Problems in Asia.

Note: Converting yen to dollars for comparison purposes for the years 1975 to 1985 would be misleading, inasmuch as the relationship of the yen to the dollar varied from 220 in 1981 to 237 in 1983 and 238 in 1985.

The percentage of monthly wage can be a means of comparison with union dues paid by U.S construction workers. In the writer's home state of Connecticut, a laborer's union dues amount to about 1.85% of monthly wages, based on a 40 hours week. Members of mason's unions pay almost 3.3% of their monthly income to the union in the form of dues.

or less standard wage rates for various parts of the country to assist local union members in their collective bargaining efforts.

Carpenter's wages in Japan's major cities vary from $16.00 per hour to as high as $20.00 per hour in parts of Tokyo. Carpenters are currently in very short supply in Tokyo and the ones that are highly skilled can command high wages. Typically carpenters working in Tokyo are earning 18,000–20,000 yen per hour for an 8 hour work day, 22 days per month. Using a conversion rate of 130 yen, this equates to $17.25 to $19.25 per hour, not too much different from open shop carpenter rates in the U.S. Apprentice carpenters who have just graduated from a manual training school earn 100,000 yen per month, which converts to approximately $769 per month, or roughly $4.80 per hour. Carpenters, like other construction trades working throughout Japan, do receive premium pay for work beyond the standard eight hour work day, but the rate of premium pay is not fixed and can vary from 30% to 50% over the worker's base rate.

Methods of Negotiating Wage Adjustments

When Mr. Katoh visited the United States several years ago, he was interested in observing how trade unions were organized along craft or guild lines as opposed to the enterprise approach in Japan. He said he found American unions "rebellious." "Unions in the United States are like a corporation," he said. The only similarity he found between U.S. and Japanese unions was that fact that in both countries there were many sons who followed their father's trade and belonged to the same unions.

When asked how Zenkensoren assists its fellow union members in their efforts to obtain higher wages, the approach to the problem and the answer to the question was typically Japanese and rather difficult for someone coming from the land of "rebellious" unions to comprehend: "We will post signs on buildings where wage negotiations are going to take place and these signs will contain messages to employers and the public at large. The wording on a sign might state 'We think that carpenters in this area ought to be paid 21,000 yen per day, but they are only being paid 17,000 per day now.' "

The writer asked what would happen if the employers to whom this message was directed ignored the plea and decided that wage levels for carpenters would not be increased to 21,000 yen per day. Mr. Katoh responded in a matter-of-fact way, "Well some companies will agree to pay our carpenters the wage increase they demand and others will not. If the carpenter's request is denied, he will threaten to leave the company and find work with another subcontractor willing to pay what is considered a fair hourly wage." The Japanese penchant for avoiding confrontation is certainly an important ingredient in wage negotiations.

Lobbying Efforts by the Union Federations

Along with lobbying efforts in the classical sense, Zenkensoren performs many such efforts that are purely Japanese. They print many kinds of posters which are distributed to member unions, and these posters are meant to be tacked up on or near construction worksites to serve as public relations flyers. Some signs will promote carpentry as a worthy and honorable occupation in an attempt to improve the image of carpenters as well as to attract new workers. Flyers will be printed and sent to the wives of carpenters telling the women that their husbands are doing a good job and that they should be proud that their spouse is a carpenter. Posters will also be printed and placed in public places to educate voters to the fact that it is necessary to enact legislation which will provide better retirement benefits for older carpenters.

Some of these kinds of posters are reproduced in Figs. 8-1 through 8-4. The poster depicting a neatly dressed carpenter with chisel in hand (Fig. 8-1) is a mailer meant to be distributed to employers. The carpenter is requesting an increase in wages to 21,000 yen per day. On the reverse side there are cartoon characters representing carpenter and employer. The boss is frowning at a request for an increase in pay, and the worker is saying that he needs more money because the cost of living is so high. The last scene in the comic strip format shows carpenter and employer with arms around each other, smiling, having reached an agreement on a new wage level.

The poster containing a triangle at the top with the date 1988 (Fig. 8-2) is a request to the federal government to treat retired carpenters better and increase their retirement benefits. The vibrant, smiling young lady athlete in Fig. 8-3 is probably saying "Carpenters make better lovers—they don't get board easily." And finally, the poster depicted in Fig. 8-4 with a perspiring caricature of Prime Minister Takeshita and another official throwing a net over what appears to be consumer goods, is meant to portray that the government ought to be placed under some pressure to improve the lot of the Japanese carpenter.

Zenkensoren has created Housing Day, a day set aside each year to accomplish two goals. One is to promote the construction and sale of housing and the importance of owning one's own home. Secondarily, Housing Day, with much fanfare, extolls the carpenters needed to build these houses and therefore serves as a method of promoting carpentry as a noble and much needed trade as well as acting as a day of recruiting to attract new workers to the industry.

Fig. 8-1. Poster depicting carpenter with request to increase daily wage to 21,000 yen.

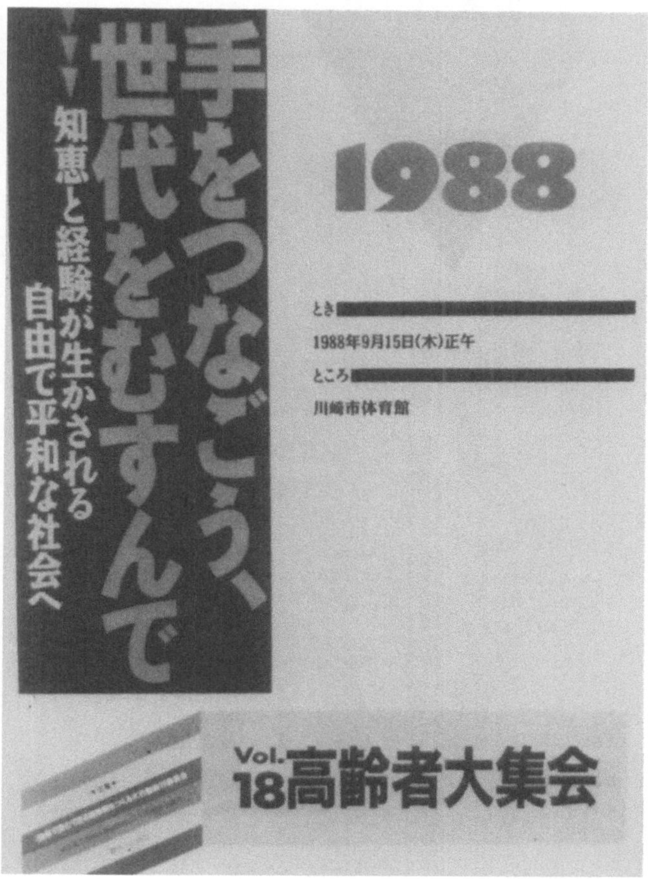

Fig. 8-2. Poster requesting that the federal government provide retired carpenters with increased pension benefits.

Fig. 8-3. Japanese equivalent of "Carpenters make better lovers, they are just plane sexy."

Fig. 8-4. Message to Prime Minister Takeshita requesting assistance in improving carpenter's social and economic position.

Disputes and the Federation's Role

According to Mr. Katoh, a strike is not an effective method for airing disputes in the construction industry in Japan. Strikes are not very effective ways to settle disputes because of the multi-tiered method of subcontracting prevalent in the country. At least in the carpentry end of the business, a carpentry subcontracting firm will probably employ a second tier firm to perform a portion of the work. This second tier subcontractor will, in turn, employ a third tier sub, who in turn, might hire a fourth tier company that may be employed just for materials distribution or cleaning purposes.

This multi-layered system which is so prevalent in Japan makes it difficult to use a strike as a threat since the workload is so subdivided and not solely dependent upon one subcontractor. If a subcontractor in one of the lower tiers on the chain decides to strike, he will be replaced by the subcontracting firm that hired him in the first place.

Mr. Katoh said that with the high level of employment and construction activity that now exists, particularly in Tokyo, along with the shortage of skilled carpenters, the threat of a worker quitting his job when demands are not met is an effective way to resolve disputes and grievances.

Table 8-4 shows statistical data regarding the number of disputes, numbers of workers involved in these disputes, and the number of working days lost during industrial disputes in the United States, the United Kingdom, West Germany, and Japan for the years 1976–1985. According to these data, the United States has registered the lowest number of disputes, averaging just about 161 per year for the ten-year period. West Germany has had the least amount of workers involved in industrial disputes, averaging only 181,000 per year. The statistics for Japan's involvement in industrial disputes reveals a dramatic decrease in all three categories beginning in 1981 with a more or less downward trend each year thereafter.

Table 8-5 lists the disputes in Japan categorized by type, i.e., attributable to wage increases, changes in work hours, shutdowns, and so forth. It appears that disputes over wage hikes accounts for most of the labor-management problems in Japan. Table 8-6 lists the reasons for employees leaving their place of employment.

THE SEWAYAKU

Individual construction workers will generally register their complaints to management through the use of a *sewayaku*, the Japanese equivalent of a trade foreman. In years gone by the *sewayaku* had the responsibility of insuring that the required size and composition of work crews were assembled each day to meet the needs of the job. If day laborers were needed on a temporary basis, it was the Sewayaku's responsibility to scour the city for indigent labor.

The *sewayaku* was a semi-direct employee of the subcontractor in the highest subcontracting tier who depended upon this experienced foreman to insure that all lower tier subcontractors also obtained the number of workers necessary to perform their tasks. It was the *sewayaku* that presented the individual worker's complaint or grievance to the boss and tried to settle any labor disputes that might occur within his company or with lower tiered subs. When a worker's grievance was not settled to the worker's satisfaction, the worker either accepted the *sewayaku's* resolution or would seek employment elsewhere.

EDUCATION, TRAINING AND TECHNOLOGY— ALL PART OF THE UNION'S CONCERN

The federations are deeply concerned with providing vocational training and other forms of education to their members as a means of upgrading the quality of life in general. Technology is viewed from a much different perspective in Japan, and advances are welcomed since they do not endanger job security due to the lifetime employment concept. The unions are very much concerned with attracting new members and improving the image of construction workers in general as new entrants into the employment arena seem to be attracted to other types of jobs.

Table 8-4. Industrial Disputes in Major Industrialized Nations.

	United States[a]			United Kingdom[b]		
Year	No. of Disputes	Workers Involved	Working Days lost	No. of Disputes	Workers Involved	Working Days lost
1976	231	1,519,000	23,962,000	2016	668,000	3,284,000
1977	298	1,212,000	21,258,000	2703	1,165,800	10,142,000
1978	219	1,006,000	23,774,000	2471	1,041,500	9,405,000
1979	235	1,021,000	20,409,000	2080	4,608,000	29,474,000
1980	187	795,000	20,844,000	1330	833,700	11,964,000
1981	145	729,000	16,908,000	1338	1,512,000	4,266,000
1982	96	656,000	9,061,000	1528	2,102,900	5,313,000
1983	81	909,000	17,461,000	1352	573,800	3,754,000
1984	62	376,000	8,498,800	1206	1,464,300	27,135,000
1985	54	323,900	7,079,100	887	643,000	6,402,000

Source: ILO Year Book of Labour Statistics—1986.

[a]Excludes disputes involving less than 1000 workers.

[b]Excludes political strikes and disputes lasting less than one day except when more than 100 working days lost.

[c]Excludes disputes lasting less than one day except when a loss of more than 100 working days is involved.

[d]Excludes disputes lasting less than half a day.

Mr. Miyoshi Sasada, Secretary of the Organizing Department of Zenkensoren said that the federation must do several things to attract younger people to the construction industry:

1. Wage levels need to be increased.
2. Robotics need to continue to be pursued and development encouraged so that machines can perform dangerous tasks and work in environments that are unpleasant or unhealthy for human workers.
3. The subcontracting system in Japan ought to be simplified so that the multitiered system is greatly reduced and follow the practices in the U.S. system more closely.

Mr. Sasada said that one of the main attractions offered by most federations is inexpensive insurance coverage for workers and by increasing the benefits, construction work might help to bring new workers into the industry.

Since Japanese unions do not function in the same manner as U.S. construction unions, skill levels need to be created, and the training necessary to attain these skill levels must be provided. Zenkensoren has the responsibility of establishing training facilities for carpenters and initiating formal skill level requirements in order to create some semblance of uniformity in classifying workers' abilities. Training centers are established at the local level and are funded by membership dues. For new workers entering the field with only a junior high school ed-

Table 8-4. Industrial Disputes in Major Industrialized Nations *(continued).*

West Germany [c]			**Japan** [d]		
No. of Disputes	**Workers Involved**	**Working Days lost**	**No. of Disputes**	**Workers Involved**	**Working days lost**
1481	169,312	533,696	2720	1,356,025	3,253,715
81	34,437	23,681	1712	691,908	1,518,476
1239	487,050	4,281,284	1517	659,966	1,357,502
40	77,326	483,083	1153	449,504	930,304
132	45,159	128,390	1133	562,921	1,001,224
297	255,334	58,398	955	247,114	553,726
40	39,981	15,106	944	215,701	538,143
114	94,070	40,842	893	224,129	506,873
1121	537,260	5,617,600	596	115,058	354,050
53	78,190	34,400	627	123,390	264,050

Table 8-5. Industrial Disputes in Japan Categorized by Type of Demand.

Year	No. of Cases	Pay Raise	Temporary Allowance	Change in Working Hours	Objection to Discontinuance/ Shutdown, Cutback of Business	Other Reasons
1965	3,051	1,586	864	60	50	72
1970	4,511	2,131	1,260	16	28	509
1975	8,435	5,304	1,971	47	32	1,723
1977	6,060	5,476	1,081	58	32	2,780
1978	5,416	3,697	1,158	42	32	2,109
1979	4,026	2,923	721	41	31	1,247
1980	4,376	3,236	722	48	26	1,914
1981	7,660	4,749	642	41	26	5,298
1982	7,477	5,951	806	17	43	4,928
1983	5,562	4,375	672	22	40	2,891
1984	4,480	3,122	552	25	42	2,681
1985	4,826	2,890	595	29	26	3,050

Source: Ministry of Labor, Tokyo, Japan.

Note: Totals of individual columns do not add up to number of cases inasmuch as some cases carried multiple demands.

Table 8–6. Reasons for Employees Leaving Place of Employment (by age group and Male and Female)

Sex	Causes	Total	15–24	25–34	35–44	45–54	55 and over
Male	Total	100.0	100.0	100.0	100.0	100.0	100.0
	Involuntary reasons	56.9	37.5	40.0	57.1	54.5	79.2
	Personnel cuts/bankruptcy	11.1	—	13.3	14.3	18.2	4.2
	Business depression/instability	13.9	12.5	13.3	21.4	18.2	8.3
	Other employer reasons	12.5	12.5	13.3	14.3	18.2	8.3
	Mandatory retirement	19.4	—	—	—	—	58.3
	Voluntary reasons	40.3	62.5	60.0	42.9	45.5	16.7
	To find better job	13.9	37.5	26.7	21.4	9.1	4.2
	Reasons of housekeeping, schooling, health	13.9	12.5	13.3	14.3	18.2	8.3
	Others	12.5	25.0	20.0	14.3	18.2	4.2
Female	Total	100.0	100.0	100.0	100.0	100.0	100.0
	Involuntary reasons	28.9	18.2	11.1	22.2	40.0	66.7
	Voluntary reasons	71.1	81.8	88.9	66.7	60.0	33.3
	To find better job	21.1	27.3	22.2	22.2	20.0	—
	Reasons of housekeeping, schooling, health	18.4	18.2	11.1	22.2	20.0	33.3
	Others	28.9	36.4	44.4	33.3	20.0	—

Source: Ministry of Labor, Bureau of Statistics, February 1986.

ucation, the carpenter training course is of three years' duration. For new high school graduates, the training course is only two years long. Training sessions are available in two schedule forms, an eight hour session each Saturday, or nightly sessions during the week, each lasting two hours. The courses offered in these training sessions range from basic carpentry skill instruction, to mathematics, to measuring and dimensional instruction, to courses in sociology. Upon completion of the course, the newly graduated carpenter candidate with his certificate of training can take a government sponsored examination to qualify as a Carpenter Second Grade. After working in the field for a specified length of time, a worker can then take the examination for Carpenter First Class.

THE COUNCIL OF JAPAN CONSTRUCTION UNIONS

Established in December 1954, the JCU, as it is known, is comprised of 39 individual unions of construction workers with a total membership of 63,000. JCU has ten regional chapters throughout Japan and coordinates labor problems with social concerns in the construction field with other industries such as the automobile industry, utility company industries and financial and commercial fields. JCU maintains contact with several foreign labor organizations so that it can be informed about the state of labor relations in other parts of the world.

The function and purpose of JCU is described in their Council brochure, reproduced on the following pages, and allowing for minor errors in translation, the Wingbest calling imparts the feeling of nurturing and the quest for harmony that is so prevalent in most Japanese organizations.

A CLOSER LOOK AT LARGE CONSTRUCTION COMPANY ENTERPRISE UNIONS

Mitsuta Miyawaki is an engineering university graduate who has been working for one of the Big Six construction companies in Japan for about eight years. He was recruited from college by the company along with a number of his classmates as well as graduates from other colleges and universities during the Spring Offensive of 1980. All of these new recruits entered the corporation much the same as a new class of college students enters the university in their freshman year. They enter as classmates and will progress through the company as such. All of these newcomers entered the corporation with a deep sense of pride in having been offered a position from such a prestigious construction company,

EXPANSION AND DEVELOPMENT OF J.C.U

"VITAL ORGANIZATION"

J.C.U.WAS ESTABLISHED IN DECEMBER 1954, WITH 16 INDIVIDUAL UNIONS, AND NOW IT GROW UP AND ORGANIZES WITH 39 INDIVIDUAL UNIONS FOR 63,000 MEMBERS.

J.C.U IS THE INDUSTRIAL ORGANIZATION CREATED BASED ON THE MUTUAL RELIANCE AND FRIENDSHIP EACH OTHERS, CONCENTRATED BY LIBERAL AND DEMOCATIC UNION ORGANIZED BY THE LABOR UNION STAFF, WHO WORKS IN THE CONSRRUCTION INDUSTRY, AND ITS RELATED FIELD IN JAPZN.

WE SET TO WORK FOR OUR SELFRELIANCE AND INDEPENDENCE, RESPECT THE CHARACTERISTIC AND THE INDEPENDENCY OF MEMBER INDIVIDUAL UNIT, PROMOTE THE SOLIDARITY AND HARMONY BETWEEN EACH INDIVIDUAL UNIT,AND TRY TO FORM VIVID, VITAL AND DEMOCRATIC ORGANIZATION.

REGIONAL ACTIVITY

WE HAVE 10 REGIONAL ORGANIZATION (CHIKYO) ACROSS THE COUNTRY.

CHIKYO IS A NUCLEUS IN THE J.C.U WORK AREA PLAYING A VERY IMPORTANT ROLE IN DEVELOPING ITS WORK EFFECTIVELY AND SMOTHLY TO ALL ORGANIZATION BY MEANS OF THE SEMINOR,RESERCH,ASSOCIATION AND SO FORTH.

TO BE A REAL KEY INDUSTRY FOR J.C.U

THERE ARE SO MANY PROBLEMS WHICH HAVE TO BE SETTLED BY US IN THE CONSTRUCTION INDUSRTY. WE BELEVE THAT WE SEE THE BRIGHT FUTURES FOR THE CONSTRUCTION INDUSTRY, AND WE ALSO UNDERSTAND THAT WE SEE THE IMPROVEMENTS OF VARIOUS LABOR CONDITIONS FOR THOSE WHO WORK IN THE CONSTRUCTION INDUSTRY, ONLY WHER WE SETTLE VARIOUS PROBLEM.

THEREFORE WE PROMOTE AGGRASSIVELY VARIOUSE INDUSTRIAL WORKS.

MODERNIZATION OF THE CONSTRUCTION INDUSTRY.

J.C.U SETS TO WORK FOR MODERNIZATION OF CONSTRUCTION INDUSTRY BASED ON THE FOLLOOWING THREE VIEW POINTS:

1 PORPOSAL FO POSITIVE MEASURE AS LABOR UNION AIMING AT THE IMPROBEMENT OF OVERALL LABOUR CONDITIONS.

2 UPLIGHT OF CONSTRUCTION INDUSTRY AS A INDUSTRIALIST WHO SEEKS FOR THE LASTING DEVELOPMENT TO THE CONSTRUCTION INDUSTRY.

3 ENHANCEMENT OF ITS SOCIAL STATUS BY FULSILLING THE SOCIAL FUNDAMENTALS.

FULFILLMENT OF ACTIVITY FOR INDUSTRIAL POLICY

J.C.U EXAMINES AND GRASPS SEVERAL PORBLEMS RELATING TO THE CONSTRUCTION INDUSTRY, AND WORKS FOR THE SUGGESTION OF POLICY PROPOSAL WHICH SHOULD BE CULTIVATED BY MEANS OF

1 EXCHANGING INFORMATION WITH RELATIVE ORGANIZATION AND WITH THE AUTHORITIES CONCERNED, AND
2 PARTICIPATING VARIOUS POLICY RELATED SEMINARS INSIDE AND OUTSIDE THE INDUSTRY.

FOR INTERNATIONALIZATION

WHILE THE INTERNATIONALIZATION IS BEING IN PROCESS, THE COMPETITION BETWEEN THE ADVANCED AND NEWLY DEVELOPING COUNTRIES INTENSIFIED OVESEAS.

BECAUSE OF THE PARTICIPATION OF FOREIGN ENTEPRISE INTO JAPANESE MARKETS, WE ARE ASKED WHAT THE JAPANESE ENTEPRISE SHOULD BE.

J.C.U WILL SEE SUCH SITUATION, AND SET TO WORK FOR SETTLING VARIOUS PORBLEMS CONCERND.

IN ORDER TO PRACTICE

<SEMINAR/GROUP ACTICITY>

- REAL ESTATE/HOUSING MEASURES SEMINAR GROUP.

  TO RESARCH THE ROLE OF CONSTRUCTION INSUSTRY AND WHAT THE URBAN REDEVELOPMENT SHOULD BE.

- MODERNIZATION OF CONSTRUCTION INDUSTRY MEASURE SMEMINAR GROUP.

  TO RESARCH THE MODERNIZATION AND STRUCTURAL REFORM OF CONSTRUCTION INDUSTRY.

<CULTIVATION OF MEASURES>

EXCHANGE OF INFORMATION WITH THE UNION INSIDE THE INDUSTRY, MANAGEMENT GORUP, GOVERMENT, LOCAL SELF-GOVERING BODY, OTHER INDUSTRY RELATED TO LABOUR UNION, AND MASS COMMUNICATION.

<DISPATCH OF A MEMBER TO THE COUCIL TO PRACTICE THE MEASURES>

- MINISTRY TO LABOUR, CENTRAL EMPLOYMENT SECURITY BEAUEAU SPECIAL INVESTIGATION COMETTEE CONSTURCUTION LABOR OFFICES.
- MINISTRY OF INTERNATIONAL RRADE AND INDUSTRY, SMALL AND MEDIUME ENTERPRISES' MODERNIZATION COUNCIL CONSTRUCTION OFFICES.

<APPROACHING TO REGIONAL MATTERS>

WE HAVE SOME CONFERANCES FOR REGIONAL MATTERS.

INTERCHANGE WITH OTHER ORGANIZATION

J.C.U REALIZES THE SOCIAL OBLIGATIONS AS THE EMPLOYEES'UNION REPREZENTATIVE OF THE CONSTRUCTION INDUSTRY, STRENGTH-ES THE INTERCHARGE SOLIDALITY WITH INS-IDE AND OUTSIDE THE LABOR NION, WHO HAS THE SAME AIM, AND DESIRES MANY PEOPLE'S PROSPERIRY THROUGH ITS ACTIVITY IN THE SPRIT OF MUTUAL COOPERATION.

J.C.U NOT ONLY HAS KEPT THE RELATION-SHIP WITH OTHER ORGANIUZATION RELATING TO THE CONSTRUCTION INDUSRRIES,BUT ALSO HAS KEPT THE RELATIONSHIP WITH SUCH IN-DUSTRIES AS AUTO,ELECTRICPOWER,COMMERC-IAL, FINANCIAL AND SO FORTH.

IN THE MEANTIME,J.C.U HAS KEPT THE REL-ATIONSHIP OVERSEAS WITH SUCH LABOR UNION AS INTERNATIONAL FEDARATION OF BILDING AND WOODWORKERS (IFBWW),INTERNATIONAL F-EDARATION OF ECONOMICAL,CLERICAL,PROFESS ION AND TECHNICAL EMPLOYEES(FIET) AND SO FORTH.

FOR SECURE A SURE MEANS OF LIVING AND PEACEFULLY LIFE

J.C.U SUPPORTS MAINTENANCE AND IMPROVEMENT OF OVEAL L LABOR CONDITIONS SUCH AS WAGES,SYSTEMS AND WELFARE, AND AT THE SAME TIME, IT MAKES EFFORTS TO RISE THE F-UNCTIONS OF DESIGINING AND PROPOSING THE MEASURES.

PREPARATION AND ENHANCEMENT OF L-ABOR CONDITIONS.

J.C.U WILL PREPARE THE PLACE WHERE PEOPLE DISPLAYS MULTIPLIER EFFECTS, MAKE EFFORTS TO PREPARE COMFORTABLE NAD ATTRACTIVE WORK PLACE,AND SUPPORT TO MAINTAIN AND IMPROVE OVERALL LABOR CONDITIONS BY EXCHANGING THE INFORMATION EF-EECTIVELY.

STABLE LIVING

WE WILL NEGOTIATE THE WAGEINCRE-ASE FOR MAINTENANCE AND IMPROVEM-ENT OF SUBSTARTIAL STANDARD OF L-IVING PROMOTE THE INDIVIDUAL WAGE SYSTEMS TO RECONSTRUCT THE WAGE SYSTEM AND PERSONEL SYSTEMS,AND TRY TO FILL UP THE RESERCH AND A-NALYSIS ACTIVETIES.

PREPARATION OF SOUND WORK CONDITIONS

IN ORDER TO SET THE UNIFORM STANDARD TO CUT THE WORKING HOURS FOR A COMFORTABLE LIVING AND WORK PLACES WORTH LIVING AND WORKING, WE HAVE COOPERATIVE SYSTEMS WITH INDIVIDUAL LAVOR UNITS TO APPROACH THE AUTHORITIES CONCERND AND EACH ENTERPIRSE.

PREPALATION AND ENHANCEMENT OF THE OVERSEAS WORKING CONDITIONS

WE ALSO INVESTUGATE THE OVERSEAS WORKIG CONDITIONS.

PREPARATION OF ENVIRONMENT FOR OVESEAS SERVICES WHEN WE SEE THE ENVIRONMENT FOR THOSE WHO WORK OVERSES,NOT ONLY THE LABOR CONDITIONS BUT MEDICAL CONDITION, MENTAL MATTERS AND THE EDUCATION OF ONE'S CHILDLEN WOULD BECOME MORE SERIOUS.

WE WILL POSITIVELY DEAL WITH SUCH MATTERS AND EXCHANGINGE INFORMATION FOR A INDIVIDUAL LABORUNIT TO IMPROVE ITS SYSTEMS.

HEALTHY SOCIAL LIFE

WE SET TO WORK TO ADOPT SUCH AS "INDIVIDUAL PENSION PLAN" FOR HEALTHY SOCIAL LIFE AS WE ARE SOON ABOUT TO FACING WITH AN AGING SOCIETY.

one of the world's largest. During the orientation period that followed, Mitsuta and his fellow newcomers were left with the impression that one day, one of them could, and would, be president of the corporation.

Upon entering the corporation, Mitsuta joined the company union. He describes his company union as more of a club; they organize company outings and picnics and union members get together one or two evenings a week after the work day to drink beer and discuss company policies and corporate philosophy.

Last year, Mitsuta said that several people in the company wanted to strike because they were unhappy with one of the corporation's personnel policies. Their complaints were discussed with the union leader who agreed to present the grievance to management along with the threat of a potential strike by some of the employees. When presented with the complaint, the company's director said he would certainly take the matter under advisement and present his decision to the union leader in short order. As far as the threat of a strike was concerned, he asked the union leader to tell the potential strikers that it was perfectly all right to strike, but their work would certainly pile up while they were striking, and after the disagreement was resolved they would be expected to work longer hours to make up for the time lost during the strike. When this message was conveyed back to the employees involved, they decided that they would not strike, and when they received word that management did not feel it would be appropriate to change the policy being challenged, the entire matter was dropped.

Mitsuta said that he was not in agreement with the corporate policy relating to unused vacation days. The company policy allows him to take 40 vacation days per year, based upon his seniority level at this time. This rather magnanimous policy is not what is appears to be, however. Since company policy stipulates that everyone works a six day work week, but Mitsuta's department is only required to work two Saturdays each month, the two other Saturdays are counted as vacation days. Subtracting two Saturdays per month, or 24 from 40 still leaves 16 actual vacation days. Mitsuta said that he occasionally works on a third Saturday of each month and when he does, he actually accrues another vacation day to his annual allotment.

The existing company policy allows for repayment of unused vacation time, but only up to twenty days per year. If an employee works every Saturday during the year and takes no vacation time at all, the company will pay him for only 20 vacation days not the full 40. Mitsuta thought that this policy was unfair and that employees should be paid for *all* unused vacation days. Last year Mitsuta only took 4 days vacation and felt that he should be fully compensated for 36 work days.

The grievance was lodged with the union leader, who discussed the matter at great length with the manager of the Engineering Group. The union leader returned several hours later and said that after much discussion, the company felt that it was not in their best interest, and therefore not in the best interest of the employee, to change the present reimbursement policy. The matter was effectively dropped at that point. When Mitsuta was questioned about this policy and asked why he just didn't take all of his vacation time, since the company would not pay for unused days beyond the twentieth, the reply was as foreign as the voice answering the question. "In four or five years, I will have enough seniority to qualify for a management position being placed in a post where I will have responsibility for possibly sixty people. If I take too many vacation days, my supervisors may feel that I am not responsible and do not display enough interest in my work or my company. This could prevent me from becoming a manager."

This same kind of attitude was displayed by another engineer from another large construction company when the subject of salary increases were being discussed. This engineer said that his company union was very friendly with corporate management and maintained close ties with the company's managers. Their union organizes a company baseball game in the spring and a beer party in the fall. When complaints arise they discuss them with their union leader to decide how and when further action ought to be taken. Several members of one work group in the company felt that their annual salary increases should be higher than they normally would anticipate, so they contacted their union leader and requested that he discuss their concerns with management. This was done, and the company manager said that he did not feel that this request was in the best interest of both parties, and the matter was quietly set aside.

Although both blue and white collar employees join the same enterprise union regardless of their job classification, educational background, or position in the company, at a certain point in the management hierarchy union membership is denied, or if the new manager is a union member at that point, he must sever his union affiliation. The level at which this happens seems to vary somewhat from corporation to corporation. One mid-level manager in a construction company slightly below the size of Big Six related his experience with his company's union and the outcome of a complaint registered with the company. Being a lawyer in the company, he was called upon to handle many of the company's contractural matters outside of Japan. It was not unusual for him to spend extended periods of time on assignment in foreign countries. While on temporary assignment, he would receive a daily allowance from the company to cover the cost of housing and meals away from home. This living expense would be adjusted according to the cost of living in the foreign country, some assignments requiring higher allowances. At one point, the lawyer was advised that his next assignment would carry him to a country with a particularly high cost of living and his allowance would be adjusted accordingly. However when he received his first subsistance check he did not feel it was adequate and through the company union leader, he requested a substantially higher per diem. The matter was discussed between union representative and management and a negotiated settlement somewhere between what the company originally offered and what the lawyer wanted was determined. The lawyer said that although the settlement should have been higher, he would accept the company's decision on the compromise.

The union leadership in these cases seems to take the form of surrogate, arguing the case for the union member, but saving the complainant the embarrassment of an outright rejection of his claim should management decide not to honor it. As one construction expert explained, when asked to summarize the company's relationship to the enterprise union, "Belonging to the enterprise union embraces the concept of family in Japan. It is just another visible sign that the employee has entered into a relationship with his company where the company and the union act as parents and watch out for their children, and do what is best for the entire family."

This is in stark contrast with the union leadership of one of America's most troubled corporations, the United States Postal Service, where productivity has been declining for the past eighteen years while the cost of a three cent stamp skyrocketed. The leadership of the American Postal Worker's Union sees its primary function as one of defending its members against the practices of management rather than attempting to form an alliance with management to jointly improve the mail system and prevent further increases in the postal rates. But we do have more mailboxes on the streetcorners in this country, and they are more brightly painted than they are in Japan!

9

Movers and Shakers—Robots and Seismics

ROBOTS

The term robot conjures up visions of Hollywood science fiction creations with claw-type appendages, traveling back and forth on wheeled feet and responding to commands with a monotone, synthesized voice. Construction robots in Japan do not fit this image. In fact most of them look like ordinary, recognizable pieces of equipment that have been slightly modified to fit their unmanned tasks.

One thing is for certain, the design and developement of construction robots is a priority item among Japanese construction companies, and they are already using unmanned equipment to perform the following tasks:

- Steel-trowel finish concrete slabs.
- Bend and place reinforcing steel bars.
- Weld shear studs in the field.
- Shot-crete walls.
- Apply spray fireproofing to structural steel framework.
- Spray paint exterior walls.
- Conduct on-site soil compaction tests.
- Test the integrity of exterior tile and masonry walls.
- Inspect clean rooms.
- Sweep floors.
- Hoist steel beams.

The impetus behind construction robotics began in 1983, when a program known as The Development Project for High Level Construction Technology Systems Using Electronics was launched. This joint effort by government and industry was created to accelerate the development of robots based upon some compelling reasons:

1. The need to attract new workers to an industry with a tarnished image. Safety records were poor, job-site injuries were high, many work tasks were unsafe, unpleasant, and physically demanding, and wage rates were lower than in other industries.
2. As part of a total quality control program in the construction industry, it would be inevitable that the development of robots would be rigorously pursued.
3. A shortage of skilled workers exists in Japan just as in many other industrialized nations, and the development and implementation of robots, could ease this shortage to a degree. (In 1986, the construction industry in Japan reported experiencing a shortage of 100,000 skilled workers, mainly electricians, equipment operators, concrete form carpenters, plumbers, and welders.)

The task of developing construction robots would naturally be placed at the doorstep of Japan's construction giants like Kajima, Ohbayashi, Shimizu, Taisei, and Takenaka Komuten, and they have responded in a positive way, producing some rather exotic yet extremely practical construction equipment.

The Concrete Floor Finishing Robot

Three companies, Shimizu, Kajima, and Takenaka Komuten have developed robots to steel-trowel finish concrete slabs with no human assistance other than the pressing of a start button.

The Kajima robot is a trackless, self-propelled troweling machine on wheels that travels over newly poured concrete slabs, guided in its mission by a microcomputer, a gyrocompass, a travel-distance sensor, and a touch sensor to avoid columns and other obstructions. This robot is a second generation creation and is dubbed the Mark II (see Figs. 9-1 and 9-2). The earlier model, the Mark I, was developed in 1984; its successor, the Mark II is a much improved version that left the laboratory for the field in 1986.

Kajima also developed a concrete surface-roughness measuring system as a companion piece to the Mark II in order to check the surface finishing accuracy of the robot. Kajima's Mark II robot meets the following specifications:

Size	Length—4′7″; Width—3′11″; Height—2′2″
Weight	Total weight, 408 pounds. The demountable trowel unit weighs 165 pounds when detached from the mainframe.

Fig. 9-1. Kajima's floor finishing robot, the Mark II.

■ Composition & Specifications

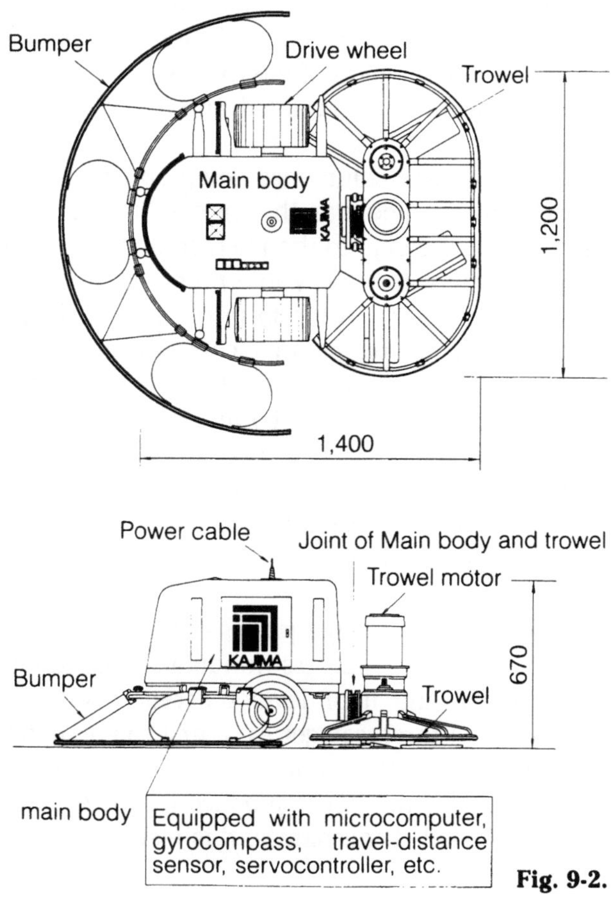

Fig. 9-2. Schematic of the Mark II robot.

Travel speed	Maximum 49 feet per minute
Finishing capacity:	60 square feet per hour lapping 20″ 93 square feet per hour lapping 8″ (According to Kajima, human productivity figures are 16 square feet per hour, therefore they figure that Mark II is 4–6 times more productive than a mason.)
Drive system	Electric motor.
Safety devices	Touch sensor to avoid obstacles. A travel stop device at end of its path.
Other detectors	A laser sensor will cause the robot to stop if it approaches an unguarded floor opening. A touch sensor will cause the robot to cease operations if the concrete it is traveling on is too soft.
Alarm device	Flashes white when operating, changing to red when an emergency is detected.

How the Mark II Operates (Fig. 9-3). The travel device section of the robot contains a self-navigation system which incorporates a gyrocompass, measuring rollers, and an 8 bit microcomputer. The robot is programmed by setting digital switches in the data entry panel and inputting the length and width of the floor to be finished. The lapping width, speed of operation, and the direction of the robot's travel as well as its first turn direction must also be programmed into the microcomputer.

By pressing a button the robot will commence going about its task. If it encounters an obstacle on the floor, as the bumper touches the obstruction, it immediately sends a signal to the microcomputer so that the robot can continue on its way without any interruptions.

The Mark II has on-board diagnostics in the form of seven LED display units, and when there is a failure, or an incorrectly performed operation, the robot's diagnostic system can assist the operator in pinpointing one of the twenty recordable codes to correct the problem.

The trowel portion of the Mark II consists of two three-bladed steel trowels. The unit's electric motor rotates the pair of trowels in opposite directions to offset the reaction forces created by rotation and to assist the machine in achieving straight-line travel.

The Performance of the Robot. Kajima judges concrete floor finishing accuracy by three standards: flatness (undulations), surface roughness (roughness of finish), and floor level differences (uneven troweling). They found that in all three categories, a first rate concrete finishing mason performed at about the same accuracy levels as the robot. Although the machine was found not to produce results superior to a human, in the absence of skilled mechanics, this electronic trow-

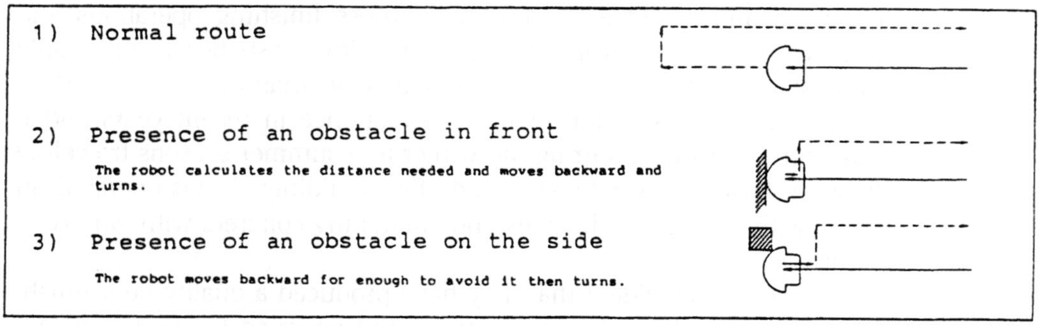

Automatic Obstacle Avoidance System

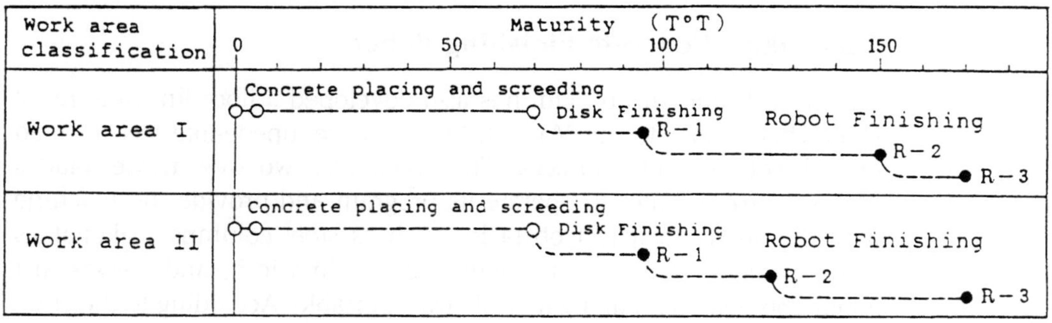

Relationship between the Maturity and Work Process

Mark II Execution Capacity

Travel speed (mm/sec)	200	
Lapping width (cm)	50	20
Execution capacity (m^2/h)	450	700
Comparison with a plasterer*	About 4 times	About 6 times

Note: *Provided that the trowel finishing capacity of a plasterer is 120 m^2/h.

Fig. 9-3. Operating parameters for the Mark II robot.

eling machine could easily take over these finishing operations and produce high quality work at more competitive costs because it is able to work faster and therefore be more productive than the normal worker.

The Mark II has gotten plenty of experience in recent years and it has been put to work during the winter and summer seasons troweling floors in factories, warehouses, and office buildings. It has operated on floors with varying thicknesses and containing concrete with variety of slumps.

Kajima feels confident that they have produced a quality floor finishing machine that can be programmed and operated by workmen, not engineers, and as they look to the future, the Mark II will be just one part of a completely automated concrete placement and finishing system.

Takenaka's Concrete Finishing Robot

Takenaka Komuten Co., Ltd. has also developed a floor finishing robot somewhat similar to the Mark II, but it has centipede-like feet to tread on still soft concrete surfaces. This robot has two steel trowel blades and a microcomputer to operate its program and provide the machine with the ability to detect obstacles such as steel columns. Takenaka's robot measures 6.5 feet in length, 3.2 feet in width, and weighs just about 286 pounds, making it easily transportable. According to the company, their machine will steel-trowel finish 3230 square feet of slab per day, and based on their figures for human productivity, they claim a 300% increase by the use of their robot.

Shimizu and Ohbayashi

Shimizu's robot is very similar, in size and weight to that of Takenaka, and it is claimed that this robot can finish 3226 square feet of slab in one eight-hour shift. Ohbayashi's floor finishing robot is gasoline engine powered and it travels on four wheels. Like the other robots, it has two sets of steel finishing trowels, but although Ohbayashi's robot is also programmed to avoid obstacles, it differs from the other robots in that lasers spotted strategically around the floor assist it in controlling levelness.

Concrete Placement Robot

As part of Kajima Corporation's quest to fully automate the concrete placement and finishing operation, they have developed a concrete "distributor" which is essentially a round hopper that is placed centrally to a large concrete pour. The hopper has the ability to rotate 360 degrees and has a roll-up discharge hose attached to it. Concrete is placed in the distributor from an outside source such as a concrete pump or an extended length concrete truck chute, and as the distributor swings in its arc, the rollup hose is either extended or retracted to place concrete unattended. It does not take much imagination to see how this simple concrete distribution machine could be fully automated and in combination with an automatic screed machine, command Kajima's Mark II to commence action at the proper time and in the proper sequence—but first, reinforcing steel might have to be placed in the forms.

Reinforcing Steel Bending and Placing Robots

Japanese construction projects consume about 5 million tons of rebars each year and virtually all of the bending is done by hand at an estimated cost of $1.2 billion. Ironworkers are scarce in Japan and it would follow that robotics would be explored in order to assist in this field.

In 1985, Aoki Corporation and Komatsu, the equipment manufacturer, introduced an automated reinforcing bar cutting and bending machine that was meant to be transported to the construction site. This machine can bend and form bars ranging in size from 0.375 inch diameter (#3 bars) to 1.6 inch diameter (#14s) and in lengths up to forty feet. Approximately 50 basic bending patterns can be programmed into the machine and it can crank out 8,000 pounds of material per hour.

When a magnetic card is inserted into the robot's microcomputer, a menu will be displayed on a CRT screen and the operator can select the length, shape, and size bar to be cut and bent. The computer will visually confirm the shape requested by the operator to verfiy that the proper bends have been selected. If nonstandard shapes are required, a program manual is provided to allow an operator in the field to create special shapes and bends.

Initially bars have to be loaded into the machine by hand, but once the microcomputer has been programmed, and the magnetic card is withdrawn and inserted into the bending unit, all subsequent operations are performed automatically. All that remains is for a nearby worker to remove the cut and bent bar so that the robot can continue to produce the next one.

According to a spokesman for Aoki, this robot will be marketed worldwide at an estimated price of $140,000. The manufacturer claims that this automated rebar bending and cutting machine will produce savings of 10–15% over previous methods of handling steel reinforcing bars.

The Rebar Placing Robot

The construction of nuclear power plants requires concrete foundations laced with large diameter reinforcing bars. Rebars with a diameter of 38 mm (half-way between a No. 11 and a No. 14 bar), 26 feet long, and weighing 220 pounds require a crew of between five and seven men to pick these bars up and walk them into place. Carrying heavy rebars like this over a field of previously installed bars is slow, costly and hazardous.

The combination of high cost, low productivity, and unsafe conditions is the perfect breeding ground for robotic development and Kajima did not miss this opportunity. They developed a playback type robot (Fig. 9-4) that can carry twenty 38 mm x 8 m long bars and place them one by one in a predesignated pattern, either parallel or zig-zag. Bar arrangements can be programmed to place the bars in patterns ranging from 1/4″ to 39″ centers. This robot has the ability to perform radial reinforcing bar placement, and with the substitution of another placement arm it can insert vertical bars in wall forms.

Fig. 9-4. Kajima Corporation's MR 38 rebar placing robot.

The Soil Compaction Test Robot

Mitsui Construction Company engineers have produced a self-propelled robot that provides real time soil compaction tests at building sites or on civil engineering projects. The MRS 400-1 robot performs contact free, nondestructive, quantitative measuring of soil compaction by means of a backscattering-type nuclear density device coupled with a moisture gauge. As this unmanned traveling robot traverses a construction site, it needs no input from workers once it has been programmed. It is possible to measure soil compaction at predetermined points or continuously along column or boundary lines. The MRS is equipped with a voice alarm, and should anyone stray into its path, it will warn the trespasser to get out of the way, and if this warning goes unheeded, an infrared sensor inside the robot will make it stop if it touches someone or something.

There are two versions of this robot in use today (Fig. 9-5). The MRS 400-1 is an electrically powered, self-propelled model, and the MRS 400-2 is a trailer type robot meant to be towed by another piece of equipment. The travel performance of the MRS 400-1 is as follows:

Maximum load weight	330.69 pounds
Travel capability	Forward, backward, left and right steering, and spin turn
Travel speed	7 feet per minute
Climbing ability	$^1/_{10}$ slope
Battery power	DC 24V, 120 Ah/5h
Emergency stop function	Contact detector, emergency stop
Alarm function	Voice approach alarm

Fig. 9-5. Mitsui's soil compaction testing robots, (left) the MRS 400-1 and (right) the MRS 400-2.

The compaction tests performed as the robot travels along on articulated wheels are simultaneously recorded with its exact position on the site. These test results can be either reported as "real time" and displayed or transmitted to a host computer, or the test results can be stored in the robot's on-board computer to be retrieved at some latter date.

The robot's on-board systems consist of:

1. A main CPU with a 64Kb memory to control the communication, movement, and measuring subsystem.
2. Sub-CPU movement control system to control the robot's movements and to control various sensor signal input boards along with a 32 channel relay board for the drive control.
3. A radioactive isotope instrument operation and display unit.
4. A wireless modem to relay data from the main control system, such as measuring specifications, travel path specification, measuring results, and present position information back to the host computer.

The Robot's Position Measuring System. On board sensors detect the robot's direction and distance traveled and assist it in grasping the present position of the machine in relation to the boundaries of the area being tested. Information mats, plastic strips with embedded wires, are installed around the perimeter of the test area to aid the robot in determining its position on the site.

There are two directional sensors installed in the robot, a vibration gyro and a magnetic azimuth sensor. Vibration gyros are not affected by magnetic noises but they are susceptible to mechanical noises such as vibrations. The opposite is true of magnetic azimuth sensors which are unaffected by mechanical noises, so by using a combination of the two, the strong points of one sensor compensate for the weak points of the other and insure that the robot's directional control will be reliable. The MRS 400 contains a magnetic proximity sensor, and as it approaches the information mats forming the boundary lines of the project, it can react accordingly and change direction.

The Soil Compaction Degree Measuring System. Radioactive isotopes provide a quick and quantitative method for the measurement of soil compaction. The nuclear density soils testing procedure has been around for years, but instead of driving radiation rods into the soil, the robot uses backscattering-type nuclear density testing coupled with a moisture meter.

Since there is an air gap between the measuring instrument and the surface of the ground, a formula has been devised by Mitsui's engineers to account for the many variables inherent in this type of test procedure.

Software loaded into the MRS 400's computer takes into account humidity density and the gap between ground and instrument, and with the variation in humidity density, moisture density, and the air gap factored in, continuous monitoring and testing becomes possible.

Detailed specifications on the measuring system is listed below.

Measuring method	Gamma ray dispersion
Density	Heat neutron dispersion
Moisture	
Measuring range	1.00–2.50 t/m
Density	0.05–0.09 t/m
Moisture	
Measuring time	15 seconds or more
Ray Source	2.3 MBq (70 Ci) Co-60
Gamma rays	1.1 MBq (30 Ci) Cf-252
Heat neutrons	
External output	16 bit parallel I/F
Measuring indication method	External signal command, panel operation

Mitsui claims that the MRS400 robot can be used to measure the soil compaction degree of low slump concrete and it can also measure the thickness and density of asphalt paving. The MRS is also capable of uncovering the location of underground cavities, and in some cases, it can detect the presence of underground water.

According to the people at Mitsui the MRS 400 is still in the development stage. They are working on a unit with tracks to allow the machine to travel on inclined surfaces and soft and irregular soils. They are striving to improve the robot's position measuring accuracy and would like to see more high speed control and communication capabilities when they install a CPU with more memory and more rapid functions.

The Wall Tile Integrity Testing Robot

Tile facings as an exterior wall treatment are popular and attractive surfacing materials in Japan. Whether factory installed on precast panels or other substrates or built in place in the field, proper adhesion of tile to substrate is critical. This type of wall requires periodic inspections, and preventive maintenance is required to keep these surfaces in tip-top shape.

Kajima Corporation had been searching for ways to inspect walls for loose tiles on the exterior of mid-rise and high-rise buildings. Workers on scaffolding would visually inspect the surfaces of a building's walls. When suspect tiles were tapped with a test hammer, a trained ear could detect loosely adhered pieces. Performing these tests manually had several obvious drawbacks:

- The judgment as to tile soundness required limiting this kind of work to experienced and highly skilled workers.
- Efficiency declines rapidly after extended periods of investigation.
- The human testing method is time consuming and thus costly.

Since 1974, Kajima had been experimenting with ways to eliminate or greatly reduce the human element in this tile testing procedure, and after years of research they perfected a robot that can check the soundness of wall tiles by using an impulse response procedure.

Kajima's robot travels vertically up the face of the wall to be tested on guide cables suspended from a permanently mounted rail system on the roof of the building (Fig. 9-6). These guide rails, similar to those used by window washing crews, allow horizontal travel, while the cables afford vertical travel for the robot.

The testing robot, riding vertically up the building on guide cables, taps each tile with a small striking hammer. A microphone picks up the sound of the hammer striking the tile and transmits it back to a data processing unit on the site, where predetermined standard sound values, established by laboratory tests, are stored in the CPU. As the repulsive power signal and the sound level signals are received at the central processing unit, they are compared with the stored standards and evaluated to determine if the tile is properly adhered or if it is in the process of delaminating.

Fig. 9-6. Kajima's Corporation's tile separation detection robot.

The data transmitted by the robot and evaluated by the CPU is transferred to a floppy disk which allows the information to be displayed graphically. The travel path of the robot is superimposed upon a drawing of the building's elevation and with a vertical and horizontal grid overlay and the ability to enlarge sections of the grid system, exact locations of peeling tiles can be plotted. Hard copies of the CRT display will be created by a printer and these printouts will be distributed to the field maintenance crews. Schematics of the system are displayed in Figs. 9-7 and 9-8. Kajima claims that based upon previously recorded and analyzed data, the robot can make the subtle distinction between tiles separating from the mortar and mortar separating from the substrate.

Not to be outdone by a competitor, Ohbayashi Corporation created *Kabedohda*, a robot that creeps up a wall on suction-cupped caterpillar treads searching for cracks in mortar, bricks, tile, and precast panels. By emitting sonic waves, *Kabedohda* can pick up imperfections in the building's skin and with its video camera it can relay the nature, extent, and location of the problem back to an operator on the ground.

Spraying Robots

Fireproofing and Paint Sprayers. Shimizu has developed a robot to apply spray fireproofing onto structural steel members in high rise buildings, always a dirty job at best. This machine, called the SSR-2, is a self-propelled, preprogrammed spraying machine that goes about its task unaffected by the airborne dust so annoying to human applicators. With its special software programming package developed by Shimizu engineers, the SSR-2 can adjust its spray patterns to comply with the structural systems that vary somewhat from building to building.

Shimizu has also created a paint spraying robot to paint the exterior walls of high rise buildings while remaining inside the building. This robot has a long articulated arm that it meant to extend out of the building before the windows are installed. A spray nozzle attached to this multi-jointed arm applies paint as the robot proceeds on its programmed travel path around the building. According to Shimizu engineers, this spray painting robot can increase productivity by as much as 200% and the ability to paint the exterior of a building without subjecting painters to the dangers of scaffolding and staging is certainly another added benefit.

Spray Lining Robots. Tunnel construction in the mountainous region of Japan makes use of the intrinsic strength of the bedrock left in place, and the use of rock bolts and concrete spray lining to keep the walls in stable condition.

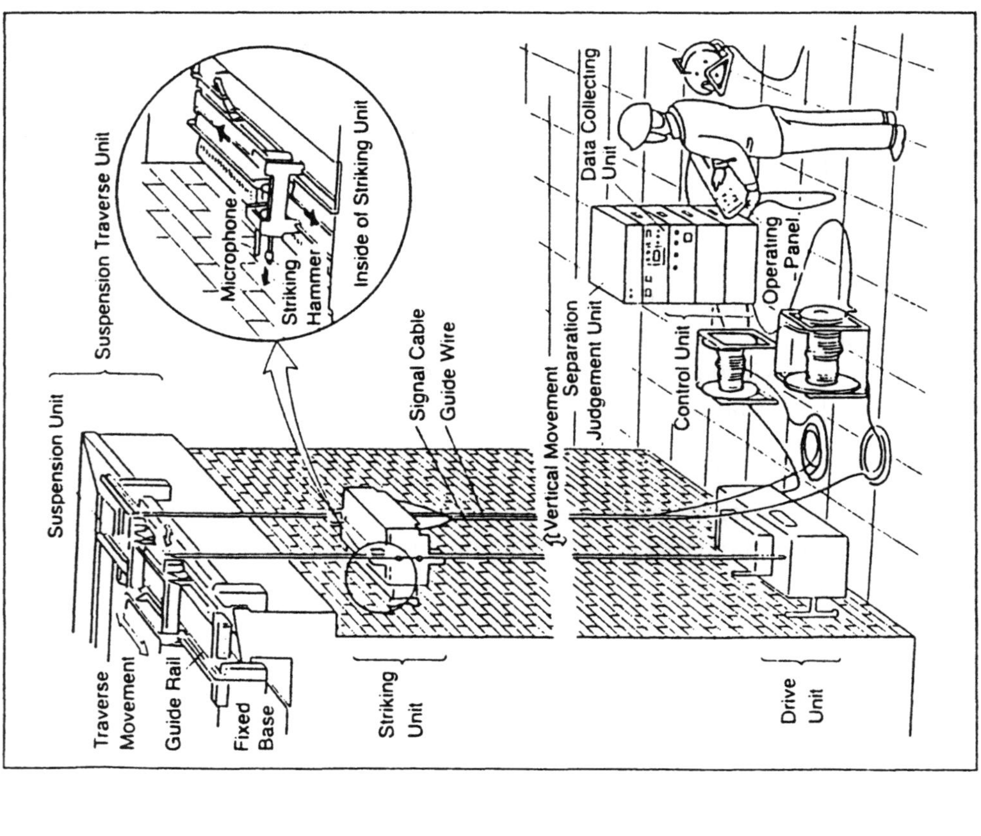

Schematic Drawing of Device (in field)

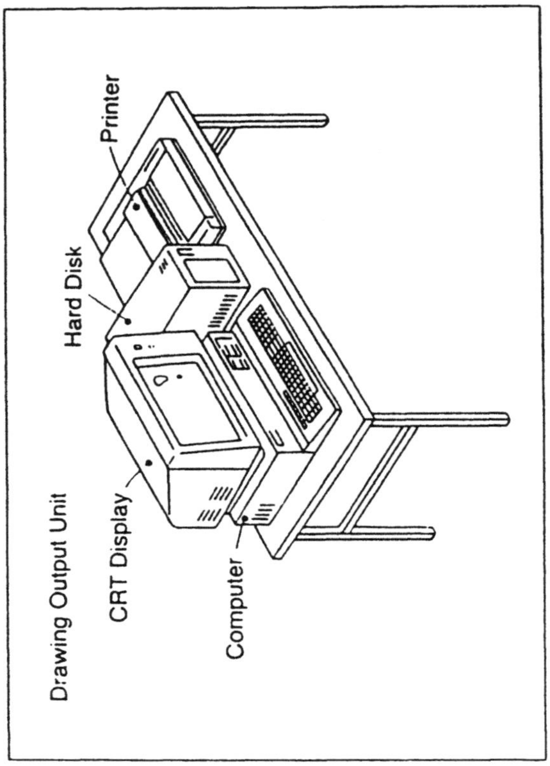

Schematic Drawing of Device (in office)

Fig. 9-7. Schematics of Kajima's tile separation detection system. (Left) Field array. (Right) CPU and printer in office receive and display field input data.

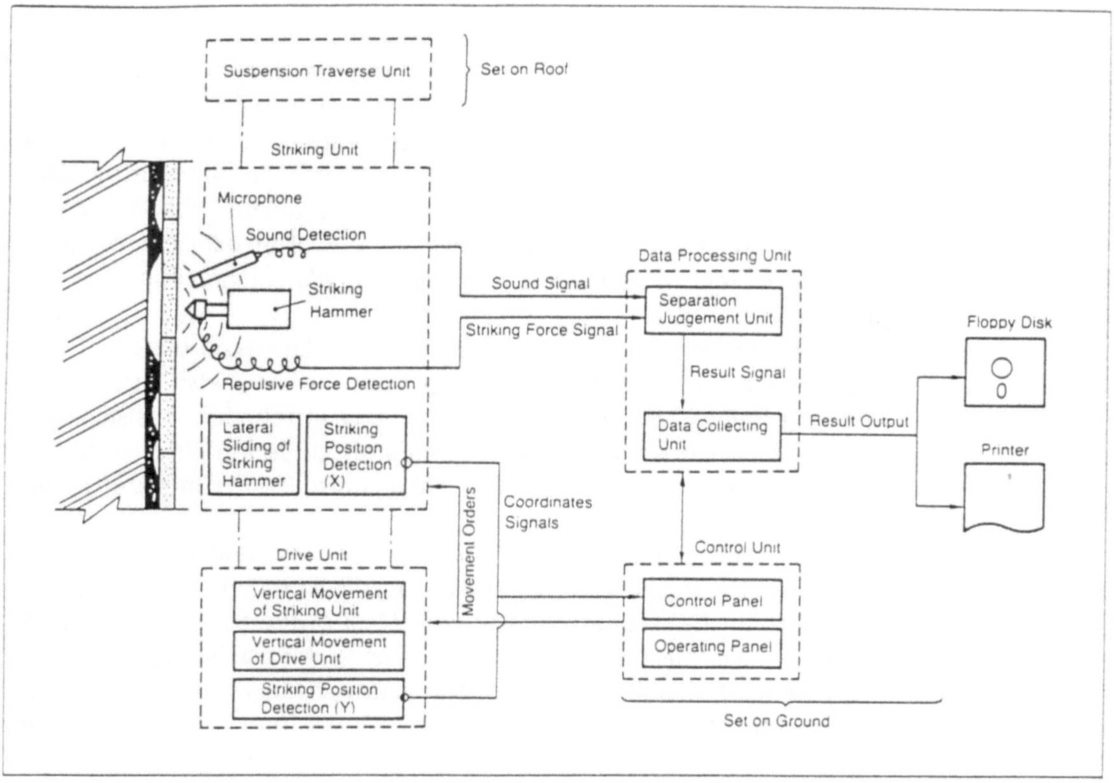

(a) **Detection System**

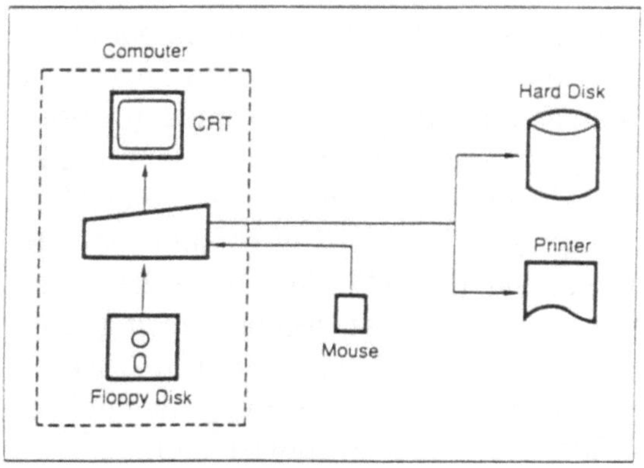

Result Output System

(b)

Fig. 9-8. Schematics of Kajima's tile separation detecting robot.

When concrete spray linings are applied by workers, it is a dirty, tiresome, and dangerous job. Usually airborne mineral dust fills the tunnel and the ever present danger of falling rocks, rebounding concrete and concrete separation adds to the potential for accidents. Just holding the heavy hose and working it back and forth becomes tiring, and efficiency drops rapidly. Recognizing all these problems, the Ministry of Construction in 1984 made the development of an efficient concrete spraying system a high priority for the nation's construction research programs.

The precursor of the spray lining robot can be traced back to 1965, when the Katsuki Tunnel was being built for the Uetsu line of the Japanese Railway System. This early attempt at robotics consisted of a self-propelled carrier suspended from a steel guide rail attached to the roof of the tunnel. Pneumatic drive parts controlled the movable boom containing a spray nozzle.

Spray lining (SL) robots evolving from this early model have now all been designed to ride on rails, tracks or wheels rather than being suspended from the ceiling of the tunnel. Most of the SL robots are built around a base machine on tracks or wheels. An oscillating nozzle remaining at right angles to the spraying surface at all times is essential, and in order to prevent the sprayed on material from coming loose and falling off, multiple passes of the nozzle are required to spray on thin coats of material and build up to the final thickness slowly.

Mitsui developed the SL-2 Robot (Fig. 9-9) to meet all of these requirements: it had a crawler base; the spray mechanism would maintain stable distance between nozzle and spraying surface; the spray mechanism would apply several thin coats; and the operation of the robot would be remote via a small and simple control panel.

The boom of the SL-2 travels freely on an arched rail and the nozzle on this boom can move horizontally, vertically, and elliptically, in extended or retracted positions. Building on its initial development of the SL-2 robot, Mitsui has produced the SL-3 which is used in large cross section tunnels. The SL-3 uses an arch rail that can be opened or closed, and the latest robot, the SL-4 (Fig. 9-10) uses a fixed rail system. Lastly, for very small cross section tunnel work, Mitsui developed the SL-Mini.

Kajima has built a shot-creting robot developed in conjunction with the New Austrian Tunneling Method (NATM). This robot has been designed to operate in a tunnel with a cross section ranging from 11½ feet to 34 feet in height, and with radii ranging from 6½ feet to 18 feet. The nozzle on this Kajima robot is a five spindle manipulator with an oscillating range of 150°. This robot is crawler mounted and can be controlled in a fully automatic mode, a circular mode, a receding mode, and a combined circular/receding mode.

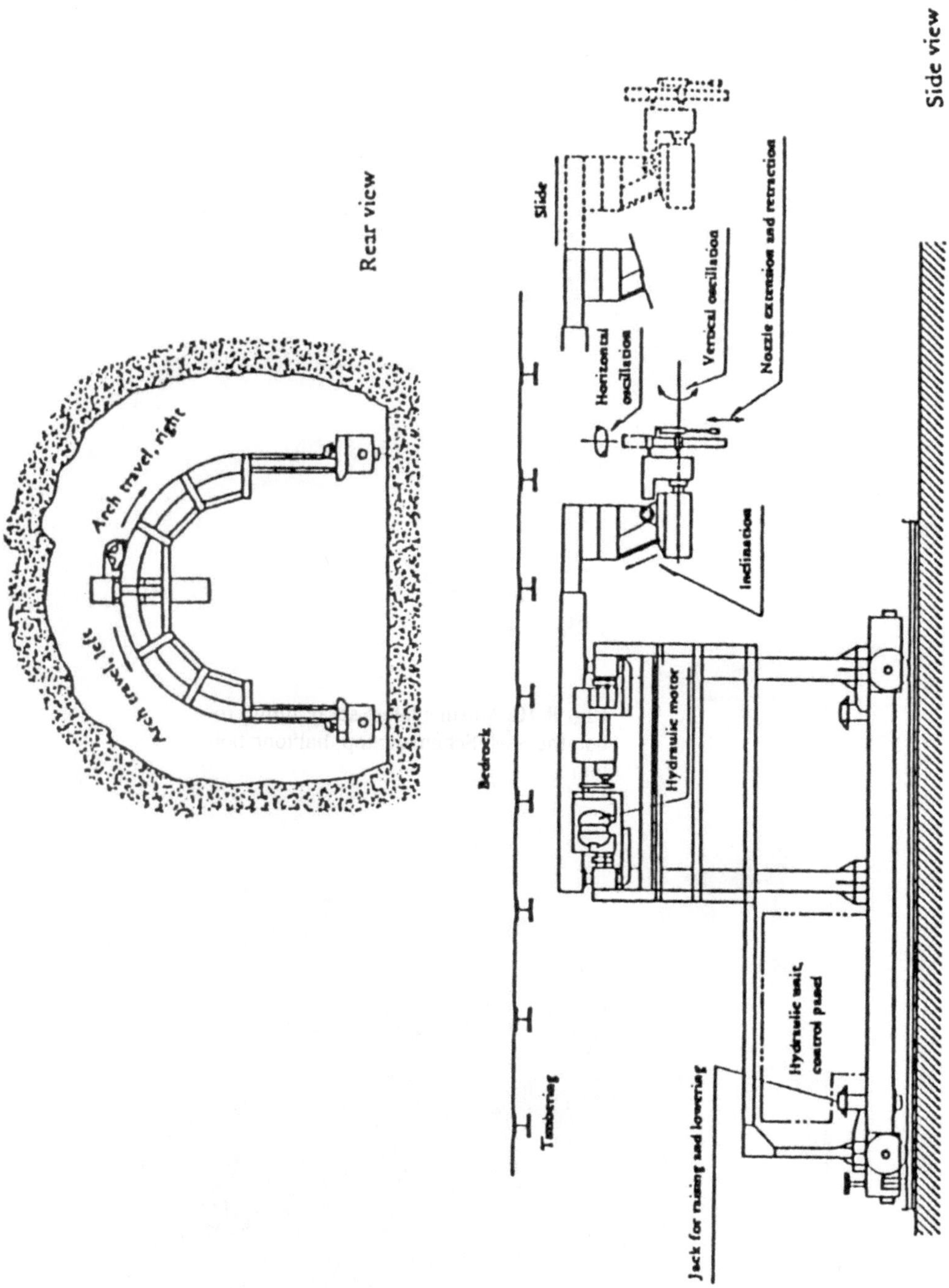

Fig. 9-9. Mitsui's SL-2 spray lining robot.

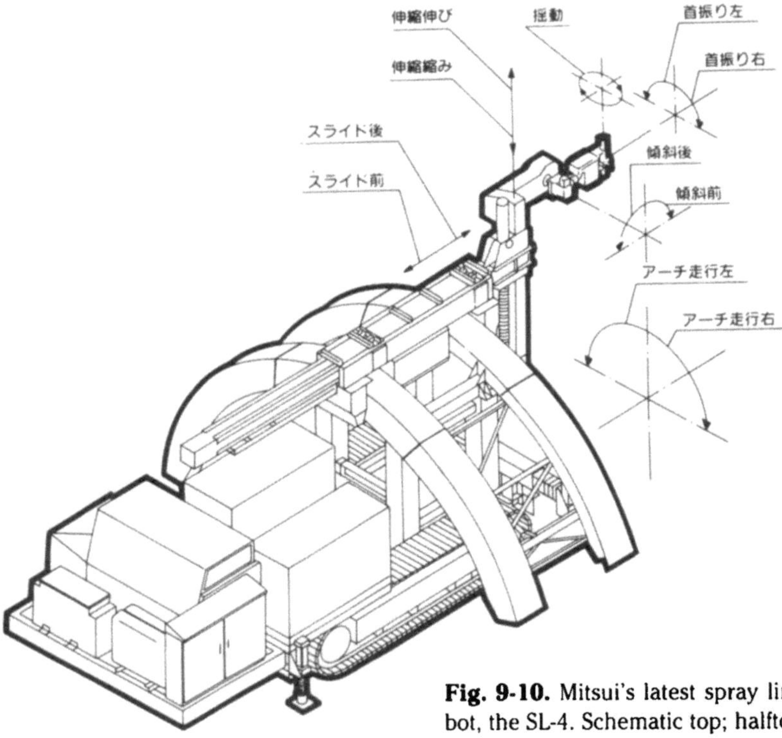

Fig. 9-10. Mitsui's latest spray lining robot, the SL-4. Schematic top; halftone bottom.

大断面用　SL-4型
建設省東北地方建設局
●月山ダム仮排水路トンネル工事

Underseas Robots

Komatsu, a leading Japanese manufacturer of construction equipment, has developed a robot to work on the ocean floor, a task that is not only dangerous when performed by divers but is also very costly.

In order to prepare the ocean floor for breakwater foundations, the surface of the ocean floor must be raked and then leveled. Komatsu developed an electrically powered robot specifically for this purpose. This robot is a 72 ton monster that rests on eight legs and is 57 feet in length and 33 feet in width. Rising 26 feet above the ocean floor, it can travel in four directions at speeds up to 83 feet per hour. This robot first rakes the surface of the ocean bottom and then compacts the seabed to a final tolerance of 1.7 inches utilizing a laser positioning system tied into a ranging unit located on land. Penta-Ocean Construction Company, Ltd., a major Japanese marine construction company claims that the Komatsu robot cleared an area of 2000 square feet in one day instead of the 111 square feet area normally cleared by divers. From the exotic to the mundane, Japanese engineers explore many areas where robotics can replace tedious and repetitive human tasks.

A Housekeeping Robot

Toshiba Corporation and Mitsui Real Estate Development Company Ltd. teamed up to develop a robot to sweep and clean floors in office buildings. Appropriately named *Auto Sweepy,* this robot, encased in a sleek white housing, can be seen in office buildings at night silently going about its job while human coworkers empty trash cans and dust desk tops. *Auto Sweepy* is 4 1/2′ long, 3′3″ high, and 2′3″ wide, and weights a hefty 1058 pounds. It contains a water dispensing system, a vacuuming device, and a revolving cleaning and buffing pad, allowing the robot to sweep an area and then proceed to wash the floor, using the vacuum to remove excess water before drying and buffing the tile or resilient floor surface. This automatic maintenance robot contains a semiconductor controlled laser which enables it to "see" obstructions and obstacles and work around them. *Auto Sweepy* does its job quietly and effectively and, according to its developers, can clean 5000 square feet of floor space in a shift.

Automated Hoisting Robots

Ohbayashi Corporation has developed several methods of hoisting objects that need no assistance from workers to either attach the load at the pickup point or unhook the load at the dropoff point.

Autoclamp is one of their systems that can be attached to a tower crane permitting the crane operator (human or electronic) to pick up steel columns or beams with a huge, clamplike device. The clamp can release its grip on the load at the dropoff point by remote control.

Ohbayashi's *Autoclaw* uses two remote controlled clamps to hoist beams and release them when required, unattended by human hands.

Shimizu has also been at work on steel structure pick-up and release robots. Their *Mighty Jack* device is a remote assembly manipulator for structural steel beams, and their *Mighty Shackle* is a radio controlled automatic release clamp for use on steel framing members.

Just like a jigsaw puzzle, the seemingly individual tasks of various robots appear to be fitting in place, and a picture emerges as the puzzle comes together. The possibility of a computer integrated building system beyond the horizon seems to be looming closer and closer, as the sun rises.

EARTHQUAKE PRONE JAPAN

The Japanese archipelago sits atop the Asian Continental Plate, which rides on the Pacific Plate in an area referred to as the Pacific Ring of Fire.

It has been estimated that Japan experiences at least 1,000 perceptible tremors each year, the worst of which occurred on September 23, 1923, when the Great Kanto Earthquake destroyed a third of Tokyo and most of Yokohama, leaving 140,000 dead in its wake. Another devasting quake hit Japan in June 1948 when the Fukui earthquake, registering 7.3 on the Richter scale, passed through an area that extended from northern Fukui prefecture to the southwest prefecture of Ishiawa. A total of 35,420 homes were destroyed when the tremors had subsided, and 3,895 people were killed. On June 12, 1978, the Miyagai-kenoiki quake shook the eastern part of Miyagai prefecture with an intensity of 7.4 on the Richter scale. The toll from this disaster was 27 dead, 10,962 injured, 1,377 homes totally destroyed, and 6,171 houses partially standing.

Seismologists are predicting that an earthquake with a magnitude of 8 on the Richter scale will occur in the Tokai region near Shizuoka prefecture, southwest of Tokyo, sometime in the near future. These areas are being "wired for sound" in order to provide the earliest possible detection of what could be a devastating earthquake.

The Measurement of Earthquake Intensity

The Richter scale was developed by Charles F. Richter working at the California Institute of Technology in 1935. It is a mathematical method of comparing the magnitude of an earthquake by looking at the logarithm of the amplitude of the waves it created and recorded on a seismograph. Numerical values were assigned with no upper limit to the range. However, earthquakes with a magnitude of 8.0 or higher are considered major quakes, and the highest recorded earthquake to date registered 8.9 on the Richter scale.

Another method of earthquake measurement was provided by Harry Wood and Frank Neumann in 1931 when they developed the modified Mercali intensity scale. This form of measurement uses 12 degrees of intensity, beginning with Level 1 (Not Felt by Anyone Except a Few People) and escalating to Level 12 (Total Damage, Existing Grades Grossly Distorted).

Acceleration, the rate of change of the velocity of a moving object, is a term associated with the force of an earthquake, and is measured in Japan in terms of the number of gallons of water set in motion by the force. Table 9-1 relates seismic intensity to the acceleration force.

Seismic Design

Earthquakes create horizontal and vertical ground movement, and when this does occur, building foundations tend to move as the ground moves, while the structure above wants to remain in its original position.

Seismic design can focus upon one of three schemes:

Table 9-1. Relations Between SEISMIC Intensity Scale and Acceleration.

Seismic Intensity Scale	Perception	Acceleration	Description
0	Unfelt earthquake	0.8 gal. or less	Unfelt by humans, but registered by seismic recording equipment
1	Slight earthquake	0.8–2.5 gal.	Felt by those in quiet areas
2	Weak earthquake	2.5–8.0 gal.	Felt by most people, sliding doors might chatter
3	Minor earthquake	8.0–25 gal.	Houses shake, doors move, hanging lights sway, water in containers moves.
4	Medium earthquake	25–80 gal.	Houses shake violently, things fall off tables, water spills from full containers
5	Strong earthquake	80–250 gals.	Walls crack, stone cornices might become dislodged and fall.
6	Violent earthquake	250–400 gals.	Houses collapse, mud slides and landslides occur, fissures created in ground
7	Disastrous earthquake	400 gal. or more	More than 30% of houses are destroyed, landslides, ground faults and fissures occur.

1 Separate the foundation from the superstructure so that deformation of the structure above is minimized.
2 Reduce deflection of the supporting members, which in turn will reduce deflection of the structure.
3 Design the supporting frame to withstand deformation.

Seismic design in Japan is concerned with providing quake-proof office buildings, factories, and homes, and in major cities such as Tokyo where super-high rise structures are becoming predominant, sophisticated seismic design is essential. Not only must the tall buildings be designed structurally to withstand strong earthquakes, but the method of attachment of exterior wall panels requires careful consideration so that they will remain attached to the building as it resists the acceleration forces of the tremors.

Seismic Design and the Nuclear Age

Japan is highly dependent upon nuclear power as an energy source. In 1985, it drew 26.9% of its energy requirements from nuclear powered generating stations. As a means of comparison, the U.S. during the same period obtained only 15% of its energy needs from nuclear sources.

These Japanese nuclear power plants, built in seismic zones, must be designed with fail-safe components in order to prevent disasters of catacylsmic proportions. The need for a centralized control point for nuclear power plant technology was the force behind the creation of the Nuclear Power Engineering Test Center (NUPEC). NUPEC was founded in 1976 by the joint efforts of the Japanese electric utility companies, equipment manufacturers, and constructors, in order to improve nuclear power plant technology.

NUPEC had two missions when it was founded: to perform tests to prove and/or confirm the safety and reliability of nuclear power plant construction; to conduct safety analysis of normal operations in nuclear plants, to include complete evaluations of incidents, failures, systems and component reliability, and human factor studies. A seismic proving ground had to be created, and NUPEC built the Tadotsu Engineering Laboratory in Kagawa prefecture to provide simulation tests for nuclear power plant components and structures.

The Tadotsu lab contains a very large vibration table installed on a concrete foundation weighing 150,000 tons. This table measures approximately 49 feet square, and is capable of creating acceleration forces up to 2.72G in a horizontal plane and 1.36G in a vertical plane.

This sophisticated facility performs seismic tests on large scale models of reactor containment vessels, primary coolant system equipment, reactor pressure vessels, and reactor core components.

Seismic Data Reporting Stations

As an adjunct to design, it is also necessary to collect information relating to the magnitude, epicentral distance, and seismic acceleration of earthquakes. Earthquake motion is greatly affected by the seismic source mechanism and the character of the propagation path from the hypocenter to the local topography, and because it is necessary to obtain earthquake data from points all along the axial lines of the disturbance, there is a need for multiple quake reporting stations in Japan. The Public Works Research Institute of the Ministry of Construction is located in Tsukuba Science City, and a series of dense instrument array observation and reporting sites have been established by this organization throughout the country. The Suruga Bay–Izu Peninsula area in the Tokai region about 150 kilometers from Tokyo has been targeted as the sixth most likely place in the world where a major earthquake could occur. In this area, the government has placed 35 observation points with 34 ground surface accelerographs and 13 underground accelerographs to pick up any advance warnings of a major occurrence.

Japan has on its books, a program called Pacific Orient Seismic Digital Observation Network (POSEIDON).

The POSEIDON program, when enacted, will conduct broadband seismological observations throughout the western Pacific region and will exchange information with seismic labs and institutes around the world in order to further the spread of seismological knowledge. Design knowledge goes hand in hand with early warning capabilities, and one of POSEIDON's goals is to further the investigation of the earth's mantle and unravel the mystery of plate tectonics.

The POSEIDON program is expected to be operational for twenty years, with the first ten year phase divided into two five year segments. Seismic stations are planned throughout the Pacific Rim to include Japan, the Izu-Marians, the Ryukyus, the Phillipines, Indonesia, the U.S.S.R, and the Showa Base in Antartica. During the initial phase of this program a data center will be built and work on a deep ocean seismic network will begin.

Base Isolation Seismic Design

Menshin is the Japanese term for base isolation, a system of isolating the base or foundation of a building from the superstructure above. In *menshin* (pronounced *men-sheen*), rubber and steel plate cushioning devices are placed between the building's foundation and structure to permit horizonatal movement to occur when an earthquake is producing acceleration forces.

This technology of separating foundation from structure is not new.

Mr. Hideo Hanakawa of the Okumura Corporation indicates that this concept of base isolation can be traced back to Japan's Meiji era of the mid-1800s. The technology was first put to use in modern times when an atomic power station was built in the southern part of France in 1983 using base isolators constructed of alternate layers of rubber and iron cushions.

At the present time there are a dozen or so buildings in the Tokyo area that have been built using the *menshin* concept of base isolation. The theory behind *menshin* is that it will protect a building from the effects of an earthquake by staggering the resonant frequency of the building from the predominant frequency of the quake and reduce the energy transmitted to the building. This special flexible connector at the base allows a building to move in a horizontal plane, and a building with these base isolation devices moves more slowly than a building with a rigid connection.

The base isolators are made of alternating layers of rubber sheets and metal disks, allowing the device to act as a vertical hard spring capable of sustaining the heavy loads imposed by the structure above (see Fig. 9-11). These isolators act as a flexible spring in the horizontal mode to allow deformity when the acceleration forces are imposed during the earthquake. The top and bottom of the isolator assembly is fitted to flangelike caps vulcanized to the core. Bolt holes in the top and bottom flanges are used to secure the isolator to the foundation and to the superstructure. Rubber is wrapped around the circumference of the isolator to protect the core from ultraviolet degradation, moisture, weather, and rodent and insect attack. These isolators are used in conjunction with dampers whose purpose is to slow down the vibration created during this period.

When *menshin* is used, special attention must be paid to all utility lines that exit the building from the superstructure. Flexible connectors will be needed for gas, water, storm and sanitary sewers, and primary and secondary electric feeders as they run from the building through the base and onto the site.

Simple precautions such as isolating the stair well leading to the basement so that neither stairs nor handrails are attached to the basement slab must be considered. Since hydraulic elevators depend upon pistons below grade for their lifting capability, traction elevators must be installed.

A Closer Look at a *Menshin* Equipped Building

Ohbayashi built a High Tech R&D Center in 1986 on the site of their Kiyose Technical Research Institute and this building contains the *menshin* base isolation system (see Fig. 9-11). A tour through the building

(a)

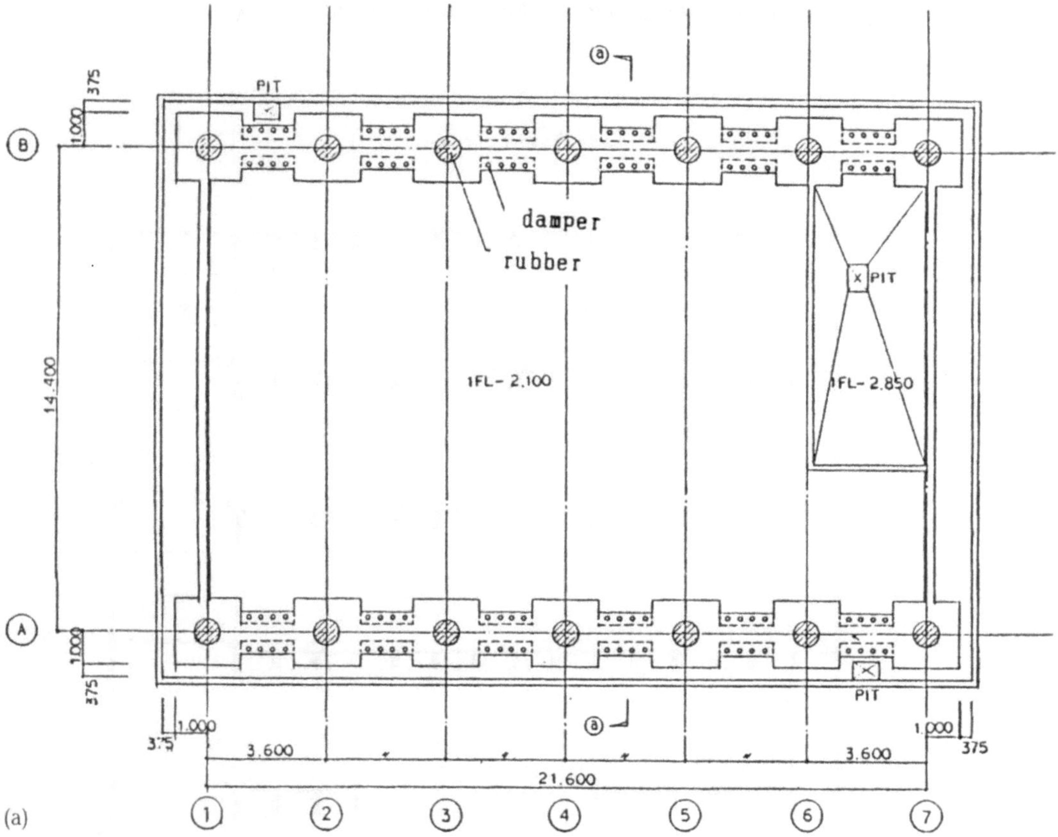

(b)

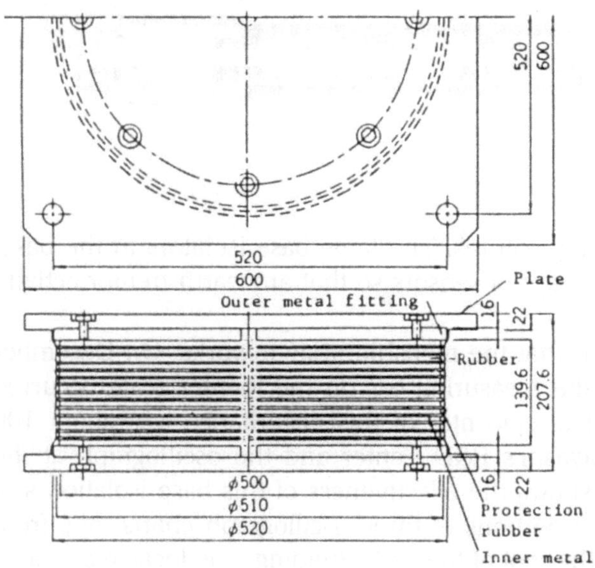

Fig. 9-11. Ohbayashi's Hi-Tech R&D Building, incorporating menshim. (a) Floor plan showing base isolators. (b) Structure and dimensions of base isolator. (c) Section through R&D building.

(c)

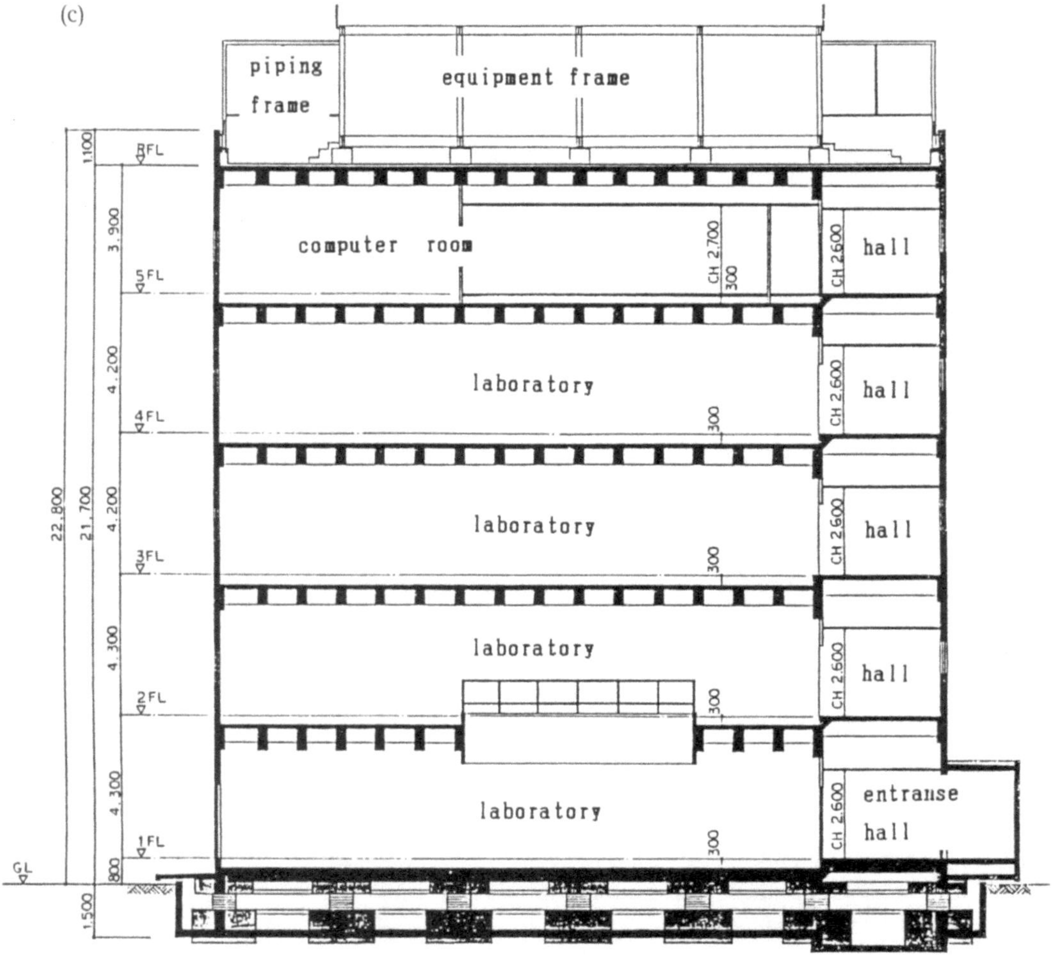

Fig. 9-11. ***(continued)***

reveals many interesting features. The large base isolators in the basement are rigged with arrays of sensors so that any earth tremor activity will be recorded.

There is no question that the menshin system works. On December 17, 1987, an earthquake measuring 6.7 on the Richter scale occurred in Chiba prefecture. The epicenter of this disturbance was about 100 kilometers from Ohbayashi's R&D Center and the oscillograph at the site in Kiyose clearly shows the effectiveness of this base isolation system. Fig. 9-12 contains portions of three oscillograph charts, one from a recorder located on the roof of the R&D Building, one located at grade, and the third located on the roof of the Institute's main building, a non-*menshin*-protected structure. Maximum acceleration at the R&D building was 10.9 gallons, while the non-base-isolated building in the

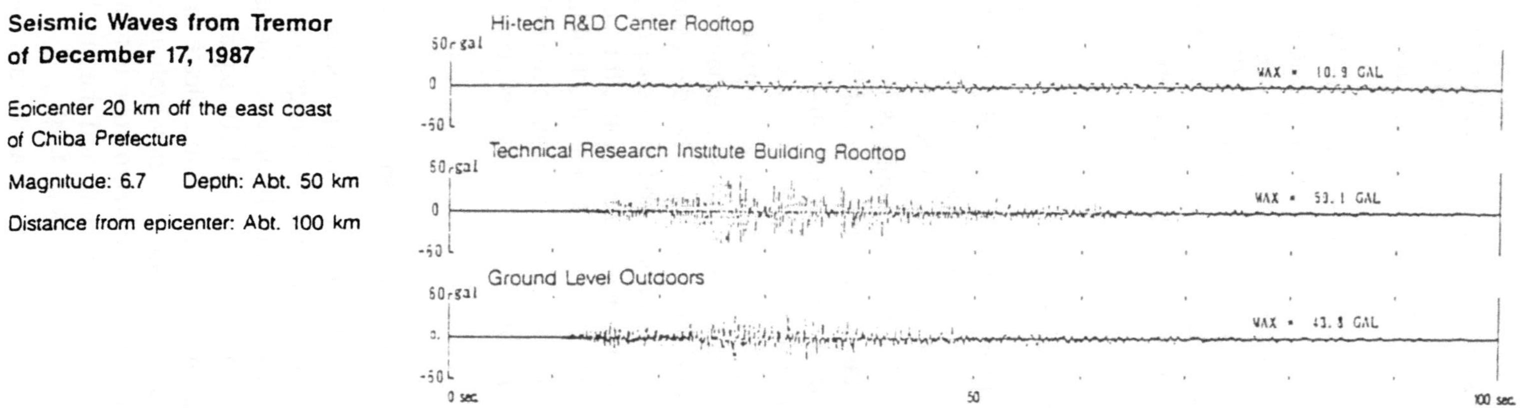

Fig. 9-12. Oscillograph readings at Ohbayashi's R&D facility during an earthquake.

complex registered 59.1 gallons at the roof level. Since the basement of the R&D building is physically separated from the structure above, there is a gap around the perimeter of the building which is covered with a flexible skirt to keep out weather and any small animals that might try to come into the building.

The traction elevator pit is suspended from the first floor and stops short of touching the basement slab by about two feet. Storm drain and other plumbing lines are suspended from the underside of the first floor slab on hangers and have hydraulic-type fittings. Where these lines exit the building, flanged stainless steel flexible connections are fitted to each pipe. Electrical cables, without rigid conduit encasement, are placed in trays hung from the slab above.

The Shibuya-Shimizu Building

The Shibuya-Shimizu office building is a "smart" building constructed with a reinforced concrete structural system and equipped with twenty units of laminated rubber bearings and 108 steel dampers in its base isolation system. This office building contains five floors above grade and one floor below grade, and the total area of the structure is 36,500 square feet. For a schematic of the building and its base isolation system, see Fig. 9.13. The use of a base isolation system permitted this building, with an average floor area of 6,000 square feet, to be shoehorned into a 9,000 square foot site and still remain economically viable, since its beams and columns could be designed much smaller because of the *menshin* design and therefore create maximum rentable space. It is expected that during an earthquake, the acceleration transmitted to this building will be one-sixth that of a building without base isolation. There is also an added benefit during non-earthquake periods, since the base isolation system drastically reduces daily vibrations caused by nearby traffic.

A *Menshin* Protected Research Laboratory

The federal government's National Institute for Research in Inorganic Materials has installed base isolation in their High Technology Research Laboratory. Although this laboratory is only a one story building of reinforced concrete with a floor area of 11,706 square feet, it contains a highly sensitive electron microscope. The delicate instrument in this laboratory must be protected against microvibration as well as earthquake ground motions. The electron microscope is mounted on an isolation table acting in a vertical direction, and the building housing the instrument is protected by *menshin* (Fig. 9-14).

The effect of a base isolation system can be dramatically displayed

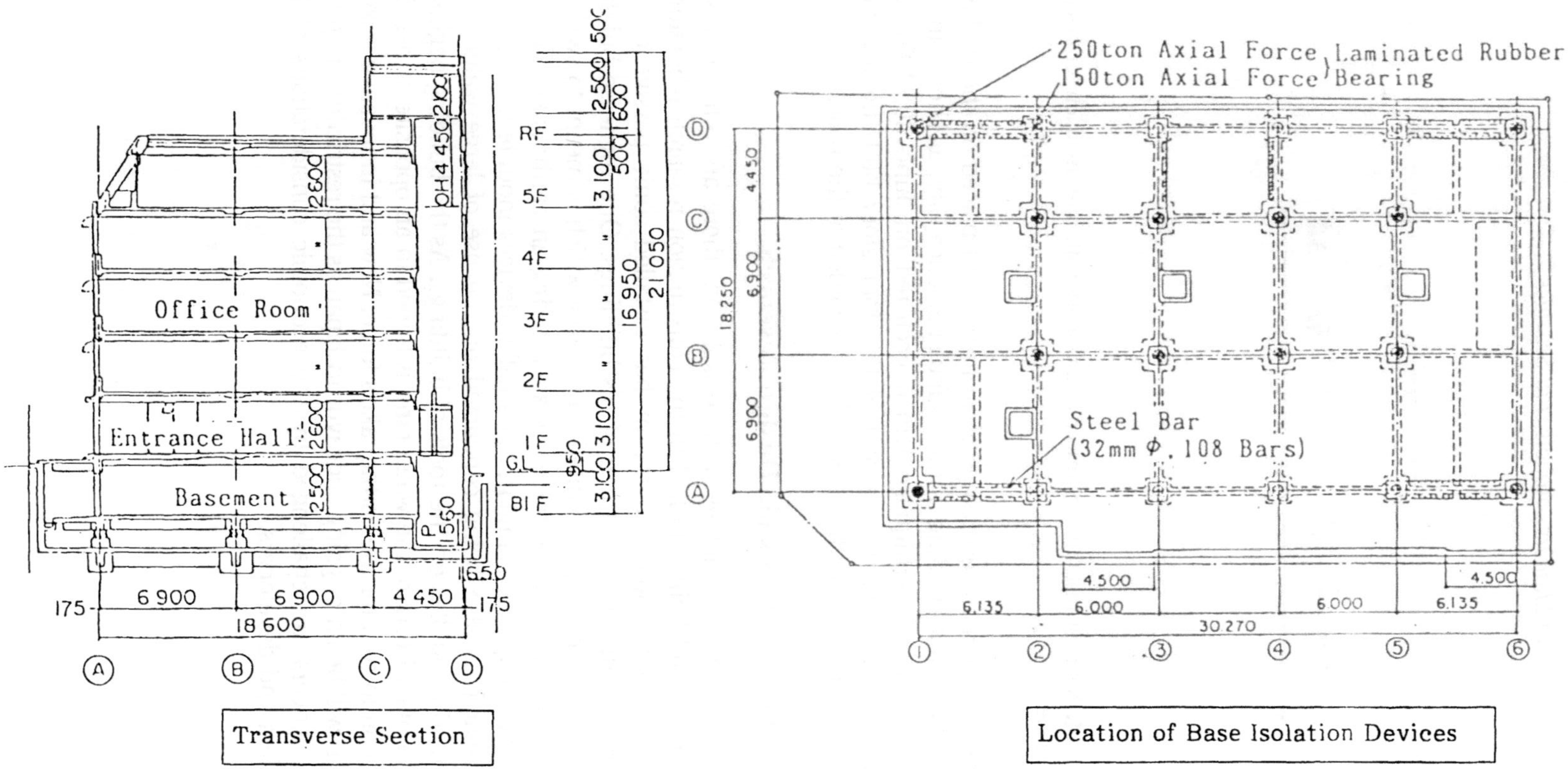

Fig. 9-13 The Shibuya-Shimizu office building with base isolation system.

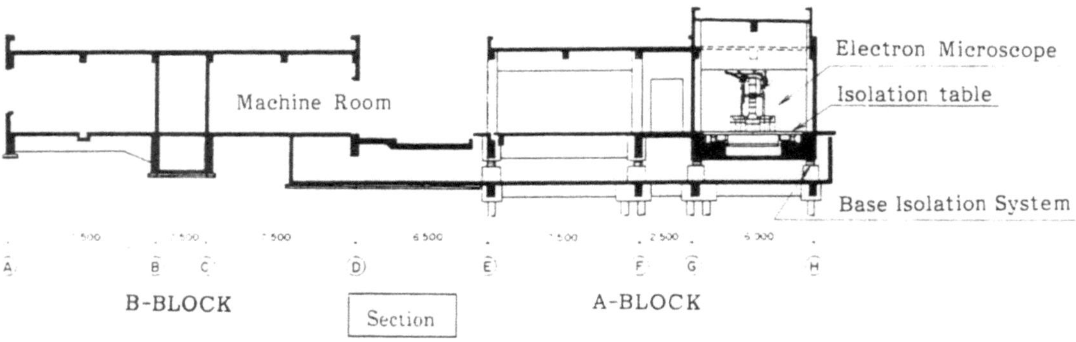

Fig. 9-14. National Institute for Research laboratory with base isolation for electron microscope.

by the oscillograph recordings of an instrument placed on the roof and on the base of a building in the Tsukuba Science City, as shown in Fig. 9-15. During an earthquake which occurred on June 30, 1987, the waveform observed on the roof of this four story building was very minimal compared with the waveform recorded at the base.

Other Forms of Seismic Design Research

Along with the technology of base isolation, there are other forms of seismic control being studied in the construction research laboratories of Japan. One leading contractor has installed some passive control systems on the roofs of three of his buildings. One high-rise building has a giant pendulum mounted on the roof which is meant to swing in reaction to the swaying forces created during an earthquake. The two other buildings have similar, but smaller devices mounted.

Research is also being conducted on the use of large water tanks mounted in cradles on the roofs of buildings. As the building begins to flex in one direction, the water tank is tipped in an opposite or reaction direction to counter the earthquake's force. New and innovative seismic design will be an ongoing endeavor in Japan as the restrictions imposed by engineers are challenged by the economic considerations of the builders and developers.

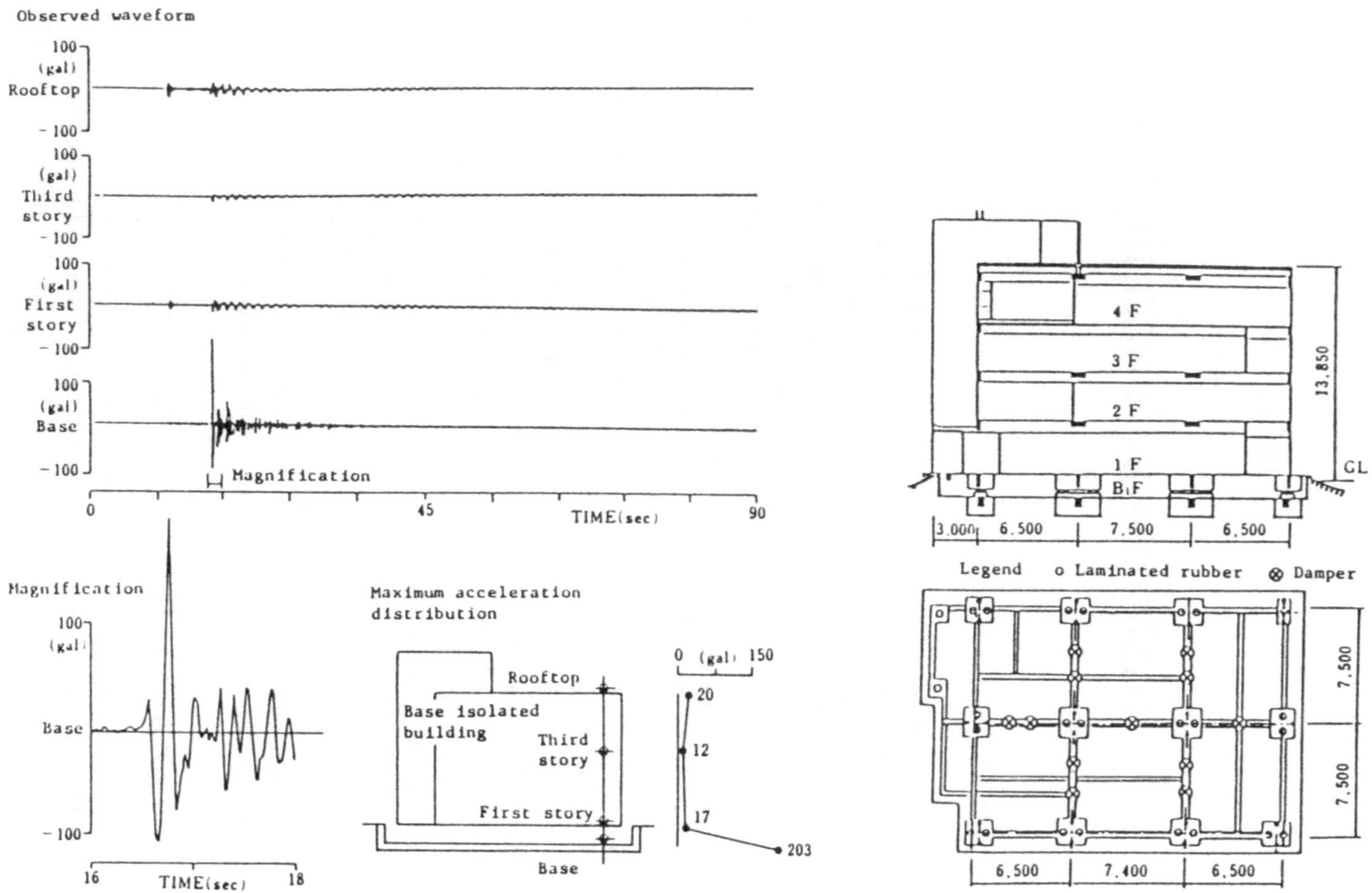

Fig. 9-15. Effect of an earthquake on a base isolation building observed by the Tsakuba Laboratory.

10

Construction Techniques

Basic construction techniques in Japan appear to be quite similar to those in everyday use at American jobsites. Structural steel framework systems contain much larger rolled sections, and cast-in-place concrete structures look like they have been beefed up, probably because of the ever present danger of severe earthquakes in Japan.

However, contractors continue to dig foundations with the same types of excavating equipment familiar to U.S. contractors, Japanese ironworkers sport the same kind of rat-tailed wrench needed to align bolted steel connections, carpenters screw sheets of drywall to steel stud frames with bugle head screws, and roofers mop hot asphalt onto layers of felt papers on the rooftops of many Japanese buildings.

There are many refinements and innovations to the construction process in Japan along with the developement of several new products and materials, but above all it is the attention to detail, the quality of the finished product, and the cleanliness of the job site that makes the Japanese contractor and his construction techniques something special.

The photographs accompanying this chapter will provide the reader with a few glimpses of the work being performed by Japanese contractors.

In some cases this same innovative builders seem to have found simple solutions to nagging construction problems.

NEW IDEAS FOR OLD URBAN PROBLEMS

It seems as though several problems are universal in nature: potholes exist on the streets of many cities throughout the world, neighborhood opposition to certain types of projects occurs in all countries, and the protection of life and property around construction sites is surely a universal concern. Japanese contractors have actively attempted to solve these problems and several others and in the main have been quite successful in doing so.

Reducing the potential for Potholes

In eastern U.S. cities like New York, city road beds are constructed of layers of compacted structural fill, overlayed with a 6–8 inch concrete slab, then capped with 3–4 inches of asphalt paving. Untopped concrete roadways in that part of the country are not common, since they would be subject to spalling and deterioration after prolonged exposure to the corrosive effects of snow melting mixtures containing salt.

When these types of concrete base–asphalt topped streets are excavated for underground utility work and repairs, the resulting cold joints in the concrete patch require maintainance, and if proper compaction of the subbase has not been achieved, concrete, weak in shear and tension, will fail and crack, creating potholes and other depressions.

In Tokyo, many of these problems seem to have been solved in a rather simple manner. Many utilities are located under asphalt sidewalks, so that street openings are not necessary when water, gas, and telephone lines have to be installed or repaired. The Japanese construct their roadbeds with full depth bituminous concrete, using 12 inches of compacted asphalt placed over a compacted structural fill base. Roads constructed in this way tend to withstand shock, heavy loadings, and the effect of severe weather much better, and when these roads require resurfacing, about 3–4 inches is milled off of the top surface, defective or soft spots are localized and repaired. When the new layers of asphalt are applied, the final surface will be as smooth and level as when the roadway was originally built.

Avoiding the NIMBY Problem

Citizen opposition to construction projects which they deem prejudicial to their neighborhood can cause disruptions and drastically increase costs as the project start is delayed, and possibly lengthy court battles have ensued. The acronym NIMBY—Not In My Back Yard—has been applied to drug rehab centers, low income housing projects, resource recovery plants, and other public projects. Lowered property values and increased traffic congestion are real concerns to residents, and Mr. Yamamoto from Osaka is not much different from Mr. Johnson from Oak Park. But in Japan, government officials seem to be more sensitive and responsive to community concerns.

An incinerator was being planned for construction in Tokyo, and the Tokyo Municipal Government (TMG) officials were expecting considerable opposition to the project from the local citizentry. The TMG officials decided that the vibrating equipment for the proposed new Suganami Incinerator should be placed underground to lessen the visual impact on the surrounding area and it would also reduce noise levels emanating from the plant. Tunnels were planned for both entry and exitways so that truck traffic in the immediate area would be reduced. TMG decided that it would be generous in its landscaping budget and they planned to plant 5,000 trees and shrubs completely around the site. The city officials proposed to build a neighborhood community building which would include a gymnasium, two heated swimming pools, an exercise room, a tea room, a senior citizens center, and a library.

When this entire plan was unveiled by the TMG the local citizens reacted in a very positive manner, and all opposition to the incinerator project disappeared. To ease the neighbors' minds about the impact on air quality in the area, the city installed a scoreboard in a prominent location so that pollutant levels of stack emissions could be displayed and monitored.

Protecting Life and Property at the Construction Site

Many busy metropolitan areas have seen an increase in construction related injuries within the past several years. Collapsing sidewalk cranes and booms, combined with dislodged and falling construction debris and the injuries and deaths that ensue make headlines all over the country. Building materials and equipment stored and parked close to public sidewalks make travel by foot either difficult or impossible. A group of New York City officials travelled to Japan several times within the past few years to exchange ideas with their sister city, and during their last visit they were very impressed with the way Japanese contractors protect their urban construction sites.

An ordinance passed in 1964 in Tokyo established standards to prevent falling objects from construction sites, and in 1967 an addendum to the ordinance added very specific requirements to the previously issued guidelines. The addendum called for steel cables to be installed completely around the perimeter of a building on all floors, and canvas supported from these cables would enclose and contain any construction debris within the building envelope. When contractors were working 32–65 feet above the foundation line, they were required to install a latticework fence extending 6 1/2 feet beyond the face of the building line, completely around the site.

Since the time of this ordinance additional measures have been enacted, and now construction sites are enclosed at grade level with a solid metal panel fence approximately 12–15 feet high. The building within this enclosure is wrapped completely on all floors with either a brightly colored heavy duty plastic netting or mesh extending from floor to floor, or if weather protection is required, a solid sheet of reinforced plastic. The New York City officials were so impressed by these cocoon wrapped structures that upon their return to the States they requested and obtained a city ordinance requiring that all new high-rise buildings be wrapped Japanese style.

Land Readjustment— A Technique for Urban Parcel Assemblage

Ownership of land is a precious commodity in Japan, not only because of its intrinsic value, but also because once purchased it is meant to be passed along to future generations.

In and around the country's major cities, many people own small parcels of land, some of which have been in the family for hundreds of years. When a developer attempts to assemble a parcel of land in the midst of these many small landowners, the difficulties that will be encountered are all to familiar to developers everywhere.

In Japan, a technique refered to as *land readjustment* will sometimes be used in order to acquire sufficient land for a specific project. Instead of merely purchasing separate parcels, some of which are not even needed in the entirety, the land readjustment concept looks to readjust property lines for the purpose of creating the optimum site.

A hypothetical situation in a downtown area will explain the concept of land readjustment more easily. Refer to Fig. 10-1. A developer interested in this area encompassing sites A–J would like to assemble enough land to create 9 buildable lots for commercial developement. Mr. A has lived in this area for 60 years, and his father and his grandfather lived in the house he now occupies. No amount of money will persuade Mr. A to sell his house and move away. Mr. J would not mind selling his land, but recognizing the ultimate potential of the combined properties he asks for an astronomical price which the developer refuses to pay. The rest of the landowners seem to be more reasonable in their demands and do not pose too much of a problem.

When the city government learns of the proposed redevelopement scheme and realizes the benefits that will accrue in the form of a higher tax base, they indicate a desire and willingness to rearrange the streets in the area, making them straighter and wider. The prefecture and federal governments, interested in seeing more open spaces in urban projects suggests creating a park in the area which they would completely fund. This is the perfect scenario for a land readjustment scheme, since all of the interested parties, the developer, Mr. A, Mr. J., the rest of the landowners, and various government agencies all want something which they could have, more or less, if individual property lines could be dissolved and each participant's needs could be addressed.

Mr. A is approached by the developer and asked whether he would sell a portion of his land in return for a fair price, having his lot re-

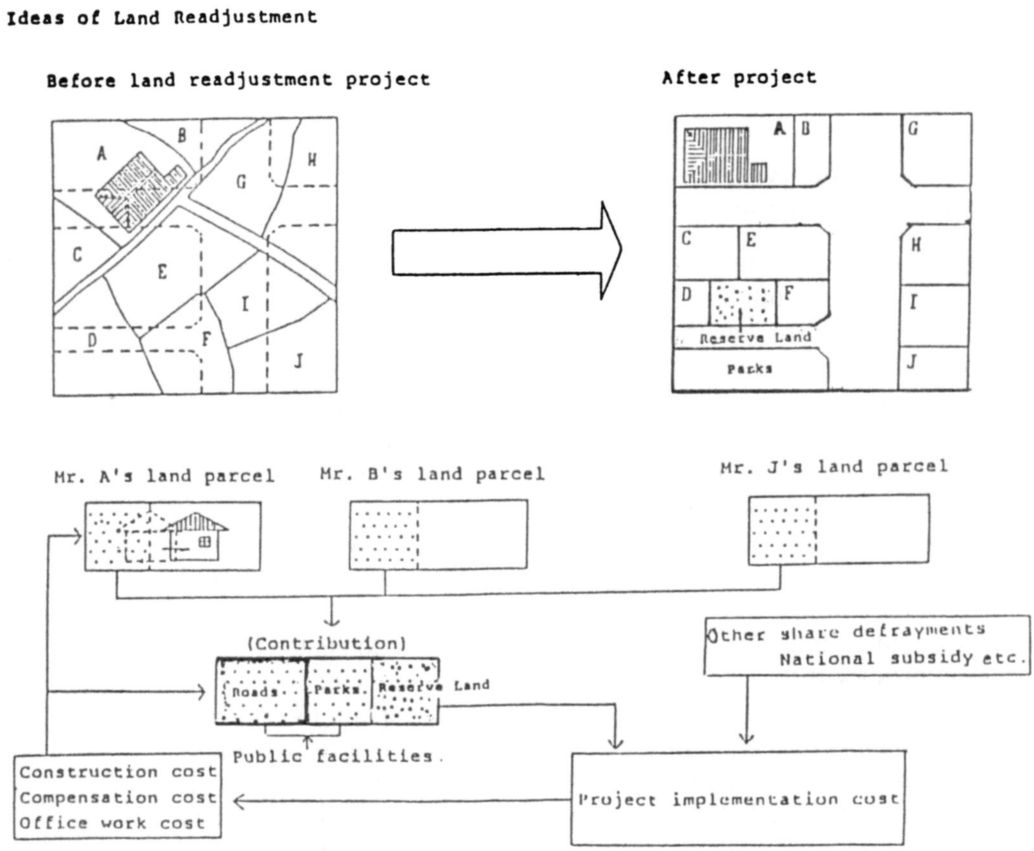

Fig. 10-1. Diagrammatic explanation of the land readjustment process practiced in Japan.

configured and having his house moved to a different spot on this revised site. Mr. J will be offered a handsome price for his entire parcel and will be permitted to retain ownership of a small portion which, if everything goes as planned, will increase in value in the coming years to such a point that its sale will give him much more than his initial exhorbitant asking price. The city, at no cost to the developer, will create boulevards through this area, making adjacent property values increase along with their tax bases. The prefecture and federal government will pump money into a local park with amenities and provide future funding for maintainance.

It appears that through the use of land readjustment both individual and collective goals can be realized, and in the end the public becomes the greatest beneficiary.

NEW CONSTRUCTION-RELATED PRODUCTS

The desire to create new construction materials in Japan satisfies a number of needs. This raw-material-deficient country has to import vast quantities of wood products such as plywood, doors, trim materials, and forming and framing lumber. It would stand to reason that a wood substitute would be vigorously pursued. Other products that could improve the quality and job site productivity of the construction process would also be much sought after by Japanese contractors. This search for new products is an important adjunct to the construction technique practiced in Japan today.

Ace-Lite—An Artificial Wood Product

Wood, one of nature's wonder products, has many special properties. Pound for pound it is stronger than steel. Because of its cellular structure, it is an excellent insulator. It can be bent, cut or shaped to almost any size and configuration.

The same cellular structure that makes wood such a good insulator allows it to absorb moisture, causing it to swell, warp, twist, and rot. Insects can attack wood and destroy it, and unless it is treated with chemicals it is a flammable material.

Onoda Cement Company, Ltd., working with Ohbayashi Corporation's Research Center, developed an artificial wood product called Ace-Lite which combines most of the desirable qualities of the natural product while eliminating the undesirable ones. Ace-Lite is made from a process using zonotolite, a substance found in calcium silicate, dissolved in water mixed with glass fibers and some high-molecular-weight organic polymers in order to create a wood substitute that can be sawed, drilled, nailed, and even whittled just like the real thing.

Ace-Lite is so highly heat resistant that it is classed as non-flammable. This product is rot resistant and will not swell in the presence of water and high humidity. It is highly weather resistant and the product is difficult to crack or split when being nailed.

The compressive strength of Ace-Lite is one of its limiting factors. At 1400 psi, it falls well below that of Douglas Fir (7240 psi), Spruce (5320 psi), or Western Pine (5040 psi), and because of this shortcoming it has found acceptance in Japan only as an interior trim material. Ace-Lite's primary use as a decorative wood trim substitute is why it is currently manufactured in thicknesses from 1/2 inch to 2 inches, and is available in widths up to 40 inches and lengths to 10 feet. One side of the product is perfectly smooth, while the other side is embossed with a wood grain finish. This material is relatively expensive, selling for about $11.61 per square foot in Japan, but research is ongoing in an effort to reduce the cost and expand the uses of this product with great potential.

Acklam—A Stone Veneer on a Honeycomb Panel

Asahi Composites Corporation working with Sekigahara Stone Company of Japan and the Ciba-Geigy Company of Switzerland, developed Acklam, an aluminum honeycomb panel with a 3–5 millimeter thick facing of natural marble or granite. The resulting panel looks to all outward appearances just like a natural stone slab, but it weighs only one-quarter as much as the usual interior decorative stone wall panels. Although the composite panel is much lighter than its natural counterpart, it is 300% stronger due to its aluminum honeycomb core.

Asahi Composites had been looking for new products to develop, and since they had already been working with the Swiss firm Ciba-Geigy, a brainstorming session led to the creation of this new composite stone panel. Asahi searched out Sekigahara Stone, who had the capability of slicing marble into sheets as thin as 0.5 millimeter, and with Sekigahara's branch office in Marina di Carrara, Italy they became a natural teammate for Asahi.

Production facilities at Asahi Composites in 1987 permit them to manufacture about 160,000 square feet of panels per month. The panels with marble facings have stone veneer thicknesses of about 4.5 mm (slightly less than 3/16″) and the panels with granite facings have 3 mm (1/8″) veneers.

Future plans at Asahi Composites includes developement of a sandwich panel for exterior wall construction.

Neoparium—Crystallized Glass Wall Panels

A demand for permanence and maintainance-free building exteriors led Nippon Electric Glass Company, Ltd. of Osaka to create a premium grade wall panel made of fused granular glass particles. Harder and stronger than granite, this new material, called Neoparium, looks like highly polished marble and is available in a number of colors, some of which have the mottled look of granite.

Neoparium is superior to natural stone in its resistance to acids and alkalies, and it does not appear to lose its original sheen when exposed to airborne pollutants. These fused glass panels have a bending strength three times greater than marble and 3 1/2 times greater than granite. Neoparium's compressive strength is 1.2–5.6 tons per square centimeter, twice as high as granite and 30% greater than marble. Because of its inherent strength, Neoparium can be manufactured in thinner sections than most natural stone products, and therefore it requires a lighter framing system for attachment.

These customized glass wall panels can be manufactured in curved sections for building radius corners or column covers, an extremely costly detail when achieved by the use of natural stone.

Neoparium panels are attached to a structural steel frame by the use of angle brackets, adjustment plates, and dowel pins located in holes predrilled in the edge of the panel. The cost of a Neoparium wall system is relatively expensive. The 15 mm thick panel sells for about $30–$40 per square foot, and the 8 mm panel is about $15–$18 per square foot, exclusive of installation costs. Although the 15 mm panel was designed for use as an exterior panel and the 8 mm is for interior use, the thinner panel can be applied to building exteriors when it is installed in a prefabricated panel system not exceeding three stories in height.

According to the U.S. sales representative for Neoparium, another $10.00 per square foot should be added for the installation of their panel onto a steel stud frame. These square foot prices, according to the sales representative, are to be used as guidelines only, since larger quantities of this material in simple, repetitive panel sizes would reduce both material and installation costs.

With 1″ thick granite panels selling for $18–$20 per square foot, exclusive of installation, Neoparium is a premium product, but when weight savings are factored into the equation, this material becomes attractive from both a quality and dollar and cents standpoint.

Lumiflon—A Fluoropolymer Coating for Field Application

For years, U.S. coatings manufacturers have been marketing long-life fluoropolymer paints and coatings such as Kynar* and Fluoropon*. These extremely durable finishes, with life expectancies of twenty years or more, have been used to protect the metal surfaces of windows, louvers, railings, flagpoles, and storefronts. These special coatings promise long life, excellent fade resistance and anti-chalking qualities, and protection from salt spray corrosion, but these paints all have to be baked onto the substrate in the factory.

The Asahi Glass Company, Ltd. of Japan has developed a fluoropolymer resin for use in paint that promises the same qualities as the baked-on product, but can be applied in the field, since it cures at ambient temperatures.

Lumiflon-based coatings can be applied by spray, brush, or roller, and the paint can be used on a variety of metals as well as on concrete

*Kynar is the registered trademark of the E.I.Dupont Company, and Fluoropon is the registered trademark of Desoto Corporation.

surfaces. The product is available in a wide range of colors from white to dark bronze, and there is a transparent coating that permits up to 90% light transmission, insuring that the color or tone of the base material will not be altered. The manufacturers of Lumiflon resin claim that their colors will retain their original appearance beyond 20 years, and the paint can be formulated to produce very high gloss coatings. Lumiflon is soluble in a number of common solvents, so it is relatively easy to apply by spray equipment.

Lumiflon fluoropolymer resins are distributed in the United States by ICI America, a division of Great Britain's Imperial Chemical Industries. As of this writing ICI's representative indicated that commercial application in this country is limited since several paint manufacturers are still evaluating the product. In Japan, the product has seen a number of interesting applications. Yasuda Insurance Company chose this coating when they decided to paint the 640,000 square feet of exterior precast concrete panels on their Shinjuku office building. TOA Paint Company applied a dark bronze Lumiflon coating to a 270,000 square foot copper roof on a building in Fukushima, 175 miles north of Tokyo. When Shimizu used Lumiflon to coat the exterior walls of a computer school in Nagoya City in 1985, the paint applied to the orange peel texture of the concrete panels gave the building's skin the look of an embossed metal panel. This unique product has numerous marine uses, and it has been applied to the hull of a luxury power boat and to the structural steel members under a bridge.

A NEW FRAMELESS GLAZING SYSTEM

Research on base isolation technology has had some spillover effect on architectural design. At the Ohbayashi R&D Center, a radically new glazing system has been designed, installed, and is currently undergoing tests in this vibration free, *menshin*-equiped building. The elevation of the first floor level of the R&D building is composed of butt glazed strip windows fitted in between the structural concrete building columns. The window installation is rather special since the lights of glass are not fixed into a frame, but merely fit into an opening framed by concrete columns on two sides and a concrete sill and head spandrel panel.

Into this concrete opening a pane of 3/8 inch plate glass is fitted by means of an aluminum clip glued to its edge and bolted securely to the concrete column. The adjacent panel is butt glazed and secured on its opposite jamb to the other concrete column with a similar glued and bolted aluminum clip (see Fig. 10.2). The appearance of this window arrangement is one of glass floating in the opening.

It appears that the vibration free nature of the rigid frame will allow

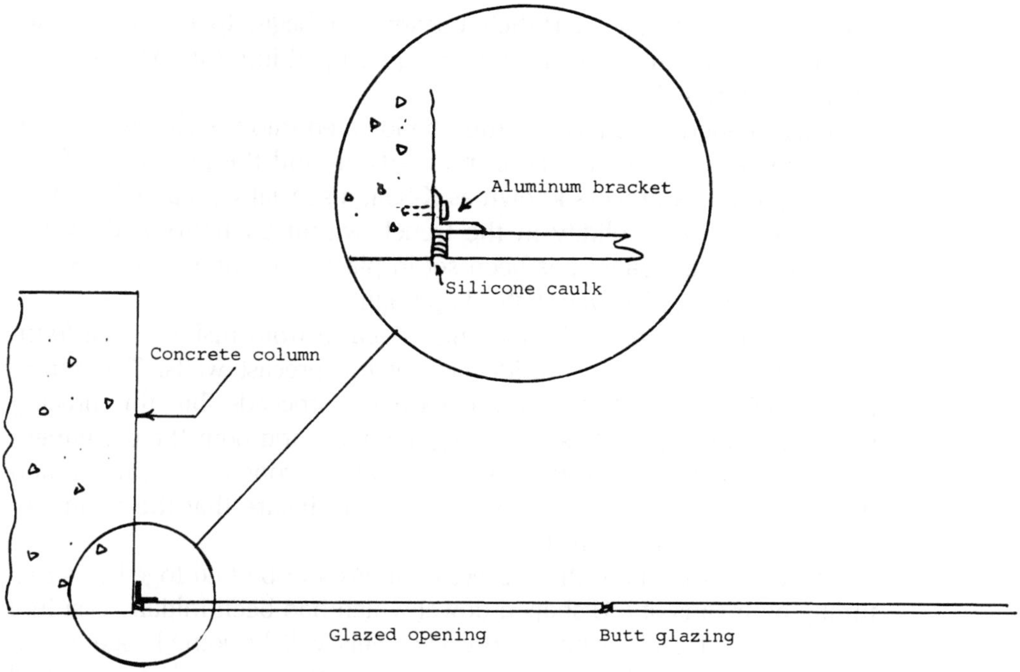

Fig. 10-2. Frameless glazed opening design at Ohbayashi High Tech Building—plan view.

the glass to expand and contract within the limits of elasticity created by the silicone adhesive and caulking, and still remain firmly in place.

A NEW METHOD OF FOUNDATION WALL CONSTRUCTION

In work on sites with zero lot line tolerances, the construction of conventional cast-in-place concrete foundation walls presents many problems, including the need to install costly sheet piling.

Kumagai Gumi has perfected a method of installing precast concrete foundation walls which serve as sheet piling initially and later become the building's permanent foundation wall system. Kumagai Gumi's PA-NE System (the name is derived from PermANEt), when used in conjunction with their K-Soil slurry trench technology, brings a new solution to an old problem.

The PA-NE process begins when a shallow trench is dug at the foundation wall line, on grade, and two L-shaped guide rails are placed in the trench. The down leg of the guide rail will serve to assist a narrow-profile clamshell bucket as it excavates the slit trenches around the perimeter of the building, and will also assist in the process of lowering the precast foundation wall panels into place. Once the guides are in place, the thin-profile clamshell bucket will begin to excavate a slit trench while a bentonite slurry is being pumped into the excavation to stabilize the trench walls.

Precast concrete panels will then be lowered into the slit trench, one after another, as the excavation proceeds around the perimeter of the building. Using a process known as K-Soil, a specially prepared bentonite slurry mix will solidify in the trench within 24 hours and by the time each precast panel has been set in place, the outer layer of K-Soil will have produced a watertight diaphragm.

Now comes the task of removing the earth from inside of the foundation and exposing the inside face of the precast walls. Temporary bracing will be installed as the excavation proceeds, but this bracing will be rapidly replaced as crews begin to form and pour the permanent structural system consisting of cast-in-place concrete columns and beams. The floor slab will also be poured to insure that the perimeter walls are locked firmly in place.

The vertical joints of these precast panels will be tied together structurally by means of a U-shaped dowel which had been initially cast into the keyway of each panel. These U-dowels will be joined together as vertical reinforcing bars are placed through them in the keyways and then grouted in securely. Where 90° corners are formed by the PA-NE walls, dowels inserted into threaded connections in the precast walls will be embedded as the cast-in-place concrete interior corner columns are being poured. The foundation construction will have now been completed.

Kumagai Gumi began experimenting with its underground diaphragm wall construction process in 1965, but it was not until 1980 that they first used it on a commercial construction project. Since that time they have installed approximately 1.5 million square feet of PA-NE wall, and it has become one of their construction mainstays.

K-Soil

A key ingredient in the PA-NE diaphragm wall construction process is the stabilized bentonite slurry which Kumagai Gumi calls K-Soil. K-Soil consists of bentonite and gelling agents added as the slurry is being agitated in place. As the bentonite slurry is being pumped into an excavation around the building's perimeter, compressed air is introduced to the bottom of the trench via a series of air delivery pipes. When the air bubbles rise in the slurry, gelling agents such as water glass and cement paste are mixed into the agitated slurry compound, and the bentonite interacts with these two additives and rapidly begins to solidify, creating K-Soil. As the agitation continues it insures that the gelling agents will be dispersed throughout the mixture and Kumagai Gumi also adds a small amount of carboxymethyl cellulose to the swelled bentonite.

Kumagai Gumi has found several other uses for K-Soil beside its PA-NE application. When sheet piling is required around a site, the introduction of K-Soil can insure that a watertight enclosure will be obtained. K-Soil has found application in shield tunneling and is sometimes used to improve the stability of soils at both ends of a tunnel excavation.

Going from the depths to the heights, Japanese ingenuity knows few boundaries.

THE NAKAGIN CAPSULE MANSION—A UNIQUE APARTMENT BUILDING

Japanese businessmen are notorious for working long hours, and since housing is so expensive in and around Japan's major cities, most executives are faced with a long commute when they leave their office late at night.

But suppose these corporate executives could have a small room or apartment in town where they could spend the night after a particularly long day, and avoid having to travel home for just a few hours' sleep? But given the continual training and education programs very much a part of a promising manager's climb to the top, a Tokyo assignment might not be long term, and if the executive is relocated to Osaka, what about the apartment? Would it be possible to have him take it with him to his next assignment?

In 1971, Japanese architect Kisho N. Kurokawa attempted to solve both problems as he set about designing the Nakagin Capsule Mansion to be built in the Ginza section of Tokyo (see Fig. 10-3).

The challege was without parallel. Could a building with small, comfortable apartments be built close to the major business district, such that when any of the company's key people was transfered to another district his apartment unit could be detached and sent along with them?

Architect Kurokawa, working with the Design and Engineering Department of Daimaru, Inc. and Taisei Corporation, developed a ferrocement living module which could be fitted to a permanent central stalk but could be plucked off of this core, if need be, and transported to another site where a similar stalk was awaiting a new attachment. The core or central stalk of the structure would contain the elevator, mechanical systems, and all of the utility risers to which each individual apartment unit would be connected.

Each capsule module would be the same size and configuration, measuring about 9 feet wide, 14 feet deep, and about 9 feet in height. One window, a circular one, would be fitted porthole fashion in the exterior 9 foot wide end wall. The living space inside the capsule is adequate

(b)

(a)

(c)

Fig. 10-3. Nakagin capsule mansion in Ginza.

(d)

(e)

Fig. 10-3. ***(continued)***

by Japanese standards and snug by Western comparison, but the interior of the unit is well planned to utilize the space to the maximum. The interior layout is U-shaped, with a bed forming the bottom loop of the U. On the left wall, at the foot of the bed, an upholstered bench extends between the bed and a small triangular bathroom at the top of the U. On the wall containing the bed's headboard there is a full length closet behind the front door to the apartment, a counter running the length of that wall containing a sink, range, and under-counter cabinets. A deep fascia extends the length of the wall at the ceiling and incorporates a televsion set, an exhaust fan, speakers for the stereo system in the head board, and two round air diffusers. The headboard over the bed also includes a reel-to-reel tape deck and a digital clock. The round porthole window is fitted with a fan-shaped blind that can be opened fully or partially by rotating the individual blades of the fan in a circular direction. A shag rug on the floor concludes the furnishing in this cozy apartment. The idea must have seemed sound at the time but apparently the concept never really caught on, and although is it now fully occupied by the general public, it stands as a testimony to one man's effort to make life easier for others.

A PENCHANT FOR PREFABRICATION

There seems to be a fascination in Japan with prefabrication of one sort or another. At various job sites, exterior wall panels can be seen joined together on the ground so that two or three of them can be hoisted into place with one pick of the crane. With the advent of lightweight carbon fiber reinforced concrete wall panels, large sections are created after they are delivered to the site and raised into place swiftly and easily with an electric hoist. At another site, interior preglazed window units are twinned together at their jamb sections so that they can be set into place as one unit. On projects with cast-in-place concrete structural systems, reinforcing bar cages for columns are fabricated on the ground with splice connectors attached to the bottom of their vertical rods so that a crane can lift them onto dowels protruding from previously poured slabs or other structural members. Factory installed tile cladded exterior wall sections are prevalent and tile faced interior wall panels can also be seen on some construction sites. Plumbing pipes with fittings attached are evident, as are prefabricated ductwork assemblies.

Prefabrication efforts extend beyond individual building components. Mitsui Construction Company has developed a prefabricated housing erector set of sorts.

The MCS System by Mitsui

Over the past ten years Mitsui Construction Company has built about 11,000 dwelling units utilizing a system of prefabricated concrete panels installed in a three-dimensional pattern which they call MCS—Mitsui Checkered System (Fig. 10-4). The precast concrete panels, manufactured off-site, are installed in the structural steel framework of the building in any number of three-dimensional configurations. These panels can be used to form exterior walls, interior walls, corridors, and shear walls. A pattern of interior walls on one floor will be reversed on the next upper floor to create a checkerboard configuration which adds to the structural integrity of the building and addresses the seismic considerations so necessary in Japan.

Prefabrication, new products and new construction techniques are a part of everyday life in the construction industry in Japan as the quest for a better way is reinforced by an industry committment to continuing research and developement.

As indicated earlier the photographs on pages 331-337 show a few glimpses of the work being performed by Japanese contractors.

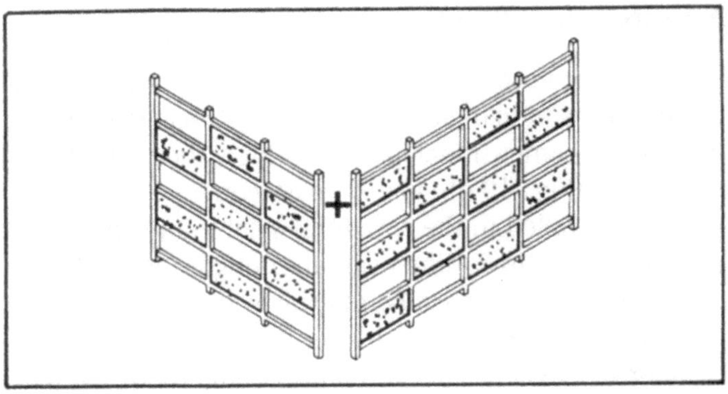

Two-dimensional checkered construction

(a)

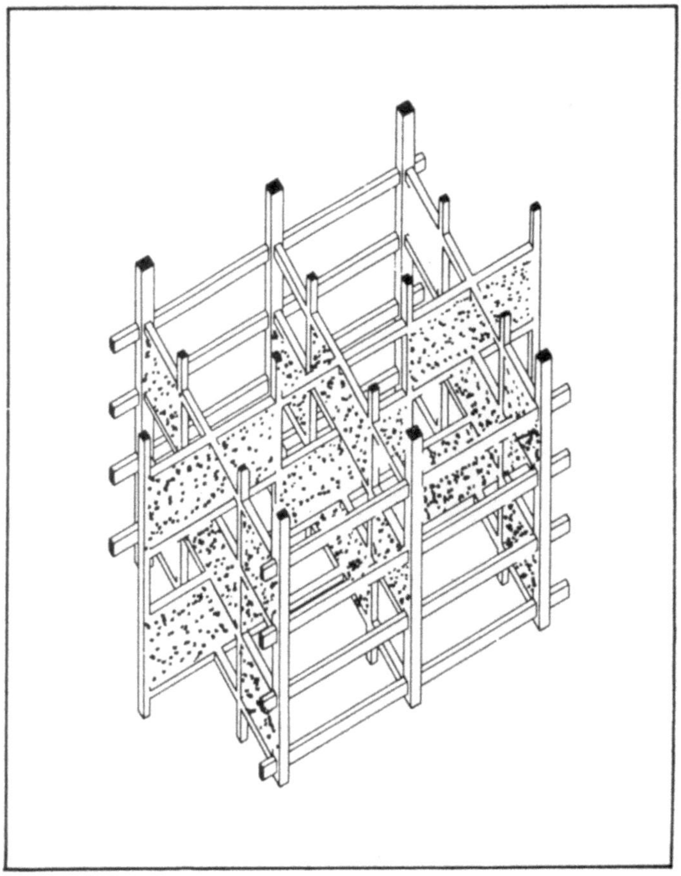

Three-dimensional checkered construction

(b)

Fig. 10-4. Mitsui MCS construction system.

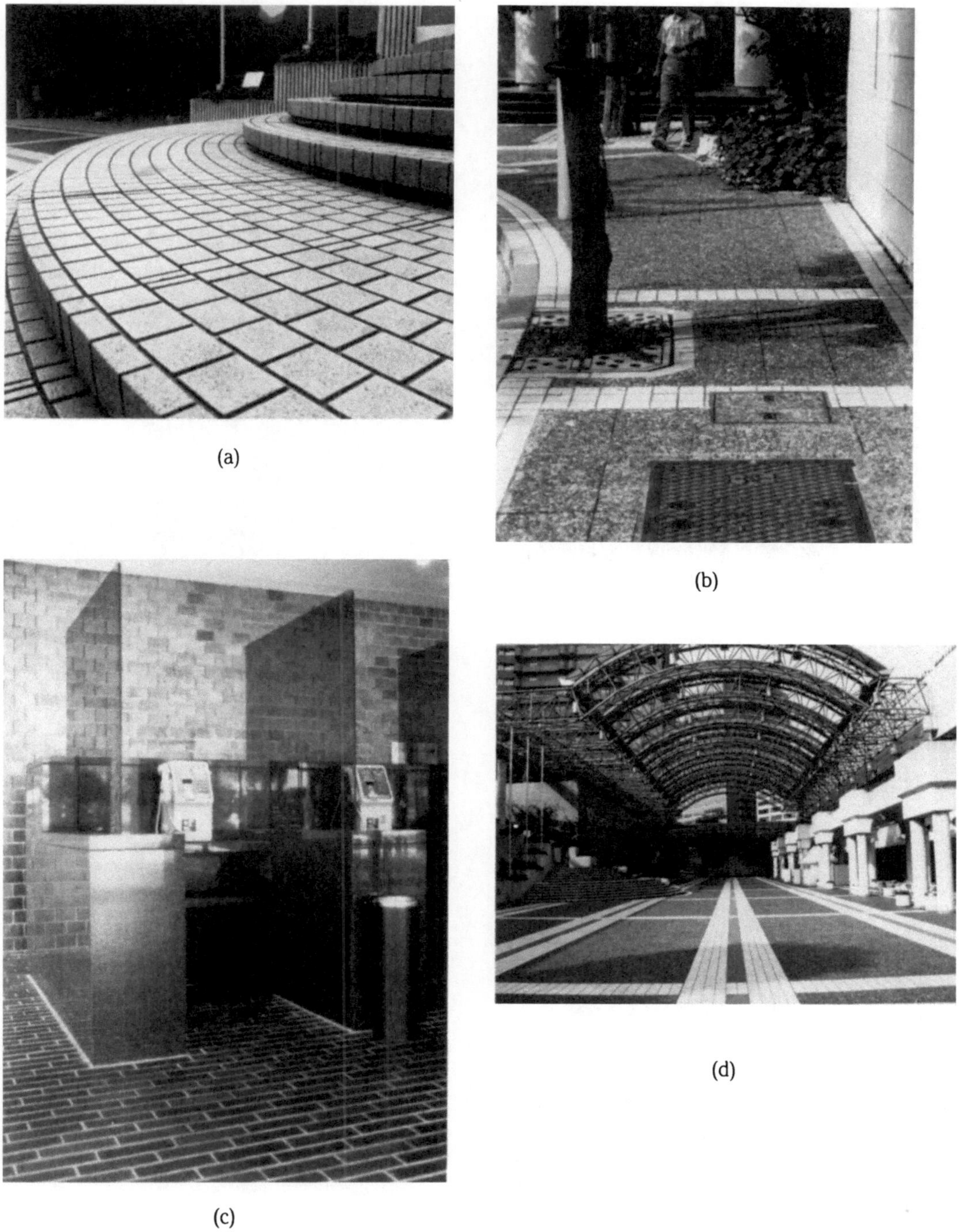

Fig. 10-5. A photographic essay—The Japanese penchant for details. (a) A curved walk constructed of pavers at Ark Hills. (b) Details show up everywhere, even in a sidewalk tree well. (c) A simple exterior telèphone booth becomes an elegant structure when polished stone and glass are used. (d) The courtyard at Ark Hils. Perhaps overdone for some tastes, but the attention to detail is beyond criticism.

Fig. 10-6. A photographic essay—Striking high rise design. (a) Kasumigaseki Building, Japan's first skyscraper, built by Kajima in 1966. (b) The Yasuda Fire Insurance Building bears a striking resemblance to 9 West 57th Street in New York City. (c) The Collins 20 building in Shinjuku. (d) The clean, classic lines of an apartment building.

(a)

(b)

(c)

Fig. 10-7. A photographic essay—The artistic abilities of the Japanese craftsmen. (a) The carved wood beams of a Buddhist temple. (b) More temple beam carvings and intricately detailed tile roof. (c) Stone column enclosures at the Ginza Saison Theater in Tokyo express the same artistic nature.

(a)

(b)

Fig. 10-8. A photographic essay—(a) A diaphragm wall under construction. (b) Another project underway in Tokyo. The absence of debris and cleanliness of the site is a Japanese trait.

(a)

(b)

(c)

Fig. 10-9. A photographic essay—Sights familiar to the American builder, but in Japan. (a) This typical scene is in Yokohama. (b) A kettle of asphalt boiling on a rooftop as vertical surfaces are being mopped before roofing felts are applied as base flashings. (c) Another sight familiar to U.S. contractors on the run.

(a)

(c)

(b)

Fig. 10-10. A photographic essay—A study in contrasts, Japanese style. The austere lines of a tiled apartment house wall (a) contrasted with a chrome bedecked Mister Donut at the Great American Carousel Plaza in Tokyo (b) and the simple symmetry of a pedestrian walkway (c).

(a)

(b)

Fig. 10-11 A photographic essay—Mastering the art of the architectural precast concrete panel. (a) A rather unusual wall panel on a Tokyo office building. (b) A closer look at the fine wood-grain texture of this superbly detailed panel.

11

Research and Development

The quality and level of research and developement in Japan is a subject of concern to both industry and government. The Japanese government has been uneasy about the quality of research in their country, feeling that their efforts lag far behind the United States and Europe, even though Japan spends 2.81% of its gross domestic expenditures on R&D, while only Germany and France come close with 2.67% and 2.26%, respectively, trailing a U.S. 2.77% gross domestic expenditure. This feeling of lagging behind might be due to Japan's propensity to feel insecure in its position among world powers.

Japan exhibits high levels of research and development as applied to mass produced consumer products, but they seem to avoid highly advanced, basic research activity. A look at the Nobel Prize recipients over the years seems to bear out this premise. In physics, the United States has dominated the field, capturing 25 medals in the last 25 years. Only two Japanese scientists have won Nobel prizes in physics. One was awarded to Leo Esaki, who shared his medal with Ivar Giaever of the U.S. and Brian Josephson of Great Britain in 1973, and the second physics medal went to Japanese scientist Shinichiro Tomonaga along with two American physicists in 1965. In the field of chemistry, the United States has again shown overwhelming domination, while Japanese chemists have won only one Nobel prize since the awards were first offered in 1981. Kenichi Fukui and the American Roald Hoffman won a Nobel prize in chemistry in 1981 for their efforts in applying quantum mechanics to predicting the outcome of chemical reactions.

R&D AND THE JAPANESE CONSTRUCTION INDUSTRY

Construction industry research in Japan, carried out by both government and private institutions, appears to have reached very high levels when compared to other countries. The government, primarily through the Ministry of Construction, operates several agencies devoted to construction research, both applied and basic, but it is in the private sector where considerable amounts of human and monetary resources are committed to the research and development of new products, new procedures, new designs, and new methods. The significant outlays for research and development in the private sector, some $300 million annually just from the Big Five, seems even more unusual in the light of the claim that competition in the domestic public works market is almost nonexistent due to the practice of *dango*. Construction research in the private sector also seems somewhat unusual, given the long term client-builder relationships that exist which could lead to a relaxing of construction innovations rather than a strengthening of them.

In the face of such alleged limited competition, why do these Japanese contractors pour millions of dollars each year into their top-flight research centers? Perhaps these builders have been using the last several years wisely, knowing that eventually their closely controlled domestic 'market would have to be opened to contractors from other countries; while the situation lasts, they would use whatever time they had left to bring their company to the cutting edge of construction technology. It is also conceivable that these major Japanese construction companies devote so much time and effort to research and development simply to make their individual companies the best they can possibly be. Research and development is as much as part of the Japanese psyche as raw fish is part of a sushi bar. Their long term, market share approach to construction goes hand in hand with their use of R&D as a tool to create more marketable products in an ever changing world economy.

Borrowing from Western technology, which they greatly admire, the Japanese have emerged as innovators in their own right. At times they have picked through the trash of U.S. technology to retrieve a few prematurely cast off gems.

U.S. TECHNOLOGY—JAPANESE OPPORTUNITIES

When Sony Corporation paid Western Electric 25,000 yen for the nonexclusive rights to manufacture transistors in 1953, who could have predicted the day when the Japanese electronics industry would dominate the world, and more recently, the manufacture of D-Ram memory chips, the basic building block of the semi-conductor industry? Although Ampex Corporation and RCA Corporation conceived the idea for a video cassette recorder in the 1960s, this $6 billion per year market today contains not one American VCR manufacturer—the industry is dominated by the Japanese. A Japanese automobile manufacturer became interested in an innovative engine design invented by a German and dismissed by an American corporation who saw little value in its development. Mazda bought the manufacturing rights to this revolutionary Wankel engine and installed it in its RX-7 series sports car, one of its hottest selling cars.

American R & D might have been fooled a couple of times in the past, but U.S. industry learns from its mistakes—Right? Wrong! High definition TV (HDTV) is being promoted as being a video system with the clarity and sharpness of 35 mm motion picture film. As HDTV comes into its own within the next few years it is expected that TV screens will grow in size to display extremely well defined pictures, which according to some experts will supplant the movie cameras and screens in motion picture theatres. American industry first produced HDTV in 1950, but now, nearly 40 years later, the technology has been "back burnered" in this country while it is cooking in Japan. HDTV is predicted to be a $20 billion market by the end of the 20th Century, and U.S. R&D might lose yet another battle to our friends from the East.

Finally, Mag-Lev—magnetic levitation—the principle of magnetic poles repelling like magnetic poles, had been applied to train travel by the scientists at Long Island's Brookhaven National Laboratories, but after twenty years not much is being done about mag-lev in the U.S. This concept, when applied to train travel, would literally suspend railroad cars in the air, and by passing AC current through the train tracks could propel trains at high speeds on a cushion of air. The current rapid advances in superconductivity bring the concept of mag-lev trains very close to commercial reality, but while the U.S. is considering conducting further research on the subject, the Japanese have a prototype mag-lev train in daily operation.

GOVERNMENT RESEARCH AT THE FEDERAL LEVEL

In Japan, the principal government agency devoted to construction matters is the Ministry of Construction, a huge bureaucracy with 26,000 employees. Among other areas of responsibility, this agency oversees all construction research and development work conducted by the government. The primary research arm of the Ministry of Construction is the Building Research Institute, established in 1946 as a research center for housing, planning, and the building sciences.

BRI—Building Research Institute

In 1979, BRI moved from its somewhat modest accommodations in Tokyo to a sprawling site in the new Tsukuba Science City, where it now occupies twenty buildings with a total floor area of 516,000 square feet. In 1986, BRI's operating budget was $11.264 million, which did not include revenues of $2.14 million received from other agencies as payment for their research studies conducted by the Institute. BRI has eight active divisions not including its Administrative Department and its Research Planning and Information Department.

The Housing Department. National and local housing problems, conditions, and needs are studied in this department. Newly developed construction materials relating to housing are analyzed here, and new housing construction techniques are sought out and evaluated in this department.

The Materials Department. This department at BRI performs extensive tests and researches building components and basic construction materials such as concrete, wood, polymerics, ceramics, metals, and composite materials. The Materials Department develops standard methods for testing construction materials similar to the functions performed by ASTM, AISC, and ACI in the United States.

Structural Engineering Department. Substructure and superstructure design and construction analysis with particular emphasis on seismic and wind loading considerations is the objective of this department.

Production Department. Building production techniques are studied by this department and advanced building methods are developed here, along with solar energy research and computer assisted design and computer assisted engineering research.

Environment, Design, and Fire Department. This is a department devoted soley to the comfort, health, and safety of building occupants, and all research conducted in this section concerns itself with these three factors.

Urban Planning Department. Urban sprawl, pollution, and urban infrastructure comes under the scrutiny of this BRI department, whose responsibility it is to make city life safe, healthy, and more efficient.

International Institute of Seismology and Earthquake Engineering. This is the only BRI department that functions on an international basis, collecting and disseminating seismic data from and to all parts of the world.

Testing and Evaluation Department. A combination UL testing laboratory and a National Fire Prevention Agency which tests and establishes standards for fire resistive materials and building design and construction.

Other Government Agencies Concerned With Research

Under the Ministry of Construction there is a Public Works Research Institute which delves into public works, civil engineering type construction research. This agency involves itself with dam construction research as well as research related to flood control and soil erosion problems. The Ministry's Geographic Survey Institute concerns itself with matters relating to geography, map management, and the operation of several geodetic observatories. The Construction College division of the Ministry of Construction is a planning and management arm which in turn oversees a General Training School in Shizuoka prefecture. The Research Institute of Construction and Economy is a nonprofit organization created in 1982 under the auspices of the Ministry of Construction which each year publishes a booklet on the state of construction in Japan, called the *White Paper.* The Research Institute, known by its Japanese name Kensetsu Keizai Kenkyusyo, also publishes a booklet on the structure and composition of the construction industry in Japan which is a compilation of statistical data very similar to information released by the U.S. Department of Commerce in their quarterly *Construction Review* periodicals.

PRIVATE CONSTRUCTION RESEARCH AND DEVELOPMENT ACTIVITY

The overwhelming percentage of private R&D work in Japan is concentrated in the hands of the upper echelon Japanese construction companies who maintain highly sophisticated facilities devoted soley to research. Many of these private R&D facilities are staffed by hundreds of engineers and scientists with yearly budgets in the millions of dollars and they are without peer anywhere in the construction industry. Hazama-Gumi developed the first private construction company R&D department in Japan in 1945, and since then a number of prominent contractors have joined in to build their own research departments:

Private Research and Development Organizations

Owner	Founding Date
Hazama-Gumi	1945
Ohbayashi Corporation	1948
Kajima Corporation	1949
Shimizu Construction Co. Ltd.	1949
Takenaka-Komuten	1953

Fudo Construction Co. Ltd.	1957
Taisei Corporation	1958
Toda Construction Co. Ltd.	1958
Aoki Corporation	1959
Fujita Corporation	1960
Nihon Kokudo Kaihatsu	1960
Sumitomo Construction Co.	1960
Tekken Laboratory	1960
Kumagai-Gumi Co. Ltd.	1962
Konoike-Gumi	1963
Maeda Construction Co. Ltd.	1963
Okumura Corporation	1964
Tobishima Corporation	1965
Mitsui Construction Co.	1967
Penta Ocean Construction	1967
Tokyo Kensetsu	1968
Tokai-Kogyo Co., Ltd.	1970
Baisue	1970
Toa-Kensetsu Kogyo	1970
Zenidaka Gumi	1972
Nissan Laboratory	1972
Sato Kogyo Co., Ltd.	1972

In order to provide insight into private construction company R&D, it is necessary to take a closer look at some of these corporate research facilities.

KICT—Kajima Institute of Construction Technology

Although Hazama-Gumi opened the first R&D facility in Japan, Kajima Corporation is credited with establishing the first research institute in Japan when it created the Kajima Institute of Construction Technology in 1949. Forty years later, KICT has matured and outgrown its original site, moving to Chofu-shi outside of Tokyo in 1956, where there are a dozen buildings with a total floor area of 219,000 square feet devoted to construction research and technology. Kajima maintains a radio isotope lab at this location along with various other research laboratories and a large structure testing laboratory where extensive testing took place in 1963 before the company embarked on the construction of Japan's first high rise building, the Kasumigaseki Building.

In 1975 Kajima built the industry's first hydraulic laboratory in Japan, and in 1984 they added another large structure lab building to their inventory of research facilities. Close by, at another location, the Nishi-Chofu site, the new environmental engineering lab was constructed in 1986, and the following year the exhibits and laboratory building was completed at this location and now acts as the focal point for all of the company's research laboratories.

The Exhibits and Laboratory Building. This modern, tastefully designed structure is a showcase in which to display many of the lab's ongoing projects in a laboratory environment, but ones that also need to be tested thoroughly in day-to-day applications before being stamped with KICT's approval. A tour through this building is like a walk into the future. Starting at the roof deck:

- An atrium skylight has been built of 6″ deep stainless steel I-beams with stainless steel cap flashings over the glazed members, resulting in a skylight that is structurally rigid and maintainance free.
- On the roof deck there are special roof drains which prevent air from mixing with rainwater, thereby allowing full flow drainage and a subsequent reduction in rain leader pipe size by one-third to one-half.
- Ceramic coated roof penthouse exterior walls and ceramic coated roof copings to provide unlimited life to those surfaces exposed to direct sunlight and airborne pollutants.
- A stainless steel roof membrane consisting of sheets of 0.4 mm thickness (about 26 gauge) with welded seams—a potential new roof membrane system with unlimited life expectancy.
- A flight of stairs from the roof to the floor below built of steel fiber reinforced concrete which is lightweight and needs only steel stringers as support between floors.
- Carbon fiber reinforced concrete raised computer floor pedestals that provide the solid feel of concrete underfoot without the extra weight associated with concrete products.
- A Vinylon fiber reinforced precast concrete floor plank, nineteen inches deep, spanning 59 feet, in which voids in the slab can be used to house recessed lighting fixtures or for the installation of ductwork.
- Transparent, thin-blade venetian blinds with a mirrored coating on the outside face of each blade, allowing the blind to deflect the sun's rays when fully closed while allowing the occupant of the room a clear view of the outside.
- A 23″ × 3/8″ thick ceramic floor tile installed over a rubber-asphalt sheet for cushioning and insulation, giving a bright, attractive, maintainance free look to the lobby floor.

With a full time staff of 244 engineers, researchers, and administrative personnel, Kajima handles a multitude of research projects daily.

In the United States, it is common practice for an architect, engineer, and contractor to rely upon a material or equipment supplier's guarantee of compliance with certain performance specifications and test data. Not so in Japan. When asked why the company does not accept a supplier's performance data as given, one of Kajima's engineers replied, "Our clients depend upon our knowledge of construction and our expertise, so it is we who must be convinced that a product meets our standards, and in order to do so we must test many products ourselves." A glance to the right of where this engineer was standing revealed a large oven in which the blackened twisted shape of a window could be seen. Kajima had considered purchasing some window units from a new supplier and they had just finished completing a two hour fire test. A scorched hollow metal door frame was being removed from the oven, having just been subjected to the same kind of two hour fire test. Just around the corner from the oven, another window was being tested, this one for water leaks and air infiltration.

At the large size structure testing laboratory (Fig. 11-1) a monstrous reaction wall, almost 40 feet high and 50 feet long, was being readied to begin a load test on a full scale, cast-in-place reinforced concrete

Fig. 11-1. The interior of the KCIT Large Structure Testing Laboratory. This building is three stories high with one basement floor below and a total floor area of 45,913 square feet. The reaction wall pictured above is 39 feet high, 52 1/2 feet long, and 9.8 feet thick, constructed of prestressed concrete. The reaction floor is 9576 square feet of 7 foot thick reinforced concrete. Two overhead cranes, each with a 20 ton capacity, travel on rails just below the roof structure in this room where large scale load tests are performed. *(Courtesy of KCIT.)*

beam and column, in order to verify a Kajima engineer's structural design calculations. Although a computer simulation had already confirmed the validity of the design, Kajima felt it necessary to build a full scale model to test to destruction, as a third method of checking the adequacy of their design. In another part of the building a biaxial seismic shaking table (Fig. 11-2) had been set up to test a model structural pier which will be used in a civil engineering project. Two of KCIT's other labs are shown in Figs. 11-3 and 11-4.

Kumagai Gumi Research and Development

Kumagai Gumi, one of Japan's most respected builders, has devoted a major portion of its R&D program to three areas of involvement—technology required to solve today's problems on current construction projects; technology needed for future planning; and technology required to develop new techniques for future dominance of the construction market. Kumagai Gumi employs 130 people at their Institute of

Fig. 11-2. Biaxial seismic shaking table and testing facilities. This shaking table can simulate 1.2G movement in the horizontal direction and 2.0G in the vertical direction with a 20 ton load specimen. Kajima has created a multitude of software programs to obtain data, precise excitation information for high accuracy analysis. *(Courtesy of KCIT.)*

Fig. 11-3. Kajima's pseudodynamic loading system test area. This computerized testing system can conduct dynamic loading tests, static load tests, and fatigue tests on various types of structural members. Simulated seismic behavior can be created using computer generated seismic response analysis. *(Courtesy of KCIT.)*

Construction Technology, and the physical plant includes a large scale structure facility, a wind tunnel, and elaborate concrete testing laboratories.

Some of the products, procedures and equipment developed by Kumagai Gumi's researcher's are very impressive. From a physical size viewpoint, *Taiyogo,* a floating dock built by Kumagai Gumi which has a lifting capacity of 29 million pounds, is probably the most awesome. *Taiyogo* is 246 feet long, 154 feet wide, and rises sixty feet out of the water. Topped by two large jib cranes, *Taiyogo* is used to fabricate, transport, and install a wide variety of marine structures such as large caissons for bridge work, mooring posts, and sea berths. With the assistance of its large on-deck cranes and four ballast pumps, large marine structures can be fabricated on its deck as *Taiyogo* travels to its destination. When the structure has been completed and the destination reached, *Taiyogo* hulls down in the water, allowing the fabricated structure to slide into place. Once this is accomplished, the huge ballast pumps empty *Taiyogo*'s 16 ballast tanks and it rises back to the surface. When operating in conjunction with a concrete mixer ship, this floating

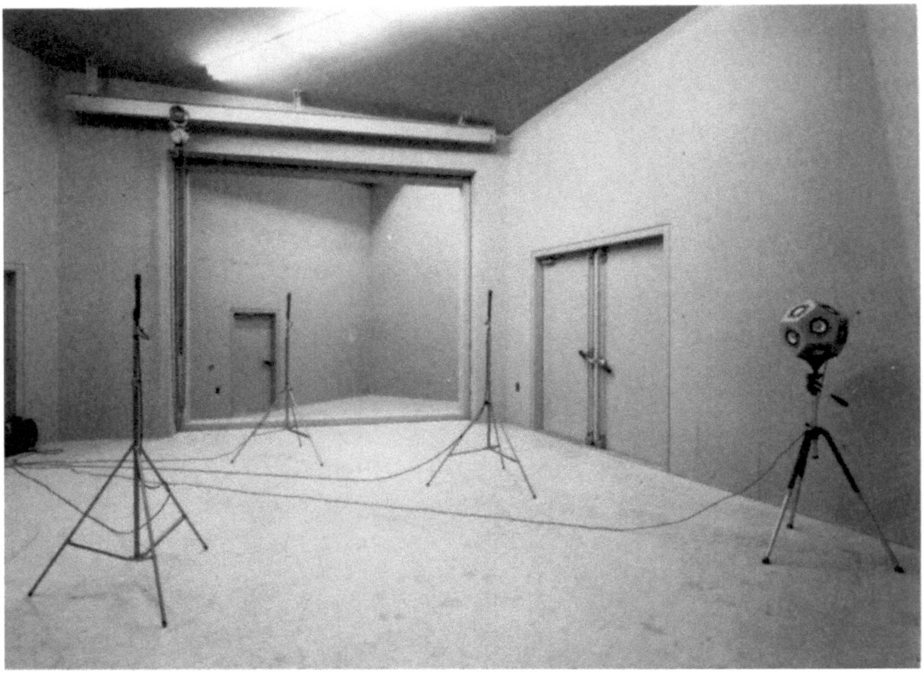

Fig. 11-4. Kajima Construction Institute of Technology's reverberation room. On the first floor of the building containing this room there is an anechoic chamber, another reverberation room, a measurement room, and an acoustic model testing room. The walls in this reverberation room are all reflective so that reflected sounds are all uniform, and this room is free from outside vibration. In a sense it is a "floating" room. *(Courtesy of KCIT.)*

dock becomes a seagoing construction site, and with an international floating dock license, she is the latest addition to Kumagai Gumi's formidable fleet of construction vessels that roam the world.

A number of other KG Institute projects have gone from the laboratory to the field:

The K-Jet System. This is a system developed to cut through steel and concrete by the use of ultra-high-pressure water mixed with abrasives. The K-Jet system is being used very effectively in bridge and tunnel work where noise, dust, and vibration must be kept to a minimum while demolition operations are taking place. K-Jet is also used in buildings with large computer operations where demolition work that can cause dust and vibration is prohibited. Hospital work is another area in which the dust free K-Jet system finds a home.

Innovative methods of concrete cutting seem to take a high priority with Kumagai Gumi engineers.

The KBB Building Breaker. The recently perfected Kumagai Gumi Building Breaker is a powerful crushing hammer-anvil that can be suspended from a mobile crane, allowing it a great deal of mobility in all planes. Powered by an electrically driven hydraulic pump, this giant clamp-shaped crusher grips a concrete wall, column, or beam in its jaws and crushes the concrete member with its tapered cylindrical main hammer and its smaller sub-hammer.

Ohbayashi's Technical Research Center

Ohbayashi Corporation established a research department at their Osaka office in 1948, but it was not until 1965 that the company constructed their Technical Research Center on an 18 acre site in Kiyose, on the outskirts of Tokyo.

By 1986, the Institute had expanded and contained a geotechnical, chemistry, and material lab, a concrete laboratory wind tunnel, acoustics engineering facilities, large structure testing lab, high tech R&D center, and a number of other support facilities. All told there are 13 buildings at the Center with a total floor area of 194,000 square feet.

Employment grew to 200 with engineers and research technicians making up the bulk of the staff.

At the research institute, concrete structures are thoroughly tested, weather and durability tests are conducted on various products, soils analyses are ongoing, and experiments are conducted on a multitude of projects from segmented membrane dome technology to laser guided slip form concrete forming methods.

With a 1987 operating budget of $34 million, Ohbayashi's Technical Research Institute is a big business that not only serves its master but conducts special research projects for many private and public agencies. In 1985 alone, the Institute earned $6 million in fees for research projects conducted for other companies.

The Technical Research Institute is divided into eight departments, some of which are engaged in both applied construction technology and also basic research.

The *civil engineering department* concentrates much of its effort on design and construction problems relating to soils with poor bearing capacities, rock formations, and seepage problems. This department is involved in tunneling technology and off-shore structure construction.

The *architectural engineering department* evaluates the performance of new nonstructural materials and construction methods. These engineers also test products and materials for durability. The architectural engineering department is charged with the responsibility to investigate industrialized construction methods and to develop robots for this purpose.

Ohbayashi's *structural engineering department* would be the envy of any major engineering college, with its large scale structural testing lab that conducts static and dynamic tests on a wide variety of structural systems. Full scale structural components undergo testing while sensors record and analyze every deformation that occurs. Models of civil engineering and building structures, including nuclear power plants, silos, natural draft cooling towers all undergo destructive testing in this facility.

The *soils and building foundation department* conducts tests on soils and foundations. Special emphasis is placed upon developing countermeasures to prevent the effect of liquefaction of soils which occurs during earthquakes.

The *seismology and building vibration department* at the Institute developed Ohbayashi's base isolation system and monitors the menshin installation at the site. Not only is building foundation isolation undergoing continuous study, but the design of machinery foundations and micro-vibration elimination in large scale integrated circuit manufacturing facilities is also a major portion of the base isolation research effort.

The Institute's *environment technology department* has a diverse area of responsibility. It conducts solar energy research, a subject critical to any country with minimal native fuel sources. The energy conservation building, completed in 1984, is a living laboratory for testing and evaluating energy conservation measures. Billed as the world's most energy efficient building (it uses only 89 Mcal of energy per square meter of floor space per year) and when it was finished, it was awarded ASHRAE's prestigious Gold Prize For Energy Conservation. This building is of flat slab construction, has a double skin on one elevation to absorb heat during the day in cold weather, and to prevent heat loss at night it lowers internally fitted, insulated plastic panels. Solar collectors and photovoltaic cells are mounted on the roof of this building and several types of thermal storage systems are installed in the basement. Building architecture, structural engineering, and mechanical technologies all joined together to incorporate over 98 energy conservation measures into this special building. The environment technology department is responsible for clean room technology and testing and also performs research on laboratory animal facilities used in the biomedical and pharmaceutical industry.

Acoustical research for concert halls, music studios and commercial buildings is handled by the *acoustic engineering department.* Studies are conducted on methods to reduce noise levels in factories, and vibration pollution is given special attention. In order to predict the acoustical impact of a new high rise building on the surrounding environment, the acoustics department will build a scale model of the proposed new building and place it among its scaled down neighbors so that the effect of wind on the shape of the new building and its acoustical implications can be closely analyzed.

The *chemical engineering department* at Kiyose investigates the chemistry of new and existing slurries, soils, rock formations, and soil stabilization chemicals. It also studies metal corrosion and concrete deterioration in order to develop measures and products to counter these destructive forces. Water treatment and waste disposal chemistry is a part of this department's day-to-day research activities. Mr. Yoshiro Ohbayashi, president of the corporation in answer to a question about the mission of the Research Institute replied, "Ohbayashi Corporation considers its mission as an engineering constructor to be that of contributing to the realization of a welfare society of a higher level in which human beings are at the center through creation of a richer, fuller environment."

Taisei and Their Research Center

Taisei Corporation's Technology Research Center follows the pattern of other Big Six contractors. They established a research department in 1958, and some 30 years later these meager beginnings have blossomed into a full blown research center in Yokohama encompassing 178,000 square feet of floor area and staffed by 354 employees, 285 of which are researchers.

Like the other major contractors, Taisei maintains a structures testing lab, geotechnical and civil engineering research facilities, acoustic, seismic, and materials testing labs. However Taisei performs considerable research on plant tissue, cell culture, and gene manipulation. Their environmental engineering function division investigates all of these fields and also the immobilization of microbe bioreactors so that commercial applications can be developed.

The pursuit of sophisticated construction research is not confined to Japan's top five or six contractors. Companies of lesser sales volumes, such as Mitsui Construction and Toda Construction, each with yearly sales volume in the $2–$2.75 billion range are also actively involved in a number of research projects at their R&D centers.

Mitsui Research and the Artificial Island

Mitsui Construction, founded in 1887, traces its origins back to Mitsui Hachiroemon Takatoshi, who first entered into business in 1673. Today the company's research and development laboratory, located in Nagareyame city in Chiba prefecture is augmented by other facilities in Kashiawa, just outside of Tokyo.

Mitsui spends a great deal of research effort on marine construction, and several years ago the company was credited with building the first offshore artificial island in Japan so that the commercial potential of a subsea coal mining operation could be pursued. This island was to be located 3¾ miles offshore from Omuta city, Fukuoka prefecture, in the southwest portion of the Japanese archipelago. Water depths at the proposed site ranged from 15 feet at low tide to 33 feet at high tide, but to complicate matters, the ocean floor was covered with a layer of muddy silt 46 feet deep. The 295 foot diameter island was built in four phases, the first of which was the removal of the existing mud silt and replacement of this material with 720,000 cubic yards of sand. A steel frame jacket was fabricated in Osaka, broken down into four pieces, and barged to the construction site where it was assembled and lowered onto the underwater sand base by 500 ton floating cranes. Steel pipe piles, 39 inches in diameter and 92 feet long were driven through guide rail brackets fabricated in the steel frame jacket, and then 480 sections of sheet piling were driven around the circumference of the soon-to-be-island. After filling the inside of the steel jacket with 131,000 cubic yards of sand, Mitsui's job was complete.

These offshore R&D activities extend to the creation of artificial reefs for fish farming, long life undersea pipe construction, prestressed concrete barges, and a multitude of offshore platforms.

Mitsui, in collaboration with Taywood Engineering, Ltd., has created an offshore floating oil production platform that rides atop a floating concrete caisson structure. Designed for use in ocean depths of 650–4900 feet, the structure has a shaft running down its entire length for drilling work, and some of its concrete buoyancy tanks are meant to be used as storage tanks. Sand ballast in the other tanks maintains buoyancy and floating stability, and the entire underwater structure is held in position by radial mooring cables attached to pile driven anchors. The floating structure can store upwards of 600,000 barrels of oil in its buoyancy tanks. Mitsui has developed an articulated underwater structure called Arcolprod—Articulated Buoyant Column System, for use in waters with severe marine conditions. A gravity base is installed initially on the ocean floor, and then the concrete platform and underwater column are lowered onto this base and connected with a hinge system that allows the underwater column to move in response to wave motions.

Mitsui's researchers have spent much time on the development of robots and have brought their soil compaction testing robot out of the laboratory where it is ready for commercial application. They plan to continue their mechatronics research efforts so that they can bring other automated construction equipment to the same development stage.

Toda Construction and Their TO-RPC Process

At Toda Construction, the completion of their Institute of Technology in Tsukuba Science City coincided with the company's centennial in 1981–1982. According to Junnosuke Toda, the company's president, the research emanating from this institute enabled the corporation to team up with France's Campenon Bernard and build several offshore concrete platforms in the North Sea and also off the coast of Brazil. Toda's research allowed them to build the Burma Biomedical Research Center in Rangoon, which garnered the company the first OCAJI prize awarded by the Overseas Construction Association of Japan to honor the transfer of Japan's construction technology to other countries.

At Toda's Tsukuba City Research Institute, the main building in the complex is a six story solar energy structure that houses two floors of research laboratories, a computer center, and a complete library. Directly adjacent to this building is a general laboratory building constructed by means of Toda's TO-RPC (Toda-Rahmen Precast Concrete) method. Using their own buildings as laboratories to test new methods of construction seems to be standard procedure among Japanese construction companies, and Toda is no exception. The TO-RPC method involves the use of factory fabricated concrete columns, beams, and floor slab sections which when used together shorten the period of construction and control costs. Precast concrete columns are erected on conventional cast-in-place foundations, and after these columns are set in place, plumbed, and guyed, reinforcing bar dowels are spliced onto foundations and grouted in. Beams are then placed upon column capitals, reinforcing bars are firmly tied together at the junction of the column and beam, and then a prefab concrete slab similar to that used in the Hambro system is set in place, acting much like a bottom concrete slab form. Ready mix concrete is then pumped onto the slab form, locking in all of the reinforcing steel and creating a very rigid, earthquake resisting structure.

In Kyobashi, Tokyo, Toda has their technology development section, which acts as liason between all departments in the company and acts as a storehouse and focal point for all corporate research activities.

CFRC—CARBON FIBER REINFORCED CONCRETE

Although Japan is notoriously devoid of raw materials, one basic building product is available in abundant supply. Thanks to many natural limestone formations, Japan has a surplus of cement and actually ranks among the world's production leaders. There are in excess of two dozen manufacturers in the country operating more than 60 factories which produce about 10% of the world supply of cement. The two leading producers are Onoda Cement and Nihon Cement, followed by Mitsubishi Mining and Cement, Sumitomo Cement, and Ube Industries. The cement industry in Japan has been substantially modernized, and productivity has increased dramatically, doubling in the period 1972–1980, to the point where worker productivity is at least 250% higher than it is in the United States, Germany, or France.

Because of the availability of cement, and because rigid concrete structures can be designed to meet strict seismic standards, concrete remains the material of choice for many Japanese designers and contractors. Not content with existing technology, most of the major construction company research institutes devote a major portion of their research effort toward developing new procedures and new materials for concrete construction.

At the forefront of their research breakthroughs is the use of new and exciting space age reinforcing materials such as carbon fibers, Kevlar®, and other super fibers.

Carbon fiber reinforced concrete is produced by adding short, pitch based or PAN carbon fibres, looking very much like black, shiny lint, to a concrete matrix in a volume of 2–4%. Concrete containing these carbon fibers has exceedingly high strength. The fibers in the concrete remain chemically stable, are not affected by the alkaline nature of cement, and resist mechanical deterioration.

Products made of CFRC have one-half the specific gravity of ordinary concrete, and with autoclave curing, shrinkage is reduced to one-third to one-half that of ordinary concrete. The relative dynamic elastic modulus of CFRC remains virtually the same after being subjected to 300 freeze-thaw cycles.

When used to manufacture architectural precast concrete wall panels, carbon fibre reinforced concrete offers many advantages:

- Due to its lighter weight, there can be a substantial reduction in the size of a building's structural steel framing members. Since the resulting light weight panels can be hoisted in place with a small crane, or ganged together for hoisting by a larger crane, erection time can be substantially reduced. Conventional precast concrete panels weight about 60–65 lb/ft^2, while CFRC panels weight about 21 lb/ft^2, and even with a stone cladding the CFRC panel will weigh in at about 41 lb/ft^2.
- CFRC panels allow greater architectural design freedom, since they can be formed into sleek, narrow shapes with slender mullions to create delicate designs; also, because of the inherent strength of the

®Registered trade mark of E.I. Dupont.

product it can be faced with stone, ceramic, or metal and still remain relatively lightweight.

- Concrete made with carbon fibers is much more weather resistant and is more durable than ordinary concrete. Resistance to water penetration is much superior with the use of CFRC products.
- Since CFRC panels are substantially thinner than conventional concrete panels, more rentable or usable interior building space is created when this material is used.

Carbon Fibers—What Are They All About?

This lightweight, man-made material was first produced in the United States, where it has been used extensively in military and space applications. Carbon fibers possess characteristics that can be immediately recognized as compatible with concrete construction use. When added to readymix concrete, carbon fibers impart superior tensile and flexural strength to the mix, creating a concrete product that is both tough and ductile, durable, and dimensionally stable, particularly when it is steam-cured.

By allowing much thinner but stronger sections to be manufactured, carbon fibers added to concrete produce a much lighter weight end product. Just one look at the comparative tensile strengths of various concrete reinforcing materials is enlightening:

½″ wire prestressing cable	36,000 lb break strength
welded wire mesh	70,000-80,000 ft-lb/inch²
billet steel (rebars) #3's to #18's	70,000-80,000 ft-lb/inch²
carbon fiber continuous filament (0.05 mm diameter)	200,000 ft-lb/inch²

There are two types of carbon fibers in general use in the construction industry today: pitch based carbon fiber, sometimes refered to as GP-CF (general purpose carbon fibre), and high modulus PAN-type material, refered to as HP-CF (high performance carbon fibre). The properties of both types of carbon fibres reveals their differences:

Fiber Type	Specific Gravity	Tensile Strength	Modulus of Elasticity	Elongation, %
Pitch	1.63	78	3.8×10^3	2.1
PAN	1.90	320	23.0×10^3	1.4

The leader in the field of carbon fiber reinforced concrete seems to be Kajima Corporation, who began extensive research on this product in 1980, and as early as 1982, found a commercial application for the product when approached by the government of Iraq. Since that time, Kajima has successfully manufactured and erected hundreds of thousands of square feet of CFRC architectural wall panels and they have even created a unique CFRC hollow section window sill to be used in conjunction with through-the-wall air conditioning units.

The CFRC Window Sill. Through-the-wall units poking through the exterior wall of a building are not looked upon too favorably by designers, who associate the look with motel construction. The ventilation unit needs to draw in outside air and a louver in the wall allows this to occur. Kajima designed and built a hollow core, CFRC window sill that allows the through-the-wall unit to take in air from an opening in the underside of the deep sill while keeping out rain. The profile of this CFRC sill is a continuous band approximately one foot deep, and is certainly more pleasing to the eye than a louver.

Al-Shaheed—A Case Study in Japanese Research and Development

Perhaps the most dramatic installation of CFRC was the very first that Kajima attempted, that of placing small panels on a segmented teardrop shaped structure in Bagdad, Iraq known as the Al-Shaheed monument. The story behind Kajima Corporation's involvement with carbon fiber reinforced concrete panels for this unusual monument might well serve as a case study in Japanese construction research and developement methods and operations.

When the Government of Iraq decided to build the Al-Shaheed Monument near Bagdad (Fig. 11-5) their original design involved a structure composed of two egg-shaped half domes, each with a height of 131 feet and a base diameter of 147 feet 6 inches. The domes were to be built with a galvanized steel structural frame onto which 103,000 square feet of gold plated copper panels were to have been attached as cladding. These gold-plated panels proved too costly at that time even for this oil-rich country, and they began to look to innovative contractors who could provide them with another type of distinctive cladding.

The structural design had been completed when the Iraqi government made the decision to switch from gold-plated copper panels to Turkish blue porcelain tiles. The galvanized steel frame had been designed to receive a cladding material weighing not more than 12 1/4 pounds per square foot, and the tiles that the Iraqis had selected weighed 6 3/4 pounds per square foot. Hence the backer panel bonding the porcelain tile to the steel frame could weigh no more than 5 1/2 pounds per square foot.

Kajima Corporation was called in to discuss the possibility of their

Fig. 11-5. Bagdad's Al-Shaheed Monument, a case study in R&D. (Top) Overall view of the monument under construction in Iraq. (Bottom) Al-Shaheed Monument completed.

becoming involved in the project since CFRC was one of the materials under consideration as a substrate. Kajima met with Iraq's engineering consultant, O.V. Earup and Partners, who advised the builder that there were other design criteria that must be considered:

- The panels and panel fastening system must withstand wind loads of 61 psi.
- The strength of the cladding should not be reduced after it has been cured in 167° water for a period of 28 days.
- Reduction of the dynamic modulus of elasticity after 300 freeze-thaw cycles must be less than 10%.
- Drying shrinkage of the backer concrete panel must be less than 6×10^{-4}.
- Bond strength of the tile must not be changed after a repeated heat cycle of 140°F × 6 hours, followed by a 16 hour period of temperatures reduced to 14°F.
- The tile panels had to be installed without using heavy equipment.

Kajima engineers, in conjunction with the Kajima Institute of Construction Technology, went to work. They checked out the mechanical and physical properties of various types of concrete products to arrive at the optimum material selection for the project. They reviewed the properties of six types of concrete product before selecting CFRC:

Type	Specific Gravity	Tensile Strength (kg/cm^2)	Elongation, $\times 10^{-6}$
Ordinary concrete	2.3	20	150
Lightweight Concrete	1.8	20	200
Ordinary concrete—steel fiber reinforced	2.4	30	300
Lightweight SFRC	1.2	15	300
Lightweight CFRC	1.0	40	4000
Glass fiber reinforced concrete	2.0	60	4000

It initially appeared that two products, CFRC and GRC, might be considered for the task, but Kajima reckoned that they would have to manufacture the concrete panels in Japan and ship them to Iraq by boat. When these panels were in transit they could warp as shrinkage occurred during shipment, and warpage would change the panels' curvature, rendering them unacceptable. Kajima would be able to prevent warpage by high-pressure steam curing at the factory, and since glass fiber reinforced concrete could not be autoclaved, the choice of CFRC became an easy one.

Now that a material had been selected, Kajima engineers and researchers back at KICT would begin a long period of testing to insure that the CFRC panels would meet, or exceed, all of the design criteria.

Since the tiles would be subjected to summer temperatures of 140–150°F in Iraq, and below freezing temperatures in the winter, Kajima would have to conduct some rather elaborate tests. Dozens of sample panels were made, some of which were ambient cured and others autoclaved. After curing, all of the panels were immersed in water for three days and removed from their water bath before being subjected to 300 cycles of temperature changes from -2°C to 50°C using the ASTM C666 standard as a test guideline.

A shrinkage test would be the next order of business, since any appreciable shrinkage of the panel after installation might cause the porcelain facing to become loose or dislodged and eventually separate and fall off. Kajima recorded dimensional changes in their CFRC panels starting at manufacturing day 1 and continuing until day 365. Continual measurements confirmed their calculations that dimensional variation in the product was barely perceptible.

Next would come the bending test to determine if the panels could stand repeated positive and negative wind loads of 61 psi. Two panel sizes were tested, an A panel measuring 19½ inches wide, 39¼ inches long, and 1½ inches thick, and a B panel measuring 39¼ inches by 78¾ inches, also 1½ inches thick. Both samples were additionally reinforced with one layer of stainless steel wire mesh with a grid pattern of 25 mm (0.9 inch) by 3 mm diameter wire. Although both A and B panels tested satisfactorily, the A size panel would be selected since it would be easier to cast, cure, and handle.

The quality of an exterior panel is only as good as the method of fastening to its structure, so Kajima began exhaustive tests on a number of types and configurations of fastening devices. They finally selected a six embedment fastener arrangement for each CFRC panel. These embedments would be securely attached to a metal furring strip, which would in turn be bolted to the galvanized framework. A two point load was applied to the test panels with this fastening system and a universal testing machine with a 100 ton capacity began to impose loads upon the test panel. Deflection at the center of the panel would be measured with a dial gauge. Extensive testing revealed that even when two embedded fasteners in the center of the panel failed, the remaining four would provide holding power that exceeded the design load bending strength by 19 times. When sufficient loads were imposed upon the panel the tiles eventually cracked in several places, but they did not become dislodged.

The CFRC panels had passed every test so far with flying colors.

Kajima research engineers expected that the interfacial bond strength between the tile and the CFRC panel would be reduced after being subject to bending stresses imposed by wind loads and thermal shock. To convince themselves that there would be an acceptable bond strength after a prolonged period, the engineers conducted a heat-cool cycle, repeated 40 times, as they subjected the panels to temperatures of 140°F for 6 hours and then cooled them to 14°F, where they remained for 10 hours. After receiving this treatment, the panels were put through a bending fatigue test 10^6 times, and when this last test proved satisfactory, the Kajima research and engineering team gave the product their stamp of approval, and the company set about the task of completing the project.

Other CFRC Projects

Since the early days of Al-Shaheed, Kajima has successfully completed a number of other projects using CFRC panels. The most visible is the 344,000 square feet of thin-spine curtain wall installed on the high rise buildings at Ark Hills, one of Tokyo's architectural showpieces, located in the Akasaka-Roppongi section of the city. These delicately finned concrete panels measure close to 5 feet in width and are 12 feet high, weighing slightly more than one ton. At the time this building was under construction the owner, Mori Building, was advised that the CFRC panels would be premium priced, but a savings in structural steel tonnage would be anticipated since a lighter frame would be required.

Kajima calculated that normal weight PC panels would demand a 26,500 ton steel structure, but with the use of CFRC, steel tonnage could be reduced to 22,000 tons. There would be a small additional savings because these lightweight concrete panels could be hoisted in place with a small electric hoist rather than a more expensive tower crane. Although the CFRC panels represented a 40% premium over conventional panels, the owner realized a $1.6 million net savings when the lighter structural steel framing system was taken into account.

Mitsui and Their Super Fibers

Mitsui Construction Company has developed a series of super fibers for concrete reinforcement ranging from polyethylene to polyester to both types of carbon fiber. Mitsui uses both long fiber and short fiber embedment depending upon the end use of the product.

Short fibers, about 5 mm long, called Fibra-Chop, when added to a concrete matrix, can impart bending strengths 150–200% higher than concrete with reinforcement. Long fiber technology, pioneered by Mitsui, involves braiding one of several exotic man-made materials into "ropes" called Fibra-Rod or Fibra-Bar. Dramatic increases in concrete strength occurs as these Fibra-Bars or Rods with tensile strengths of 57,000–500,000 foot pounds are added to the end product. Mitsui has also developed rigid reinforcing rods of Kevlar® fiber, and one of these rigid Fibra-Bars with a diameter slightly less than 3/8 inch, weighing about 1.2 ounces per lineal foot, has the same tensile strength as a conventional No. 5 deformed steel billet rod having a 5/8 inch diameter and weighing 1.043 pounds per lineal foot. Similarly, a No. 3 steel reinforcing rod, with a 3/8 inch diameter and weighing 6 ounces per lineal foot can be replaced by a Fibra-Bar with an 0.0230 inch diameter (about the same thickness as 25 gauge metal) that weighs somewhat less than one-third ounce per foot.

Mitsui uses these Fibra-Bars in the manufacture of precast underground utility pipe, among other uses. They form a cage of Fibra-Bars and place it in a centrifigal concrete molding machine. Pipe manufactured by this method is very dense, light in weight with a thin cross section but having superior corrosion resistance properties. Fibra-rod is being used in prestress tensioning systems, and Mitsui, working with Shinko Wire, has developed several types of anchors when Fibra-Rods are used as tendons. A wedge shaped anchor and a resin injection type anchor have both been tested by Mitsui and anchoring efficiencies of 100% have been indicated for each device.

U.S. Concrete Technology in Comparison

American concrete technology seems to be geared toward the developement of highstrength and super-high strength concrete mixtures with compressive strengths ranging from 5,000 psi to 14,000 psi, reaching 20,000 psi when microsilica was added to the concrete used in a Seattle, Washington project in 1987.

As of mid-1988, neither the American Concrete Institute nor the Precast Concrete Institute exhibited any serious concern for the developement of carbon fiber reinforced concrete in the United States. There are a few manufacturers of synthetic fiber concrete reinforcement in this country, but for reasons of building code requirements or other limiting factors CFRC seems to be of little interest to American engineers and builders.

®Registered trade mark of E.I. Dupont.

WHERE IS JAPANESE CONSTRUCTION R&D HEADING?

Japanese construction research and developement spans the entire spectrum of building technology—from soils analysis to substructure and superstructure design testing; from concern for the environment to maximum utilization of the environment; from refinements of time-honored procedures to dramatically new techniques; from basic building materials to space-age exotics. These research and development efforts all seem to be heading in the same direction—toward nothing less than excellence in their field of endeavor.

The Ministry of Construction's Public Works Research Institute recently unveiled a prototype bulldozer that may become the first completely automated piece of excavating equipment. Produced in collaboration with Komatsu, Tokyo Construction, Fudo Construction, and Mitsui Construction, this revolutionary machine has been designed to be guided by lasers which control its position on the site and raise or lower its blade in order to grade to a tolerance of 1/2 inch. Although the Komatsu bulldozer's blade is laser controlled, it still needs a human operator—at least for a while.

When one looks at the strides that have been made by Japanese contractors through their R&D efforts so far, the future direction of construction might just be unfolding before our very eyes.

We have already seen Japanese technology control tower cranes and various types of automatic coupling and uncoupling devices, which could portend completely automated hoisting in the near future. Automatic concrete placement is just around the corner, and automatic concrete floor finishing is already here. Now robotic bulldozers trailing soil compaction robots seem to be rumbling into view. It may be increasingly difficult to compete in the world market with a construction company that can pack up its equipment and electronic work force and be ready to travel to any part of the world on a moment's notice.

12

Competing with the Japanese

THE IMPACT OF JAPANESE CONSTRUCTION ON THE U.S. MARKET

Except for a handful of American construction giants, Shimizu, Kajima, Ohbayashi, Kumagai Gumi, and the rest of the Japanese Big 6, will not pose much of a threat to the hundreds of thousands of small and medium size contractors in this country. Although the 1988 *Engineering News Record* list of Top 400 contractors in the U.S. lists Number 400 with an annual sales volume of $43.7 million, there are about 455,000 other general contractors out there with yearly sales under a million dollars. The comings and goings of the world's building behemoths probably will not touch any of these 455,000 plus contractors in any significant way, but those U.S. construction companies breathing the air of multi-million dollar contracts might see these competitors from a different perspective.

The Big Six already operating in this country and abroad can give them a run for their money, or the major American builders can learn to coexist peacefully with their distant relations from the East.

When Alfred T. McNeil, President and Chief Operating Officer of Turner Construction Company, was interviewed in 1987, he was asked what impact the Japanese would have on the U.S. construction market. Mr. McNeil said, "From our standpoint, it has been positive. The Japanese are becoming substantial investors in the U.S., especially in the major cities that have direct links to Tokyo. Since they have different investment criteria than American investors, they may continue to support markets after U.S. investors have dropped out. This is one of the reasons the construction business continues to be good in many major cities."

In mid-1988 Turner Construction announced that they had formed an association with Kumagai Gumi, with whom they had worked before, and together they are going to pursue selective construction projects in the United States. Turner's association with Kumagai Gumi will undoubtedly enhance their ability to obtain construction contracts from Japanese clients in this country, and this new found collaboration is expected to open doors in the domestic Japanese construction market. One can be certain that Kumagai Gumi's substantial financing sources will be put to good use as they walk across America, hand-in-hand with their new found partner.

But big or small, everyone in the construction industry can learn something from these builders from Japan, and depending upon what we learn and how well we implement what we learn, the Japanese construction industry may have a substantial impact upon the way we conduct business in this country.

If Japanese contractors can show American builders and their clients that by working together and establishing long term relationships, ultimately both parties will benefit, then these Japanese contractors will have had a major impact on the U.S. construction industry.

If Japanese contractors can teach U.S. builders the value of developing long term, harmonious relationships with subcontractors in a nurturing environment rather than an adversarial one, and a free exchange of information is created between them then the Japanese contractors will have had a major impact on the U.S. construction industry. If Japanese contractors and their designers can teach American designers to say "No" to clients demanding a fast turnaround on design documents without sufficient time for careful review and coordination, we will have learned another valuable lesson with considerable impact on the U.S. construction industry.

If the Top 100 U.S. contractors dismiss the presence of the Big Six in this country as very little threat to their livelihood and future, then the Japanese contractors may have a major impact on the U.S. construction industry.

And finally, if the legacy of the Japanese contractors in America is to shake U.S. builders out of their lethargy and cause them to renew and revitalize their assessment of the country's entire industry, then, most definitely the Japanese contractors will have made a major impact on the U.S. construction industry.

WE HAVE MET THE ENEMY AND HE IS US

If and when a complete self-analysis of the U.S. construction industry comes about, the picture that will emerge will not be a very comforting one: an industry where productivity has been declining for the past twenty-three years; an industry producing a less cost effective product today than at almost any time within the past two decades. What will be revealed is an industry with a too-little, too-late management philosophy that has been in a state of almost constant confrontation with labor unions; an industry hobbled by a maze of government regulations in dire need of complete overhaul. To its credit, the construction industry is not made up entirely of Doom and Gloom mavens, standing on the sideline wringing their hands. There have been many intelligent thorough, and in-depth studies conducted by several industry groups within the past several years, and a few of these comprehensive surveys have highlighted the problems facing the industry. Sound solutions to many of the construction industry's problems have been offered, but nobody listens or seems to care!

THE BUSINESS ROUNDTABLE

In 1969, when inflation in the construction industry was reaching very high levels, Roger Blough, then Chief Executive Officer of U.S. Steel, was instrumental in setting up the Construction Users Anti-Inflation Roundtable, an organization that became a part of the Business Roundtable in 1972. Composed of 200 CEOs from all fields of American industry, this organization met periodically to discuss and develop positions on social and economic problems facing the United States.

In 1973, the Business Roundtable's Construction Committee studied the state of the construction industry in this country and issued the Construction Industry Cost Effectiveness Report. Twenty-three separate reports were finally prepared by CICE covering such topics as Measuring Productivity in Construction, Construction Labor Motivation, Modern Management Systems, Technological Progress in the Construction Industry, Constraints Imposed by Collective Bargaining Agreements, and other pragmatic approaches to industry problems and concerns. A great deal of thought and effort went into these studies, but they were not readily embraced by the construction industry.

Mr. Charles D. Brown, a DuPont engineer and *ENR*'s 1983 Construction Man of the Year, chosen for his leadership in preparing these reports, was interviewed at a National Construction Council Meeting in 1987. He said that in spite of the overwhelming evidence of problems in the industry, little more than a "dent" had been made in controlling construction costs since the CICE report had been published four years earlier.

THE CONSTRUCTION INDUSTRY COST EFFECTIVENESS REPORT—CICE

Construction productivity, almost a national priority in Japan, has been grossly neglected in this country. The CICE report indicates that construction productivity in the U.S. dropped from an index of 100 in 1972 to 82.9 in 1979—a decline of almost 20%. Since 1985, construction has been the only industry in the country with a consistently negative productivity growth, and CICE attributes more than 50% of the wasted time on construction sites to poor management, with organized labor sharing considerably in the problem.

One of the reasons for construction inefficiency may be the fragmented nature of the business and the multitude of organizations formed around it. Although there are about one million contractors in this country, 455,000 of them, more or less, are small concerns, either sole proprietorships or small working partnerships. There are about 70 national contractor organizations, more than 10,000 local and national union organizations, about 5 million construction workers and clients of every size, shape and distinction.

Company competes against company in a fiercely competitive marketplace; unions square off against management, open shop contractors vie with union contractors, clients war with contractors, and contractors war with clients, and if there is one word in the English language that best describes the nature of U.S. construction, it is ADVERSARIAL. A closer look at some of CICE's findings, and their suggested solutions, would be of value to any company concerned about losing business to a competitor from another part of town or from another part of the world.

MANAGEMENT OR LACK THEREOF

When King Cheops directed the construction of the world's first high rise in Egypt in 3000 B.C., his management concerns with pyramid construction were easily solved. When productivity slips, behead the foreman, and beat the laborers a little harder. In some cases, today's management is no more enlightened than Cheops.

Modern management techniques have been adopted by construction companies very slowly, and industry critics feel that the ever-present delays in completing a project, and the cost overruns that follow, are directly attributable to poor management techniques. And the obverse is true, sound management practicies can have a profound effect on many of the problems that normally plague contractors:

- Safety—That old bugaboo sometimes receives only half-hearted lip service from management, followed by a flood of What-Not-To-Do posters.
- Scheduling—This critical process often gets filed away once the first requisition for payment has been approved.
- Training—Can important decisions be made by personnel so poorly trained to do so?

- Motivation—This is the key to higher productivity; conversely, demotivation is the key to lower productivity.
- Labor—Management must work closely with unions or merit shop organizations in order to obtain quality oriented, skilled, and productive work crews.

The Construction Industry Cost Effectiveness Report takes a close look at each one of the factors affecting productivity, the number one concern of the industry today. The race to achieve greater productivity, already in progress in Japan, certainly needs the attention of the construction industry in America, and CICE's observations bear some attention.

SAFETY

Construction is one of America's most hazardous industries. For the period 1980–1984, the National Institute for Occupational Safety and Health (NIOSH) lists construction as having the second highest fatality rate in the country, reporting 23.1 deaths per 100,000 workers. Only the mining industry with 30.1 deaths per 100,000 workers ranks higher than the building business in on-the-job fatalities. In New York City alone, between 1979 and 1986, 81 construction workers lost their lives on the job. In nearby Bridgeport, Connecticut, the collapse of the L'Ambiance Plaza project took the lives of 28 construction workers in 1987.

Construction accounts for about 10% of all occupational injuries, and if the actual annual direct and indirect costs of these injuries, constituting 6.5% of all yearly industrial, commercial, and utility construction volume, were totalled, poor safety performance in 1988 alone would have a $10 billion price tag.

Accidents will happen because of carelessness, clumsiness, or just plain human error, but the frequency and severity of accidents can be controlled by proper management involvement, as witness the dramatic decline in Japanese construction accident rates once safety became a national priority in that country.

There are many sound safety programs available for contractors and they all contain the same basic principles:

1. Establish management committment by assigning top managers to head up the program.
2. Establish a job and safety program for both new and long term employees.
3. Audit all accidents promptly to determine the cause and possible elimination of unsafe conditions.

4. Provide proper safety equipment, and monitor its proper use.
5. Make safety a key ingredient in the construction process and create and maintain safety awareness.

SCHEDULING

The preparation of a construction schedule is a complex task that begins life on the estimating table and completes its job when the Certificate of Occupancy has been obtained, but it takes many strange side trips along the way. Construction schedules are dynamic, and sometimes major midpoint corrections must be made as landmarks are updated or shifted during the course of a project.

Any number of forces impact on a schedule and some of them are unavoidable. Some changes are due to the complex nature of the construction process, but some are created by crisis management and poor control over subcontractors and material suppliers. Effective scheduling needs to include the following considerations:

- Landmark dates established for key building component completions.
- Actual versus estimated progress monitoring.
- "What-if" scenarios so that any unexpected delays might be easier to cope with.
- Frequent and realistic schedule revisions to determine how project completion is being affected.

Scheduling is also a contractor state of mind. Among Japanese contractors, maintaining a project schedule is of prime importance because the builder has in effect staked his professional reputation on the ability to meet committment dates given to the client at the outset of construction. Failure to adhere to the schedule is viewed as a serious breakdown in the construction process, and in the case of public projects may even affect the contractor's annual performance rating.

TRAINING

Would you fly with a pilot who had been taught the rudiments of flight but had never actually flown a plane? You bet you wouldn't, but thousands of contractors, every day, entrust their family fortunes to managers that are not adequately trained for their job. Executives in the construction industry, particularly project managers, generally come from an engineering background, and rightly so. However, construction is a *business*, run by businessmen in a competitive environment, in which key business decisions play a critical role in the day-to-day operations of a company.

How well do these executives and project managers understand basic accounting practices, management theory, and decision-making procedures? Do any of them have a basic knowledge of construction law and how best to avoid construction claims? Do they know how to prepare for potential disputes? The answer, in most cases, is probably No! Productivity, or lack of it, is a major problem in the construction industry, but how many managers have had training in productivity measurement and improvement? The answer again, is probably None!

As we have seen from our brief look at Japanese construction companies, they put great emphasis on management training and increasing management skills, but we Americans don't seem to see the same value in providing our people with broad-based knowledge of the building business.

MOTIVATION

Motivation is a key ingredient in better productivity, and the need to motivate managers and workers alike becomes more critical each year. Motivation is what puts fire in the belly and an impetus to meet and exceed goals, and motivation involves rewarding those individuals who deal effectively with these goals.

Motivated people can be rewarded in any number of ways, from promotions and wage increases to bonuses, shared employee stock option plans, or sometimes just a plain "Thank You." People feel part of a program when they are allowed to voice opinions, and the Japanese do this very effectively by encouraging employee suggestions and forming quality circles to discuss way and means to increase productivity and quality. By allowing employees to identify and solve problems through employee participation, the motivational process is greatly enhanced.

The University of Texas conducted a survey among 1,000 construction workers employed on twelve large industrial construction sites in 1978, and although somewhat dated, it appears that the same things that turn workers on and, conversely, turn them off are still applicable today. (See Table 12-1)

LABOR AND THE COST OF RESTRICTIVE WORK PRACTICES

In 1980, the Business Roundtable retained the services of the Construction Labor Research Council to conduct a survey of excessive cost provisions in local building trades agreements. About 20% of all U.S. construction collective bargaining agreements in effect at the end of 1979 were examined, and certain wage and hour assumptions were

Table 12-1. Construction Worker Motivators and Demotivators. ***(Source: Construction Industry Institute.)***

Survey Results of Motivators on Twelve Construction Projects

	PROJECTS											
MOTIVATORS	A	B	C	D	E	F	G	H	I	J	K	L
Good craft relations			1*			2	2		1	1		
Good orientation program	3	3	1		1	1	1	2				2
Good safety program		1*	1	1*	2*	3		2*	2			
Work itself	2		2	2	2		3		2			2
Overtime					3		3					
Pay			2	1	1	2	2	2	2		2	2
Recognition				1*	1*							
Goals defined			1	1	1		1					
Open house & project tour	2	1		2								
Well-planned project						2*	2*	1*				
Suggestions solicited			1*		1*	1		2*	1			

KEY: 1 = somewhat important
2 = major importance
3 = extremely important

* = demotivator and motivator

Survey Results of Demotivators in Twelve Construction Projects

	PROJECTS											
DEMOTIVATORS	A	B	C	D	E	F	G	H	I	J	K	L
Disrespectful treatment		3	3	2		1			3	3		2
Little accomplishment			1		2					2		
Material availability	2	2	2		3	1	1	2	1	2	1	2
Tool availability	2	1	2		2			1	1	1	1	2
Redoing work	1	1	1	1	1				1	1	2	2
Crew discontinuity	1		2		1		2					2
Project confusion	1	1	1	1	1	1	1	1	1	1	1	1
Lack of recognition	2	2		1*	1*		1	1			1	
Productivity urged but no one cares	1	1	1								2	2
Ineffective utilization of skills	1		1				2	1			2	
Incompetent personnel	1	1	1	1	1		1	1	2	2	2	2
Lack of cooperation among crafts	1		1*					1			3	3
Overcrowding		1			1							
Poor inspection programs		1		1					1	1		
Communications breakdown	1	1	1	1	1	1		2	2	2	2	2
Unsafe conditions	1	1*		1*	2*		2	1*		2		2
Lack of participation in decision making	1	1	2*	1	1*			1				

KEY: 1 = somewhat important
2 = major importance
3 = extremely important

* = demotivator and motivator

made. An hourly wage of $14.50 was assumed along with an average annual work year of 1600 hours (200 days') duration. Overtime was assumed to have been worked at one time or another during the year and was factored into the equation.

The study revealed eight labor provisions that were responsible for adding substantial costs to construction work, and they listed each one of them with an estimate of their annual cost:

Cost Impact of Eight Restrictive Work Practices

Overtime premiums	37% of the labor contracts examined called for double time premium pay for daily work after 8 hours, while 63% stipulated time-and-one-half pay for overtime. 48% of these contracts required double time pay for Saturday work. Annual excess cost: $76 million (daily overtime) $51 million (Saturday overtime)
Time paid, not worked	Cleanup is part of the construction workday, as is clean-up for personal reasons (i.e., Painters). Coffee break time in the morning and sometimes afternoon is as American as Apple Pie, but does it have to be written into contracts? It has been. Annual excess cost: $42 million (Job site pickup time) $16 million (Cleanup time) $17 million (Coffee breaks over 10 minutes)
Subsistance-travel pay	Subsistance pay, pay provided as compensation for working away from home, and travel pay from an arbitrary point to the job site have all been abused in many contracts. Annual excess cost: $42 million (subsistance pay) $29 million (travel pay)
Shift provisions	75% of the contracts reviewed contained provisions for premium pay for second and third shifts. Annual excess cost: $41 million (second shift) $29 million (third shift)
Hours of work	Although an 8 hour day seems to be universally accepted, some unions negotiated less than 8 hour days, creating gaps between craft working hours and requiring standby work. Annual excess cost: $58 million
Crew size restrictions	20% of all contracts reviewed contained specific work crew sizes for certain tasks. It was estimated that if management could control crew sizes, there would be 10% fewer members. Annual excess cost: $42 million

Off-site fabrication restrictions	Restrictions were placed upon bringing certain prefabricated assemblies to the job site so that work for on-site tradesmen would not be jeopardized. Annual excess cost: $30 million.
Showup pay	87% of all contracts reviewed contained provisions for paying workers who are not put to work when they show up on the job awaiting a change in weather, etc. Although most open shop contractors do not have showup pay, they do try to provide at least two hours work, minimum, for those showup workers. Annual Excess Cost: $27 million

These restrictive work practices imposed by collective bargaining agreements have proven costly to contractors and unions alike. The cost to contractors is obvious; the cost to unions has been steadily declining membership rolls, while open shop and merit shop operations have grown by leaps and bounds.

There are a number of other inefficient labor practices which were discovered by the Construction Labor Research Council when they conducted their study. Many of these inefficiencies were not specifically negotiated at the time the collective bargaining agreements were being discussed, but many came about by pressure exerted by local unions and allowed to occur and continue in practice by the permissiveness of lazy management. Many of these practices prevail in the collective bargaining atmosphere, and many could also be creeping into the open shop environment.

The following checklist of 57 inefficient labor practices might be a good starting point for a management awareness session. The next time a job-site visit is planned, this list ought to be carried along and any of the inefficient practices observed duly noted.

INEFFICIENT LOCAL LABOR PRACTICES

Nonproductive Work Time

1. Late starts.
2. Early quits.
3. Excessive time for washup and putting away tools.
4. Unauthorized breaks.
5. Place of work is the change shack, or company property lines.
6. Abuse of visits for medical aid.
7. False weather excuses.

Additional Time Payments

8. Guaranteed overtime for specialty craftsmen.
9. Overtime for total craft if anyone works overtime.
10. Welder qualifications pay beyond time spent testing.
11. 40 hour guarantee for general foreman, foremen, or steward.

Premiums, Travel Pay, Incentives, etc.

12. Wage payments above basic rate for welders, instrument men, or foremen.
13. Contract welding rigs.
14. Apprentice payments above classification.
15. High time hazard pay, clothes repayment.
16. Pipe sketchmen at foreman rate of pay.
17. Overtime lunches and eating on company time.
18. Travel pay and/or subsistence.

Employment Practices

19. Restrictions on contractor's choice of foremen and general foremen.
20. Exclusive use of union hiring hall.
21. Pressure to lay off nonlocal personnel first.

Work Restrictions

22. Restrictions on equipment and materials obtained from nonunion sources.
23. Restrictions on repairs by specialty firms.
24. Unnecessary limitation of work within crafts.
25. Deliberate slowdown of work.
26. Maximum number of welds per day, or similar productivity restrictions.
27. Limitation of type of work by apprentices.
28. Restrictions on use, erection and dismantling of scaffolding.
29. Use of more equipment than needed.

Jurisdictional Disputes

30. Composite crews with more men than necessary.
31. Payment for work not performed.
32. Failure to proceed with work while waiting resolution.

Steward

33. Nonworking stewards.
34. Remain for overtime work when not essential.
35. Remain after rainout.
36. Building trade stewards on large projects.
37. Involvement in hiring and terminations.
38. Payment over scale, or other favors.
39. Contact with union for manpower requirements.
40. Last one in layoff.
41. Involvement in disputes with other than employer.

Manning Requirements and Standby Time

42. Teamster drivers on job site pickups.
43. Nonessential material checking and receiving.
44. Journeyman pipefitter to work with welder.
45. Welder refusing to work without helper.
46. Use of survey crew for layout when workload does not justify.
47. Foreman pay where there are manning cutbacks.
48. Minimum numbers on rigging crews.
49. Crews on elevators.
50. Master mechanics by ratio system.
51. Lack of apprentices.
52. Crew sizes demanded outside of labor agreement.
53. Excessive numbers of supervisors.
54. Excessive levels of supervisors.
55. Unneeded personnel for cleaning, testing, and startup.
56. Observing work of technicians or client personnel.
57. Temporary heating and lighting.

On average, according to CICE, these practices inflate labor costs in unionized construction by as much as 15%. Many of these inefficient labor practices came about because of union pressures, but blind acquiescence by contractors can also be held accountable for many others. Labor peace at any cost can be expensive.

THE PRODUCTIVITY ISSUE

There are many voices being raised in America about the decline in productivity, the existing inefficiences in the construction industry and the lack of concern for higher quality standards.

Mr. Duke Nielsen is one of those concerned individuals. Mr. Nielsen is president of Performance Systems, Inc., a Littleton, Colorado firm that specializes in advising contractors on construction productivity improvement. Duke was the former Director of Education and Training for the Colorado chapter of AGC, and in that capacity, developed a number of workshops for construction personnel.

Since establishing his own company, he has been conducting workshops around the country which stress the need and methods by which productivity goals can be effectively communicated between managers, first and second line supervisors, and the field worker. In his handbook "Performance Engineering," Nielsen indicates that thousands of productivity studies have been conducted over the past 15 years. The data for these studies had been collected by observers on construction job sites who tracked the activities of workers during an 8 hour shift to determine how much time is being spent on each activity. Duke divides a normal work day into three segments: (1) direct work; (2) waiting time (waiting for instructions, other crew members, and waiting for no specific reason); and (3) moving, but not in direct work (material handling, personal reasons, going after things).

The results of these time studies, according to Duke Nielsen, reveals that in an eight hour work day these three daily work segments break down as follows:

Direct Work	31.9%	of an eight hour day		2.55 hours
Waiting Time	31.1%	of an eight hour day		2.49 hours
Moving	37.0%	of an eight hour day		2.96 hours
	100.0%		Total	8.00 hours

Nielsen cites a number of reasons why, in his opinion, productivity has declined in recent years:

1. More administrative functions have been added to the daily routine of field supervisors which leaves less time to plan activities, organize, and assign productive work and monitor job progress.
2. The trend among general contractors to subcontract more of their work removes the general contractor's superintendent from direct control of these subcontracted work crews.
3. There have been changes in the worker's responsiveness to a superintendent's authority. Previous autocratic leadership produced results, but this form of supervision turns workers off, and now supers and other first and second line supervisors need to be retrained in new methods of motivating today's new breed of workers.

LILCO

If Duke Nielsen's direct work productivity figures seem to be an eye opener, consider another productivity study conducted at the Shoreham Nuclear Power Plant site located on Long Island's North Shore, about 55 miles east of Manhattan. The Long Island Lighting Company announced plans to build a nuclear power plant on Long Island in 1966, but construction did not start until 1973 because of years of hearings, appeals, more hearings, and more appeals by concerned citizens and various government groups. By the time construction actually started, the initial cost estimate of $70 million had risen to $506 million and the project completion had been set for 1977.

In September of 1977, when it appeared that construction costs would exceed $2.1 billion and completion might be extended until 1980, Lilco directors conducted a survey to determine why costs had increased so dramatically and why the completion time had been extended. The study revealed that although Lilco had hired an experienced, nationally known, and well respected construction management company to build the facility, the average worker at the Shoreham plant spent only 20% of a typical seven hour day actually performing productive work. Could this have been one of the major reasons for constantly updated cost estimates to the point that when finally completed and ready for low level testing in 1985, the project carried a price tag of $4 billion?

The Shoreham plant never did reach functional operation and due to tremendous public safety concerns, the nuclear power plant was never opened. In the summer of 1988, plans were announced to dismantle the entire project.

THE NEED TO ADDRESS THE ISSUE OF QUALITY

On his way to the office in Manhattan, a top executive from one of Japan's Big Six construction companies told the writer that he is always watchful for signs of quality construction as he walks in the city. While waiting at a stop light on a busy corner near his New York office, he glanced up at a polished granite curtain wall on a relatively new skyscraper. With his trained eye, he drew an imaginary plumb line up the side of the building to quickly assess the quality of the stone panel installation. Some of the panels, in his opinion, were protruding at least 3/4″ beyond what he felt was an acceptable level. As he related the story, he smiled and said, "That's not the Japanese way."

The quest for quality is something most people strive for—everyone wants to be told that they have done a good job—but over the years, there may have been a number of roadblocks thrown up in the path of the pursuit of quality. Although contractors share in the quality problem, they are by no means the sole culprit. In the sophisticated economy in which we live, the use of money has become an ever increasingly complex science. Real estate projects are highly leveraged and are targeted to enter the market with surgical precision. The tendency is to wait until the right moment, accelerate the design development, and produce bid documents or working drawings as rapidly as possible. Fast track is the name of the game as missing spaces on the design drawings will be filled in as construction proceeds, and added costs due to design "creep" are either traded off or absorbed in some form or fashion. General contractors are pressured to hold the line on budget estimates, and they in turn pass on these same restrictions to their suppliers and subcontractors, until the rule of the day is "Get it up and get it up fast!"

But quality in construction is becoming a buzz word these days as people in all fields of endeavor begin to realize that quality is not synonymous with cost, it is a state of mind.

THE MANUAL OF PROFESSIONAL PRACTICE FOR QUALITY

Mr. James W. Poirot, chairman of Denver based CH2M Hill Cos., Ltd., one of the nation's largest design firms, was designated Man of the Year in 1988 by McGraw-Hill's *Engineering News Record* magazine because of the role he played as the guiding force behind the American Society of Civil Engineers's "Manual of Professional Practice For Quality in the Construction Project." This somewhat controversial book is meant to show the way toward injecting quality back into the construction project. Mr. Poirot has had his share of detractors along the way, but he and the ASCE have taken a positive step toward the promotion of quality in the field of construction.

In the "Preliminary Edition for Trial Use and Comment," distributed to professionals throughout the country in 1987, *quality* is defined as meeting the user's requirements as to "functional adequacy, completion on time, and within budget, life-cycle costs and operation and maintenance." Quality from the design professional's viewpoint is explained as the ability to produce a set of construction documents that create a well defined scope of work by qualified an experienced staff, incorporating adequate field information. The preliminary edition of the ASCE manual states that "Providing quality in the design, construction and operation of a project costs less, is safer to the public and site personnel, and minimizes disagreements and subsequent litigation. Quality in design and construction results in significant benefits for the owner, design professional and constructor."

There is at least one general contractor out there who wholeheartedly believes in the Poirot Doctrine, and that is the J.E. Dunn Construction Company of Kansas City, Missouri, a company that created their own quality control program some ten years ago.

THE DUNN CONSTRUCTION COMPANY QC PROGRAM

Quality control has been a tradition at the Dunn Construction Company in Missouri. Mr. Lou Hilker, Senior Vice President of the company wants every employee to feel responsible for creating the best possible building allowed within the budget constraints. Mr. Hilker and CEO William H. Dunn, Senior feel that this policy would also result in increased business as the company's reputation for quality construction spread throughout its operating area.

Apparently this decision was a correct one, since Dunn's annual sales from 1985 to 1988 exceed $200 million and 70% of this volume has been negotiated work. J.E. Dunn believes that quality construction will actually lower costs over the years, based upon the simple premise that it is cheaper to do it right the first time. In the seven years prior to Dunn's formal adoption of a quality control program, the company experienced one callback for every four completed projects, but after their QC program got off the ground their rate of callbacks decreased by 65%. Dunn executives found that quality control brought another unexpected bonus—their insurance rates dropped. General liability insurance rates were lowered to $.12 per thousand as compared to an average of $.40 per thousand for contractors with $50 million or more annual volume. Dunn's excess liability and completed operations insurance premiums were both 70% below the average rates for corporations of comparable size.

When Dunn's quality control program was conceived, the executives at the company made the following decisions:

- The QC program would be controlled from headquarters, not from the site.
- The program would be headed by a full time, experienced person, whose sole corporate function would be QC, and this person would report directly to top management.
- QC responsibility would involve working with company employees, suppliers, subcontractors, design consultants, and owners.
- The cost of the QC program would be considered an item of corporate overhead.

- If it appeared that the company could become involved with a project loaded with problems, they would turn down the job rather than compromise their standards.

Once the program's objectives had been defined, the actual formulation had to be started and procedures had to be created to deal with design review, value engineering, general construction planning, material acquisitions, equipment management, field and weather conditions, workmanship, safety, inspections, and employee attitudes. Policies had to be established for every step along the way, starting with design review of the new project under consideration.

Dunn set up procedures so that a very detailed review will be made of a project's plans and specification before the work begins. Structural designs would be reviewed by outside consultants to detect any potential problems that may arise during construction. In some quarters this might raise the hackles of the original designer until the consultants begin to realize the value of a peer review.

J.E. Dunn established a set of minimum standards for most components of construction: paving, concrete work, interior and exterior wall construction, roofing, waterproofing, HVAC, and other mechanical and electrical systems. These standards are included in procedures manuals and are used when various phases of work are being inspected by their quality control manager.

Dunn credits the use of full scale building mockups as a surefire method of pinpointing potential problems in complex or difficult construction components before they are actually put into place. On one occasion, a full size mockup of a glazed panel section showed the subcontractor how to properly and effectively caulk the glass to prevent leaks. During construction of a hotel they were building in Oklahoma, Dunn built a complete bathroom mockup which even included the tiled walls. Unexpectedly, this full scale model revealed a potential problem with the ceiling height and the wall tile joining details, which was corrected on the drawing board instead of in the field. A full scale mockup of a complicated exterior masonry wall at the Vista International Hotel project in Kansas City (See Figure 12-1) helped to clarify a complex and repetitive assembly.

The interaction that Dunn creates between contractor and subcontractor creates a strong bond, even though they may from time to time require their subs to redo portions of work that fail to meet acceptable quality levels. These subcontractors soon come to realize that they too have fewer callbacks, and that once the job is done properly, there is little need to return to the site at a later date to repair shoddy work.

Although J.E. Dunn is large enough to absorb the cost of a full time quality control manager into their corporate overhead, V.P. Hilker sug-

Fig. 12-1. J. E. Dunn's full-scale mockup of a masonry exterior wall at the Vista International Hotel project in Kansas City. *(Photo furnished by J.E. Dunn Corporation.)*

gests that much smaller firms can get involved in a QC program by assigning someone to the quality control function, even though it may be only on a part time basis. But Hilker suggests that this QC responsibility be placed upon the most experienced, dependable person in the organization, and that this QC manager is provided with the time and flexibility to get the job done right. The end result of J.E. Dunn's effort, according to Hilker, is that their reputation for quality has been extremely helpful in attracting clients with quality projects. This philosophy mirrors very closely that of the major Japanese construction companies and shows that quality knows no international boundaries, both in its recognition and its practice.

Although major U.S. construction companies may not make the same commitment to investment in research and development as the Big Six Japanese builders have done, in one shape or form construction R&D needs to become a national priority in this country. In the world of international construction it may determine who will be the future Top 50 contractors.

CONSTRUCTION RESEARCH AND DEVELOPMENT IN THE U.S.

Research and development in the construction industry in this country is being conducted at many levels by private and public organizations alike, but what seems to be lacking is a central focal point for all of the work being conducted by dozens of groups throughout the country. The federal government's role in furthering construction research has been minimal at best. In 1980, for instance, the government funneled $620 million in federal funds to universities conducting engineering research, but only $1 million found its way into construction research projects. Private donations during the same year yielded another $800,000 for academic research at the college level, but altogether this seems a rather paltry sum from one of the country's leading industries.

There have been deterents to technological advances in construction in this country as Table 12-2 below indicates. Some of these barriers are formidable, but they should not be looked upon as being immovable.

Encouraged by the growing awareness of contractors, designers, material and equipment suppliers, and also owners, several prominent organizations have responded to the call to increase levels of construction research and technology.

THE CONSTRUCTION INDUSTRY INSTITUTE

The Construction Industry Institute, founded in 1983, is an officially designated research unit of the College of Engineering at the University of Texas in Austin, Texas. CII was founded in part in response to several recommendations made in the Construction Industry Cost Effectiveness Project sponsored by the Business Roundtable, CII obtains funding from its members, sixty five major contractors and industrial companies such as Bechtel, Guy Atkinson, Brown and Root, Fluor Daniel, A.T.& T.,

Table 12-2. Major Barriers to Technological Advances. *(Source: The Business Roundtable.)*

	Standards & Building Codes	Labor Agreements	Liability & Legal	No Profit Incentive	Adversarial Contractual Relations	Government Reguilations
Commercial Construction	2	3	3	3	2	3
Industrial Construction-Process	3	2	1	1	3	2
Industrial Construction-Other	3	2	2	2	2	3
Utility Construction	1	2	1	1	2	1

Key: 1 - Major barrier
2 - Sizeable barrier
3 - Minimal barrier

Dupont, Dow, Exxon, Monsanto, Mobil, General Motors, and General Electric. With membership dues of $25,000 annually, and a pledge to actively participate in CII committees and task forces, this is an organization with considerable strength. CII concerns itself with identifying problems in the construction industry, researching solutions to the problem via the creation of task forces, and then after their findings are published the organization aids in the implementation of their recommendations. The types of research projects being undertaken by their task forces can best be described by listing many of their projects, either recently completed or ongoing. See Table 12-3 (a), (b), (c). CII makes their research reports available to members and nonmembers alike, at very reasonable costs.

The Construction Industry Institute's efforts to find ways to make the U.S. construction industry more competitive recently prompted their executive committee to provide funding to cosponsor the Sixth Annual International Construction Robotic Symposium to be held in San Francisco in mid-1989. Attendance at CII's annual meetings indicates a growing interest in the organization. At their August 1988 meeting held in Beaver Creek, Colorado, attendance reached 300, an increase of 20% over their 1987 meeting. With their recently created Construction 2000 Task Force, CII is one of the bright spots on the horizon, whose mission it is to make American contractors more competitive in the world market.

THE CONSTRUCTION RESEARCH CENTER

The Georgia Institute of Technology, better known as Georgia Tech, is home to the Construction Research Center, an affiliate of the Washington, D.C. based Building Futures Council, a nonprofit group involved in improving the quality of design, construction, and facilities management.

The Construction Research Center (CRC) conducts basic research on materials and systems and provides means of communicating and teaching the construction industry the results of their research efforts. CRC will undertake specific research projects when requested to do so by their corporate sponsors, and activities at the Center involve interaction between designers, contractors, real estate developers, and industry. The Center addresses the following problems facing the construction industry today:

- Materials composition and performance, ceramic applications in construction.
- Design and development of new materials in construction.
- System performance evaluation.

Table 12-3. Construction Industry Institute Research Projects. ***(Source: Construction Industry Institute, The University of Texas at Austin.)***

PROJECT STATUS REPORT

TASK FORCE NO.	TASK FORCE NAME	PROJECT NO.	PROJECT DESCRIPTION	RESEARCH ORGANIZATION	COST	SCHEDULE 1987 / 1988 / 1989
	Special Projects		Construction Executive Program Curricula	Stanford	$ 50,000	(complete)
			Labor Motivation	Michigan	$11,000	(complete)
83-1	CICE Impact Evaluation	83-1-1	Exposure	Maryland	$ 22,000	(complete)
		83-1-2	Awareness	Oregon State	$ 19,311	(complete)
				Purdue	$ 29,306	(complete)
		83-1-3	Degree of Implementation	Maryland	$ 30,000	(complete)
				Oregon State	$ 20,000	(complete)
		83-1-4	Awareness II	Oregon State	$ 15,000	
83-2	Productivity Measurements	83-2-1	Overtime and Shiftwork	UT	$ 77,490	(complete)
		83-2-2	Alternate Techniques	Penn State	$ 42,000	(complete)
		83-2-3	Model Plant	UT	$105,000	
		83-2-4	Std. Productivity Measurements	Penn State	$ 47,000	(complete)
		83-2-5	Overtime and Shiftwork II	UT	$ 35,669	(complete)
		83-2-6	Std. Productivity Measurements II	Penn State	$ 20,000	(complete)
		83-2-7	Overtime and Shiftwork III	UT	$ 30,000	(complete)
83-3	Constructability	83-3-1	Planning	Stanford	$ 78,984	(complete)
		83-3-2	Design	UT	$ 68,000	(complete)
		83-3-3	Field	UT	$ 75,600	
		83-3-4	Modularization	Stanford	$ 80,000	(complete)

(a)

TASK FORCE NO.	TASK FORCE NAME	PROJECT NO.	PROJECT DESCRIPTION	RESEARCH ORGANIZATION	COST	SCHEDULE 1987 / 1988 / 1989
83-4	Industry Data and Statistics	83-4-1	Statistics Program	Construction Labor Research Council	$ 5,000	(complete)
		83-4-2	Industry Database Development	Construction Labor Research Council	$ 69,500	
		83-4-3	Processing System	Construction Labor Research Council	$ 22,000	
		83-4-4	Preliminary Planning		$ 15,000	
83-5	Contracts	83-5-1	Types and Clauses	Illinois	$ 64,220	(complete)
				UT	$ 54,203	(complete)
		83-5-2	Risk Allocation	UT	$ 94,000	
		83-5-3	Incentives	Illinois	$ 57,000	
83-6	Cost/Scheduling	83-6-1	Engineering Control	Colorado	$ 49,999	(complete)
		83-6-2	Planning, Cost Scheduling	Georgia Tech	$ 65,000	(complete)
		83-6-3	Contractor Strategies	A&M	$ 29,989	(complete)
		83-6-4	Scope Definition	UT	$ 35,000	(complete)
		83-6-5	Methods for Schedule Compression	Colorado State	$ 40,000	
		83-6-6	Change Order Impact	Cal-Berkeley	$ 40,000	
		83-6-7	Cost & Schedule Contingencies	Colorado	$ 40,000	
		83-6-8	Work Packages	Maryland	$ 40,000	(complete)
83-7	Materials Management	83-7-1	Attributes	Auburn	$ 23,308	(complete)
				A&M	$ 19,950	(complete)
		83-7-2	Cost-Effective Materials Managemt.	Auburn	$ 59,466	(complete)
				A&M	$ 50,000	(complete)
		83-7-3	Comprehensive Volume	A&M	$ 45,000	(complete)
		83-7-4	Microcomputer Tutorial	Auburn	$ 39,000	(complete)
		83-7-5	Comprehensive Volume II	A&M	$ 50,000	
		83-7-6	Bar Code Utilization	Auburn	$ 30,000	
		83-7-8	Pilot Projects & Bar Code Implement.	Auburn	$ 46,000	
		83-7-9	Bar Code Standardiz.	A&M	$ 30,000	

(b)

Table 12-3. ***(continued)***

TASK FORCE NO.	TASK FORCE NAME	NO.	PROJECT DESCRIPTION	RESEARCH ORGANIZATION	COST	SCHEDULE 1987 / 1988 / 1989
84-8	Design	84-8-1	Design Inputs	Worcester Poly	$ 50,000	(complete)
		84-8-2	Design Outputs	UT	$ 60,000	(complete)
		84-8-3	Inputs II	Worcester Poly	$ 50,000	(complete)
		84-8-4	Outputs II	UT	$ 50,000	(complete)
		84-8-5	CAE/CAD Effectiveness	Cal-Berkeley	$ 58,000	
84-9	Technology	84-9-1	Pipe Bending	UT	$ 50,000	
		84-9-2	Advanced Building Matrix	Michigan	$ 60,000	
		84-9-3	Anchor Bolt Installations	Iowa State	$ 30,000	
		84-9-4	Electrical Terminations	Drexel	$ 45,000	
84-10	Quality Management	84-10-1	Quality Management Problems	Clemson	$ 95,000	(complete)
		84-10-2	Quality Tracking Systems	A&M, Clemson	$ 35,000 $ 45,000	(complete)
		84-10-3	Field Trials	Clemson	$ 90,000	(complete)
		84-10-4	Organization & Techniques	Clemson	$ 40,000	
84-11	Employee Effectiveness	84-11-1	Employee Effectiveness Factors	Michigan	$ 4,500	(complete)
		84-11-2	Substance Abuse	Michigan	$ 55,000	(complete)
		84-11-3	Organizational Culture	Michigan	$ 40,000	
84-12	Project Organization	84-12-1	Objective-Setting	Iowa State	$ 50,000	
		84-12-2	Project Planning		$ 10,000	
84-13	Safety	84-13-1	Subcontractor Safety	Washington	$ 78,200	
84-14	Education and Training					

(c)

- Intelligent buildings and energy management systems development.
- Interactive video and educational programs.
- Textile applications in construction and mineral fiber reinforcement in concrete.
- Computer applications—design and construction.
- Cost effective means and methods of construction.
- Construction worker motivation, productivity, and labor issues.
- Light source impact on employee health.
- Facility diagnosis.
- Policy and code research.
- Construction information transfer.

THE CONSTRUCTION PRODUCTIVITY ADVANCEMENT RESEARCH PROGRAM

The U.S. Army Corps of Engineers announced in mid-1988 that they too were ready to enter the research and development arena. Major General Henry J. Hatch, Chief of the Corps, stated that the decline in productivity in the construction community has weakened an industry vital to this country. General Hatch recognized the fact that low productivity increases costs and therefore the Corps is paying more money and receiving less value in its construction projects. The creation of their Construction Productivity Advancement Research program will put the Corps of Engineers four-square in the middle of R&D work. The research efforts of this federal agency will be in the form of cooperative ventures with the private sector, and General Hatch indicated that the Corps' partners in these various research projects will share the costs of the projects. Research partners are expected to come from industry, trade groups, academia, and other government agencies, and the Corps anticipates an initial entry into four broad areas of research and development—design, construction site productivity, new construction materials, and technology transfer management.

The CPAR program will have an interesting twist to it; private sector partners will be able to obtain exclusive licensing agreements from the Corps to allow the participating company to market the product it helped to develop.

THE NEED FOR CENTRALIZATION

The creation of research groups around the country is an indication of what is right, and at the same time what is wrong with the state of research and development in this country. The effort, dedication, and money expended by all of the R & D organizations in this country is commendable and heartening, but the duplication of research efforts performed by many of these centers points to a crying need for an overall nationwide repository and locus for construction research.

Just a glance at some of the research programs being simultaneously pursued by the Construction Industry Institute and the Construction Research Center is a case in point:

Studies Conducted by CII	Studies Conducted by CRC
Productivity Measurement	Productivity and Labor Issues
Labor Motivation	Construction Worker Motivation
Costs/Scheduling (8 studies underway)	Cost Effective Means and Methods
Design (5 studies including CAD, CAE)	Computer Application in Design/ Construction
Industry Data and Statistics (4 studies)	Construction Information Transfer

The Corps of Engineers' new program also includes some of the topics being investigated by both CII and CRC, and while new methods of reinforcing concrete are being studied at these facilities, further north at the prestigious structural engineering labs at Cornell University, considerable effort is also being applied to multifaceted concrete research.

The future of construction research and development in this country might be in the form of one organization which could act as a focal point for all construction related information. It could be a privately funded facility, with possibly a one-time federal grant to establish and equip the physical plant and to provide the first year's operating expenses. Thereafter the Institute could be funded by private means, grants from industry, and revenues creating by selling its services. Membership would be open to trade groups, industry groups, contractor and labor organizations, individual contractors and subcontractors, and any other group interested in promoting the interests of the U.S. construction industry. An appropriate name might be The Construction Research Institute of America.

THE CONSTRUCTION RESEARCH INSTITUTE OF AMERICA

If such an organization was formed, it could act as a central clearing house for all construction related research and development under way in the United States. This Institute might consist of various divisions, each one responsive to and responsible for a certain segment of the entire, integrated R&D effort.

Research division	A section of the institute which would conduct basic research and would confirm test results obtained from equipment and material manufacturers.
Materials and systems testing division	A construction industry bureau of standards to which other national testing organizations would belong. Underwriter's Laboratories, National Fire Protection Agency, ASTM, ACI, and similar groups would have their testing procedures and results confirmed here and evaluations of entire construction systems would be made by this division.
Data bank division	A central database and repository for all construction related information, somewhat like a giant library card catalog system with an electronic index to assist in locating and retrieving requests for information.
Labor-management section	A division to collect, evaluate, and disseminate statistical data on labor negotiations, trends, demographics. One subdivision would be a productivity center.
Training and education division	This division would operate in-house and touring management and vocational training centers for supervisors, managers, and tradesmen. Mobile training centers would tour the country, bringing management and craft training to local contractors.

Legal and accounting division	Accounting principles and construction law education would provided. Bookkeeping techniques, data processing programs would be offered for sale, and a data bank of court decisions and legal interpretations relating to construction claims would be available to subscribers.

It has been estimated that the Japanese spend three times as much for construction research than we do in this country, and if we plan to compete with them, we need to devote more resources to the advancement of industry technology.

COMPETING WITH THE JAPANESE IN JAPAN

Much has been written about the closed nature of the Japanese construction market, which has been fueled by their country's huge trade surpluses—$83 billion in 1988 alone. As other world construction markets dry up, U.S. construction companies working overseas are clamoring to break down these barriers, real or imagined, and enter this market of multibillion dollar projects.

Open access to Japan's domestic market is viewed by the U.S. government as a means to redress the Japanese-American trade imbalance, and open access to this vast construction market may be a way for some U.S. contractors to keep their yearly sales volumes intact. Through the concerted efforts of the U.S. Department of Commerce things are beginning to change, and requirements for the licensing of foreign builders in Japan has been eased. Although the modified licensing requirements for contractors came about because of pressures exerted by the U.S., another group of contractors has unexpectedly become the first beneficiary of Japan's relaxed method of registration requirements.

The Japanese and Koreans have had a long history of conflict, fighting over territory has now been replaced by fighting over markets as lower-wage-scale Korean workers have made serious inroads into the automobile and electronics industries. Japanese contractors talk about Korean builders with a degree of apprehension. They feel that the Koreans are good builders, but they chop costs and profits to the bone to obtain contracts. Japanese contractors respect Korean contractors but feel they fall short of the mark when complicated, sophisticated construction projects need to be built.

But shortly after Japan announced that it was modifying its requirements for licensing applications, four major Korean contractors received permission from the Ministry of Construction to apply for licenses to do business in Japan. Construction Minister Ihei Ochi granted permission to Hyundai Construction, Samsung Construction, Daewoo Japan, and Samhwan Corporation and of the 11 South Korean contractors that have applied for special ministerial permission, the remaining seven will probably be given the go-ahead shortly.

In the meantime, Schal Associates of Chicago, Illinois was awarded a contractor's license in September 1988, and at that time Turner and Tishman were close to obtaining their licenses as well. The Parsons Corporation of California opened an office in Tokyo in September 1988 in order to concentrate on 14 projects being opened to American contractors. Parsons had recently formed a joint venture with Shimizu called Parsons Polytect, Inc., which will pursue engineering work in Japan in the private sector. However, this relationship with Shimizu should make the process of obtaining work by Parsons a little easier. American firms are now being allowed to bid on 14 projects in Japan totalling $16.9 billion.

JAPAN OPENS ITS CONSTRUCTION MARKET—KANSAI, K-1, K-2, K-3

The Kansai Airport Project

The $6.7 billion contract to build a state-of-the-art international airport on a man-made island in Osaka Bay is probably what started the Japan-bashing construction ball rolling. Kansai was one of the tantalizingly large construction projects that American contractors knew they could not realistically participate in due to many of Japan's restrictive bidding requirements.

The Kansai District

Near the geographical center of the Japanese archipelego there is a district known as the *Kinki district,* a name that refers to the area contiguous to the seat of the Imperial Palace. This area, formed by the triangle bounded by Osaka, Kyoto, and Kobe is also known as *Kansai,* which at one time was the most culturally advanced area in Japan. (See Fig. 12-2)

The strategic location of this district was one of the reasons for locating a new international airport facility there to help relieve the tremendous traffic serviced by Tokyo's Narita Airport.

The Kansai District is the world's third largest urban area, surpassed only by Tokyo and New York City, and is the hub of concentrated economic activity. Twenty percent of all of the corporations listed on the first section of the Tokyo Stock Exchange have their headquarters in Osaka, the city at the southeastern leg of the triangle. Just outside of

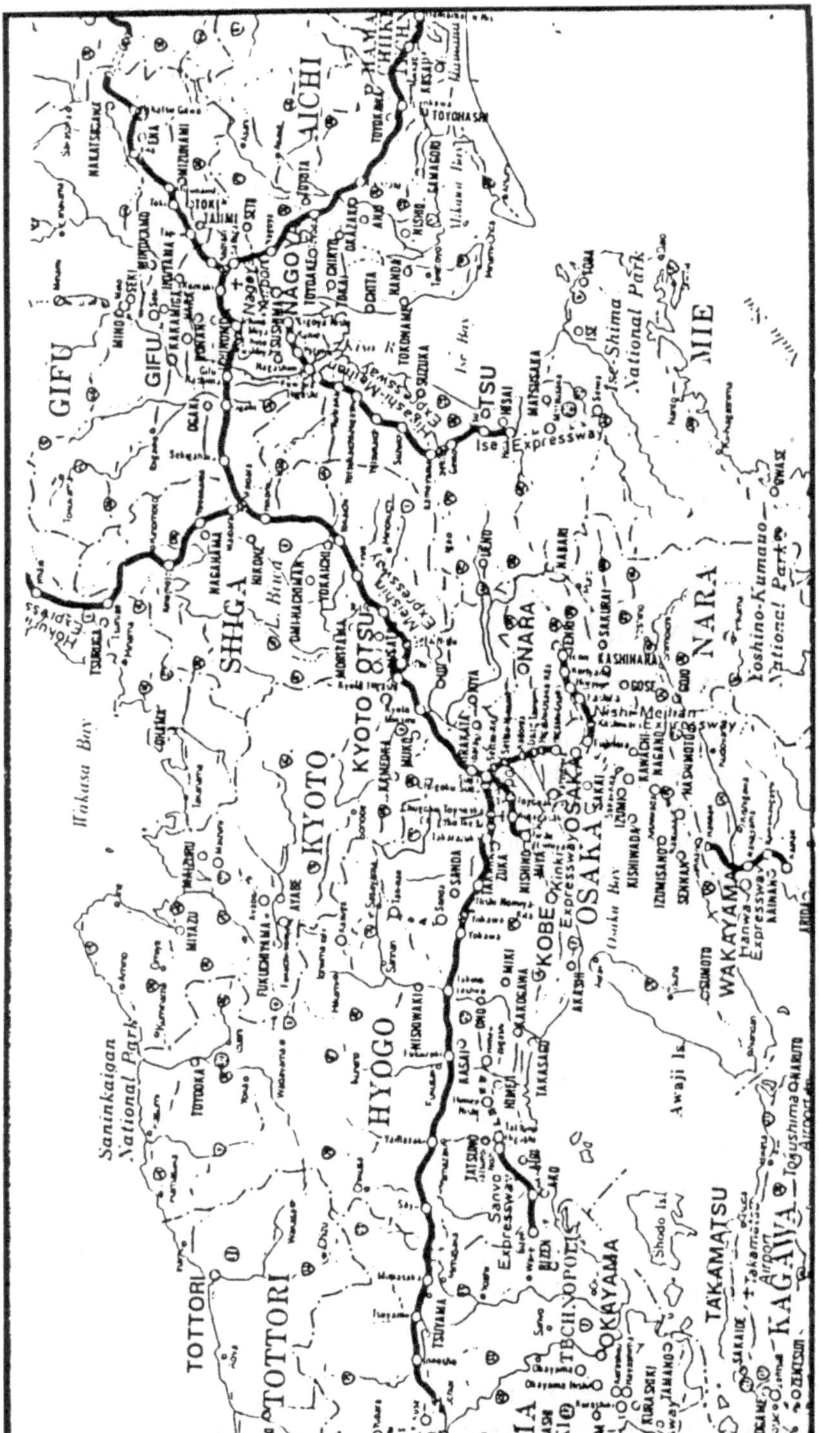

Fig. 12-2. Map of Kansai District.

Kyoto, at the northern end of the district, a new high tech town of 180,000 is planned which will contain national and public advanced science institutes to prepare the area for the challenges of the 21st Century. It is fitting that this futuristic city is to be located next to Kyoto, a thousand year old city.

Kobe, the other major metropolitan center in the Kansai District, was once Japan's largest trading port, and it is now the center of a massive project known as the Waterfront Project. Combining the creation of a new city built on an artificial island, additional harbor facilities, and housing for 30,000 people, this $9.6 billion renaissance will restore Kobe's preeminence as a major port city.

The Kansai International Airport, with its 11,482-foot-long runway, will be situated on an island of 1262 acres located approximately 3 miles offshore from Izumisano City on Osaka Bay. Phase Two of the Airport Development Plan will include the expansion of the island to 2965 acres. Two additional runways will be built to allow 190,000 passengers to be serviced daily.

Opening the Kansai to Foreign Participation

The United States government expended a great deal of effort, backed up by lots of political pressure, to get the Japanese to agree to open up the bidding procedures for the remaining portions of design, construction, and equipment contracts required by this mammoth project. If Kansai could be truly opened to U.S. construction participation, then it would be reasonable to assume that other major construction projects in Japan such as the $13.3 billion Yokohama Port Expansion program and the $7.7 billion Tokyo Bay Bridge job would also permit American participation.

U.S. Secretary of Commerce, C. William Verity and Nobuo Matsunaga, the Japanese Ambassador to the United States, culminated months and months of tough negotiations with an historic exchange of letters in late 1987. These letters formed an agreement to open up heretofore "closed" Japanese construction projects. A more formal agreement than that contained in the Verity-Matsunaga letters was finalized when Ichiro Ozaawa, Deputy Chief Secretary of the Japanese Cabinet arrived in Washington, D.C. in March, 1988 and stated that Prime Minister Takeshita had made a commitment to enforce nondiscriminatory and open bidding procedures on 14 public projects worth in total $16.9 billion.

The letter agreement reached by Secretary Verity and Ambassador Matsunaga covered three areas of participation in Japanese construction projects:

1. The Kansai Airport Project, the Tokyo Bay Bridge construction work, and a $200 million NTT Intelligent Building job would be included

in one group of public works projects that collectively would be known as K-1
2. Other major construction projects such as the Haneda Airport Expansion—Phase II, the New Hiroshima Airport, Tokyo Port Development, Ise Bay Highway, Akashi Straits Bridge, a portion of Yokohama's Minato Mirai 21 Project, and the Kansai Science City work would also be included in the agreement, and collectively these projects would be refered to as K-2 or special measures procedures.
3. A series of third sector work (local government and private sector projects) is also planned for the future and will create the same type of environment as the K-1 and K-2 agreements. Since the federal government of Japan cannot enforce participation by third sector interests in the U.S.–Japan handshake agreement, it can only request that local officials and corporate CEOs pledge to follow nondiscriminatory procurement procedures before awarding construction contracts. These K-3 projects will include terminals at Haneda, Narita, Hiroshima, and Kitakyushu airports, a conference center at Minato Mirai, a Teleport in Tokyo, a Technoport at Osaka, and creation of Rokko Island.

The Kansai International Airport Company—KIAC

The Japanese government, after the letter agreement was established, set up a series of guidelines to be followed if foreign contractors, including Americans, wished to submit bids on portions of KIAC work which remained after some initial contracts had been let. Prior to the letter agreement which changed the rules, KIAC had issued awards to foreign firms, as of October 31, 1987, as set forth below:

Firms Receiving Orders		Dollar Value	Type of Award
Bechtel Civil, Inc.	U.S.A.	$193,550	Consulting
Six U.S. and European Airport Operators		754,839	Consulting
AT&T International	U.S.A.	258,065	Data Communication Research
Del Norte Corp.	U.S.A.	161,290	Trisponder Radio Positioning System
GM	U.S.A.	225,800	Ship Engine
TRW	U.S.A.	316,000	Research on Security
Bechtel Civil Inc.	U.S.A.	96,775	Research on Security
Del Norte Corp.	U.S.A.	851,000	Trisponder Radio Positioning System
Hewlett Packard and IBM	U.S.A.	154,839	Computer
Geotronics	Sweden	38,700	Geodimeter
AGA	Sweden	38,700	Geodimeter

Wavin Repox	Holland	458,000	Wavi Float
Raytheon	U.S.A.	70,968	Digital echo sounder
Sinco	U.S.A.	6,451	Bentonite pellets
Wild	Swiss	6,451	Level OL-13
United Air Specialists	U.S.A.	19,355	Electrostatic precipitator
Arcair	U.S.A.	38,710	Automatic arc gouging machine
Hercules	U.S.A.	1,006,000 (800,000)	Concrete admixtures (To be shipped later)
Caterpillar/Mitsubishi	U.S.A.?	6,000,000	Excavating equipment
Rexnord	U.S.A.	258,000	License fee for rock crusher
Allis Chalmers	U.S.A.	3,300,000	Rock crusher
MTS	U.S.A.	645,000	Two dimensional irregular wave generator
Total U.S.A. Firms		$15,698,493	

Dollar conversion based upon yen valued at 155 in October 1987

Awards to American firms, as of the third quarter of 1987, totalled $15.698 million, and represented only 0.2% of Kansai's total construction budget, but with this new offer to open the market it was now possible for substantial increases in participation. United States firms and other foreign construction companies would now be allowed to submit an application for participation in a K-1 project without feeling that the dice were loaded.

A valid complaint from U.S. contractors in the past was that the Japanese government required that contractors bidding on major public projects had to list their previous experience in building similar projects in that country. Obviously this was a Catch 22 situation—a company needs experience in Japan prior to being placed on a public works selected bidder's list, but in effect is prevented from gaining this experience.

THE NEW K-1 RULES

Under the new K-1 procedures, there is no longer a requirement to present a history of work on similar projects to show experience levels needed to qualify as an acceptable bidder.

The Japanese government established a new set of guidelines for bidders interested in tendering offers on K-1 contract work.

1. A Master Plan will be provided for all companies that request one.
2. Procurement procedures were established for three types of solicitations:
   - Goods valued at 30 million yen or more ($240,000 at 125 yen).

- Consulting requests valued at 25 million yen or more ($200,000 at 125 yen).
- Construction requests for work valued at 700 million yen or more ($5.6 million).

3. Bid solicitation ads would be run for 30 days in *Kampo,* the official bid offering newspaper.
4. A prebid conference will be held prior to the selection of bidders on construction projects only.
5. The following time for bid preparation was established:
   - For goods—40 days.
   - For consulting work—40 days.
   - For construction work—60 days.
6. Rebidding (if all bids exceed owner's estimate or budget): three working days, if a written request is submitted, otherwise new proposals are to be resubmitted immediately.
7. A mechanism was established to respond to complaints filed by bidders:
   - An explanation for rejection will be given to nonselected bidders, if requested.
   - Complaints will be formally filed with the International Affairs Office of KIAC.
   - Unresolved disputes will be refered to the Ministry of Transport's KIA Office by the U.S. Embassy, if the dispute is lodged by an American firm.
   - Outstanding disputes will be monitored to insure that they are being actively pursued.
8. Bid documents will be performance related and foreign test data will be accepted as equal to similar data in Japan, if all other things are equal.

K-2 Rules and Procedures

Most of the rules and regulations contained in the K-1 procedures package remains in effect for K-2 projects with only a few modifications. Master plans will be made available within 60 days for those projects already under way. For projects in the planning stage, the Master Plan will be made available to those requesting it at least 12 months prior to the fiscal year in which the project is being planned. In April of each year, an annual list of proposed bid announcements will be issued so that interested parties can be made aware of projects on the drawing board. Table 12-4 contains a flowchart showing the steps required to bid on the Kansai Airport construction work.

Table 12-4. Flowchart of Steps Required to Bid on the Kansai Airport Construction Work.

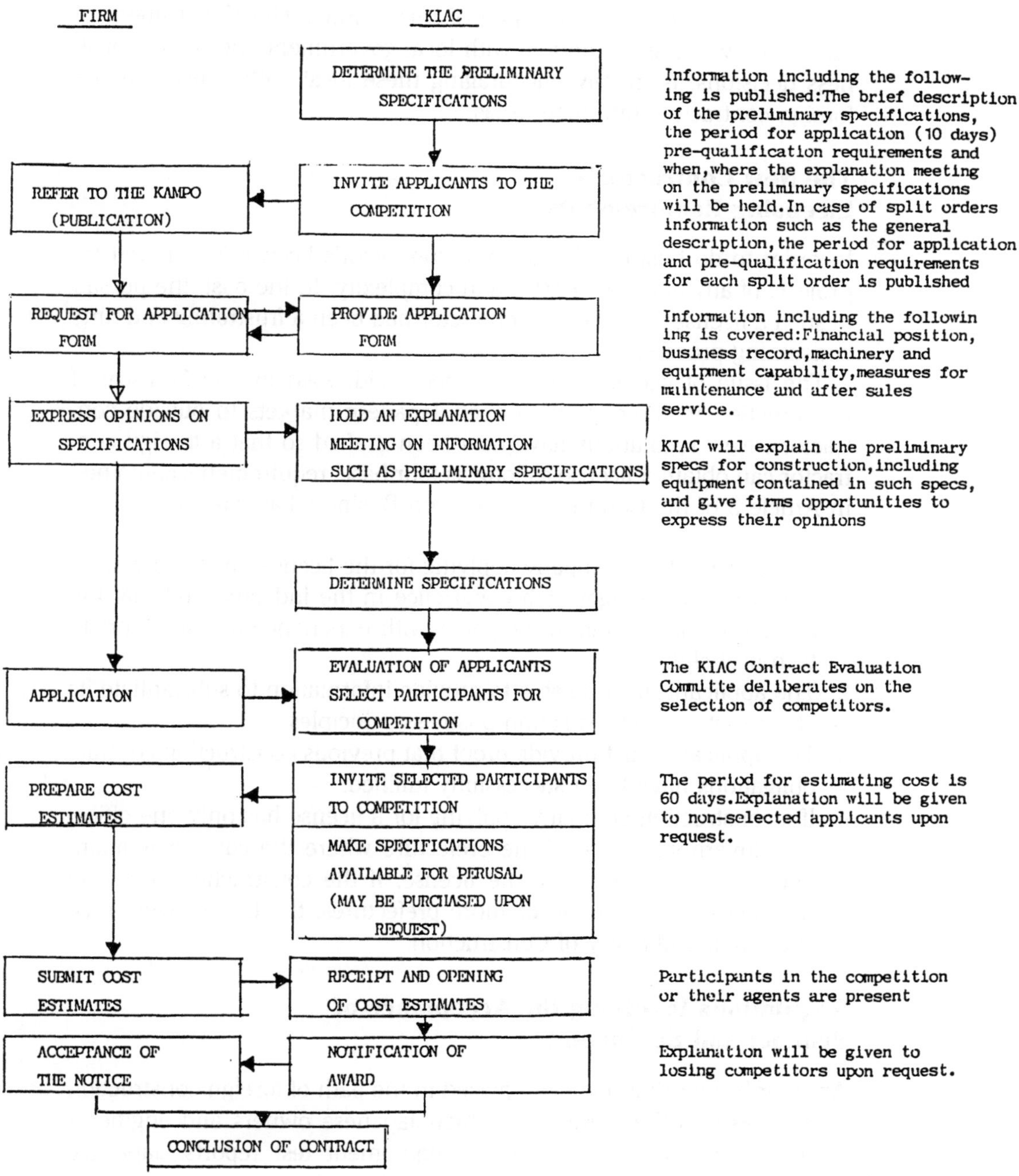

K-3 Project Procedures

Procedures for all works within the purview of third sector projects have not been worked out at the time of this writing. The Government of Japan, not wishing to interfere with local government and private business decisions, is merely encouraging these local authorities to invite foreign contractors into their areas.

The New Contractor's Licensing Requirements

In order to do business in Japan, a contractor's license is required for projects of any appreciable size and complexity. In the past, the pursuit of such a license by a foreign contractor had been a frustrating and time consuming process.

The issue of licensing has now been addressed by the Ministry of Construction in their quest to open Japanese markets to outside contractors. New regulations have now been drafted so that a foreign contractor can obtain a license as long as four basic requirements contained in Article 3 of the Contract Construction Business Law are met:

1. The construction company applying for the license must show proof that it has a manager with experience in the industry, and that the company has on staff an engineer with experience and appropriate technical skills.
2. The applicant must be able to provide information to substantiate its claim that it practices sound business principles.
3. The applicant must provide proof that previous construction contract obligations have been successfully fulfilled.
4. If a construction company applying for a license has only one office, the Governor's Office of the prefecture where the contractor maintains his office will issue the license. If the construction company maintains offices in two or more prefectures, the license will be issued by the Ministry of Construction.

Regulations Involving the Acceptance of Product and Equipment

Another barrier that had been placed in the path of foreign construction companies was the manner in which Japanese owners and engineers reviewed product compliance and equipment test reports. Japanese technical specifications were pretty much exclusionary, and the approval of Western "or equal" equipment and/or materials was almost impossible to obtain. Under the new "rules" established by KIAC, performance specifications, whenever possible, will be used. When a Japanese owner or engineer reviews a request by a foreign contractor for product approval, foreign test data and performance specifications are to be directly compared with Japanese test and performance data for conformance. If a product or piece of equipment is rejected, a written explanation of the reasons for rejection will be issued by KIAC before any contract award is made.

The KIAC Rating System for Foreign Contractors

Prior to the pressure being applied by Secretary Verity, the Ministry of Construction had established an annual rating system for contractors bidding on public works projects. This rating system favored the big Japanese construction companies who, once entrenched on the selected bidders list, would obstensibly eliminate many other contractors wanting to get on the list. As far as foreign contractors were concerned this method of rating builders was obviously exclusionary.

The new KIAC K-1 agreement addressed this evaluation or rating process in a very straightforward manner. In the evaluation of foreign contractors business history, KIAC will address such matters as credit standing, completed projects, managerial and engineering staff capabilities, and previous work experience in countries outside of the Japan just as though the company had been operating *in* Japan. If KIAC decides to reject any qualification information submitted by a foreign contractor, it is obliged to do so in writing and state the reasons why some, or all of the qualification data presented is not satisfactory.

Joint Venture Opportunities

Japanese firms entering the U.S. construction market have long since recognized the value of entering into joint venture projects with American partners. By joining forces with a local firm who has the contacts, working history and knowledge of the area as well as familiarity with building codes and ordinances, a joint venture project can be a learning experience. Joint venturing is a good way to enter an unfamiliar construction market and limit one's risks substantially.

The Government of Japan has openly announced its desire to assist American general contractors and subcontractors desirous of forming joint ventures with Japanese firms. All joint venture proposals submitted by Japanese and American participants will be reviewed by KIAC as though it was an all-Japanese proposal. KIAC offers its services to introduce American general contractors or subcontractors to Japanese Prime or subcontractors to further promote joint venture possibilities on K projects. These sincere efforts on the part of the Japanese government will provide U.S. firms with an opportunity to establish a foothold in Japan, if they so desire.

RECENT DEVELOPMENTS IN U.S.– JAPAN TALKS

Two years of negotiations, plus lots of external pressure exerted by banning Japanese participation in federally funded U.S. projects, resulted in an improvement on the 1987 Verity-Matsunaga Agreement, and created a further opening in the Japanese domestic construction market.

The May 25, 1988 Verity-Matsunaga Update

On May 25, 1988, Secretary Verity and Ambassador Matsunaga exchanged another group of letters which formed the basis for a continued effort by the government of Japan to pursue nondiscriminatory procurement policies relating to their public works projects. The text of this letter by the Ambassador from Japan is reproduced on the following pages.

The critical concessions contained in Ambassador Matsunaga's letter revolve around the relaxation of the requirement for "sufficient exposure" by foreign firms wishing to bid on Japanese public works projects.

A subsequent letter from the Japanese ambassador, also dated May 28, 1988, stated that the other letter waiving the prior experience requirement and all of the other concessions would be suspended "should limits on Japanese participation in U.S. public works be continued after September 30, 1988. It is the wish of the Government of Japan that these limits will be lifted before that time." *Sayonara* Brooks-Murkowski!

The Continuing Role of the Department of Commerce

The Department of Commerce has been designated the government agency entrusted with the responsibility to collect and disseminate information about the Japanese construction market. Executive level trade missions to Japan will be organized by the Department to arrange meetings with top level construction company executives in Japan. The Department's Office of International Major Projects will maintain a list of master plans for each upcoming project and will also obtain the annual plan of orders from the Japanese government each fiscal year. OIMP will have copies of registration forms needed by U.S. contractors and will have a list of possible contact points, if contractors wish to

EMBASSY OF JAPAN
WASHINGTON, D. C.

May 25, 1988

Dear Secretary Verity:

I am writing to you with regard to procurement procedures for the KIA (Kansai International Airport) and other major projects undertaken by similar private commissioning entities, and for public works (works to be commissioned by the government or governmental agencies), as well as procurement policies of those private and third sector entities (entities to be jointly funded by private and local government) which relate to the above public works in carrying out their major projects. These have been a matter of consultation between the Governments of Japan and of the United States with a view to facilitating access to the Japanese major projects construction market.

It is the policy of the Government of Japan that all major public works projects are to be conducted in accordance with Japanese laws and regulations under non-discriminatory procurement procedures. The Government of Japan believes that international competition should be facilitated in major projects procurement by all entities, public and private. The Government of Japan on its part welcomes the entry of foreign firms into the Japanese market and will, as a general principle, undertake positive steps to promote and facilitate the entry of foreign firms into the Japanese construction market.

With regard to obtaining construction licenses, prior experience in Japan is not required when firms apply for them. The ordinary processing period for business licenses for foreign applicants from the Minister of Construction of the Government of Japan will be three to four months. As for licenses to be issued by prefectural governors, the ordinary processing period will be one to two months. After 90 days, upon request, the applicant will receive an explanation of the status of the application from the Construction Contractors Division, the Ministry of Construction.

It is confirmed by the two Governments that, regarding KIA and other major projects undertaken by similar private commissioning entities, the arrangements described by my letter to you of November 4, 1987, are acceptable. The list of the projects to which the measures described in my letter of November 4, 1987, will initially apply is attached as Appendix I.

Regarding public works, lack of sufficient experience by foreign companies in the Japanese private sector has made it difficult for them to enter into Japanese public works procurement. The Government of Japan will take the special measures described in Appendix II so that foreign companies may be designated as qualified to bid on any of the major projects listed in Appendix III without regard to their prior experience in Japan. Registration procedures for public works, as explained in Appendix II, will not limit participation of or present a barrier to foreign companies.

The major projects listed in Appendix III are intended to provide sufficient expanded opportunities to foreign companies so that they may become familiar with the bidding system of public works in Japan. The listing of these projects does not limit or prejudice application by foreign companies to participate in projects not listed. The Government of Japan hopes that foreign firms' participation in public works will be developed beyond the projects listed herein as those firms gain experience.

In view of further enhancing international competition in the construction market, the Government of Japan believes that non-discriminatory procurement policies should also be followed by those private and third sector entities listed in Appendix V which relate to the above major public works projects. In this regard, the relevant policy of the Government of Japan and two policy statements by presidents of entities concerned are included in Appendix IV.

A monitoring committee consisting of the Ministry of Foreign Affairs, the Ministry of Transport, and the Ministry of Construction will meet with the U.S. Embassy in Japan at the request of either side to monitor the implementation of all measures relevant to U.S. interests covered by this document and to solve problems where possible. Relevant information will be made available on request. In addition to these sessions in Tokyo, officials of the two governments will meet at the request of either party to resolve any issues with regard to these measures.

The Government of Japan may decide to add additional components of those projects listed in Appendices I, III, and V and to provide additional information on these components in the future through monitoring.

The Government of Japan will supply information upon request to the monitoring committee regarding the status of Governmant of Japan-approved, and yet to be approved, projects as well as undecided components of projects currently included in Appendices I, III, and V.

A Government-to-Government review will be held after two years to examine whether these arrangements are serving the expected purpose of facilitating sufficient access of foreign firms into the Japanese construction market. Depending on the results of this review, the measures covered by this document may be adjusted, if necessary.

I would like tó confirm that the arrangements described in this letter are acceptable to our two governments. The Government of Japan will implement the measures described in this letter and its appendices in a mutually beneficial manner.

Sincerely,

Nobuo Matsunaga
Ambassador of Japan

The Honorable
C. William Verity
Secretary of Commerce
Washington, D.C. 20230

Appendices:

I. Projects subject to measures described in Matsunaga-Verity letter of November 4, 1987

II. Special Measures applicable to Major Public Works Projects

III. Major Public Works Projects Subject to Special Measures in Appendix II

IV. Policy Statement regarding procurement by private entities and third sector entities

V. Projects commissioned by private and third sector entities

obtain registration forms directly from the Japanese entity issuing them. The U.S. government has made an exemplary start in furthering the opportunities for U.S. contractors to enter the Japanese construction market.

It now remains to be seen if this "window of opportunity" will be opened by American builders or whether the short-sightedness they have exhibited in the past will prevent them from viewing the potential for new construction work on the other side of the glass.

Japanese Projects Included in This New Beginning

Projects being offered to foreign contractors, and more specifically American companies are quite substantial in both scope and dollar value.

- Haneda Airport Expansion—Phase II, 1990–1995. This project involves construction of a new runway, taxiways, parking and loading aprons, roads, water tank and garage for fire fighting equipment, CIQ building, navigational aids, electric power facility and equipment. Total estimated value 110 billion yen ($880 million at 125 yen to the dollar).
- New Hiroshima Airport—1986–1993. Bypass road, waterway, parking and land improvements, runways, taxiways, aprons, garage for fire-fighting official building, navigational aids, radar and air traffic control facilities, electric power facility and equipment. Total value not disclosed, but opportunities for U.S. investment could reach $356 million.
- Tokyo Port Development—Commenced 1984.Initially included the Metropolitan Expressway 12 and has recently been expanded to include construction of a teleport communications center. Total value of the project is estimated at 3.4 trillion yen (about $27.2 billion at 125 yen) with the teleport portion consuming 1.9 trillion yen of this total.
- Akashi Kaikyo Bridge—1986–1998. A suspension bridge across the Akashi Straits valued at 383 billion yen.
- Ise Bay Highway—1987–1996. Construction of cable stayed bridges (Meiko Central and Meiko East), viaduct dredging, paving, and highway lighting valued at 88 billion yen.
- Minato Mirai 21 Project—1988–1993. Land consolidation, redevelopment in Yokohama to include port construction and major international conference and convention facilities. Total value 230.8 billion yen.

- Kansai Science City—1988–. Land consolidation, Second Keihan Expressway, Center for Graduate Studies in Advanced Science and Technology. Total value 713 billion yen.

THE OPENING OF THE JAPANESE MARKET TO AMERICAN DESIGN FIRMS

It seems rather strange at first glance that the Japanese construction market has opened its portals to U.S. design companies while it was the U.S. construction community that started banging on their front door demanding admittance. Although the Kansai International Airport Company had already selected a design concept for the airport from Aeroport de Paris, as of this writing a final design for the passenger terminal had not yet been awarded.

The final list of 15 bidders for the design work, chosen from a field of 48 applicants in late May 1988, included a number of U.S.–Japanese joint ventures. I.M.Pei–Kumagai Gumi; Ellerbe Becket and Parsons Overseas–Shimizu Corporation; Hellmuth, Obata and Kassabaum, Inc.–Kajima Corporation; Cesar Pelli Associates–Takenaka Komuten. Strangely absent from the selected bidder's list was prestigious Skidmore, Owings and Merrill, who had teamed up with Taisei Corporation but was not chosen to continue in the competition. Each of the bidders will receive payment of $80,000 for their final design proposal work, and after selection of the successful bidder, a joint venture will be formed with Aeroport de Paris.

The Japanese interest in American design firms has extended beyond the Kansai Airport project. A Japanese development company, Fukuoko Jishu has retained three U.S. firms to design two substantial projects on the island of Kyushu, the southernmost isle in the Japanese archipelago. Jerde Partners, a Los Angeles architecture firm, will be working with Chicago architect Stanley Tigerman, along with Michael Graves of Princeton, New Jersey, on a shopping center project in downtown Fukuoka and also on an apartment complex in the same city. Simo Stone, the $80 million apartment building complex, will be designed by Michael Graves and Stanley Tigerman, each working on a separate building, while a Japanese architect, Arata Isozaki, will provide design schemes for the other two buildings in the development. Could it be that the Japanese interest in collaborating with U.S. architects has a more far reaching goal?

The reluctance of Japanese developers and contractors to enter the U.S. market as primes may be due to a number of understandable concerns. One of the puzzlements to Japanese contractors upon first entering the U.S. market is the varied and complex sets of local building codes, planning and zoning requirements, and the myriad ordinances and regulations emanating from local and federal agencies. American builders seem to have considerable trouble complying with all of these confusing rules and regulations; to the Japanese it must seem to be a nightmare. And just when the local laws have become somewhat familiar a construction project in another part of the country will start bewilderment all over again.

Architects and engineers in this country must be intimately familiar with all of the various building code requirements as well as other government restrictions in order to create acceptable working drawings and specifications. By hiring American architects and engineers to work in Japan, their developers and contractors can begin to develop a relationship with these "foreign" design groups that could be used at some future date when these same Japanese developers and contractors decide to enter the American market. Long term goals are an acknowledged Japanese trait and the alliance with U.S. design firms could be a calculated relationship that is part of their matrix for future American market share.

WHAT DOES THE FUTURE HOLD?

Asia, the Pacific Rim, and Eastern Europe appear to be the markets of the future. These areas of vast land resources and large populations are emerging as immense pools of industrial potential that are just beginning to flex their international economic muscles. The Soviet Union, China, and India together, cast a long shadow over the rest of the world, if just by the enormity of their combined land masses and population count. These three countries represent 24% of the earth's 13.57 million square mile land area and 42.6% of the entire world population. As Russia and China emerge from eras of repression and India learns how to cope with its divisive political groups, these three countries will be on the threshold of enormous industrial expansion that will probably be followed by a fast paced demand to improve each country's standard of living.

If America's future is looking outward, then it ought to look to the East where the largest consumer goods market in the history of the world may someday emerge. Recent statements by Russia's Secretary General Mikhail S. Gorbachev appear to be directed toward the creation of a more capitalist society, almost a necessity if the U.S.S.R. does not want to slip back to a second rate power in the next century. As arms limitation agreements are ratified by the U.S.A. and Russia and each country's military budgets are reduced, the 15% of the U.S.S.R's gross national product now devoted to defense will probably be diverted to infrastructure or consumerism. China awakened to the joys of capitalism during the period 1973 to 1986 and their gross national product increased by 130%. In January of 1989, Beijing had been considering ending the state ownership of industry but just six months later, in June, gunfire echoed through that city as the government began a crackdown on student protesters, and foreign investors became uneasy about the future of China's economy. But when any significant portion of China's 1.05 billion people become industrial wage earners, it boggles the mind to think of their needs that will require fulfillment.

What direction will the construction industry take in these three countries as demand for commercial, industrial, and residential work begins to accelerate? And what about the vast sums of money that governments in these countries will have to spend to provide roads, airports, waste treatment facilities, bridges, hydroelectric and nuclear power generating plants, and all of the other kinds of infrastructure needed by an industrialized society?

Will these countries welcome foreign contractors into their borders to provide the management skills and technical knowhow so necessary to build these projects quickly? If U.S. firms can successfully export Coca-Cola, Pepsi-Cola, and McDonalds, why can't we sell construction expertise as well? American contractors in the international marketplace will have to become more mature in their approach to construction work in foreign countries, and Japan might just have been our Asian learning experience. The knowledge gained in dealing with an exclusionary market ought to stand U.S. contractors in good stead as other dynamic markets unfold in the Pacific Rim and Eastern Europe. The global construction giants of tomorrow will need to have a strong presence in these areas as the world becomes one megamarket. For those companies in the big leagues of international construction the time for learning is rapidly coming to a close and the slow learners will have to take a back seat to the ones that have done their homework properly.

For those smaller companies that constitute the backbone of the U.S. construction industry, there are many lessons to be learned as they strive to survive and excel in their individual business endeavors. Planning, the establishment of long term goals, a committment to practice the highest standards of the trade, and the ability to know one's market thoroughly are universal qualities to be pursued by the Bechtels of the world as well as that small general contractor from Maple Valley, Vermont.

HARLEY DAVIDSON—A STORY FOR NOT ONLY MOTORCYCLE BUFFS

There are few products more readily identifiable with American culture than the big, powerful Harley-Davidson motorcycles—whether one conjures up a vision of a group of Hell's Angels riding en masse down the highway on Harley Choppers, or Mom and Pop out for a Sunday ride on a luxurious Harley ElectraGlide or the black and white police Harleys of the famed California Highway Patrol racing up the freeway. And who can forget Marlon Brando's portrayal of a rogue biker in *The Wild One,* riding, of course, a Harley Hog.

This Pennsylvania motorcycle manufacturer contracted much the same illnesses and afflictions as many American companies when international markets changed in the 1960s and 1970s, and they failed to fully understand or take notice of some of these profound changes. Complacency and the switch in emphasis from product orientation to financial orientation brought many basic industries to their knees. In America in the 1960s, conglomerates were the name of the game. Corporations bought disparate companies, on the premise that if management could effectively and profitably manage, say, a vacuum cleaner company, it could also effectively manage a chicken parts processor, a chain of discount shoe stores, and a lingerie manufacturer. American Machine and Foundry Company became AMF in an era when it was considered fashionable or forward thinking to streamline the corporate name; U.S. Steel Corporation became USX; and Minnesota Mining and Manufacturing became, simply, 3M. AMF was going to devote its considerable corporate energies to the fast growing leisure time industry. They already owned a bowling alley equipment company and a boat manufacturer, and they bought Harley-Davidson in 1969.

AMF directed Harley to increase its production, and by the early 1970s the company annual 15,000 bike quota of previous years had been upped to 37,000 motorcycles. While AMF pushed for higher production rates, it took no steps in improve Harley's manufacturing capability. As a result, product quality took a nose dive and labor-management relations became severely strained. It was about this time that the venerable California State Highway Patrol announced that they would no longer purchase Harley-Davidson motorcycles because they deemed them to be unstable. AMF brought a new manager, Vaughn Beals, on board in 1975 to help straighten out Harley's problems, but Mr. Beals found that most of his suggestions were falling on deaf corporate ears, and he finally assembled a group of Harley-Davidson people who bought the company back from AMF.

Japanese motorcycle manufacturers, particularly Honda, were making serious inroads in America, thanks to their introduction of small engined bikes and because of their practice of dumping excess inventory into this country at very low prices. Now Honda introduced its Gold Wing, a big 1,000 cc bike that would directly challenge the large machines being manufactured by Harley-Davidson.

Vaughn Beals appealed to Washington to stop the Japanese from "dumping," and soon the U.S. motorcycle manufacturers received tariff protection. Mr. Beals then took on the task of getting his company back on its feet. Harley managers toured Japanese motorcycle plants and took note of their manufacturing procedures and such things as just-in-time inventory practices. Beals was relentless in his pursuit of the highest production quality levels, and by 1984 the California Highway Patrol was again riding high on Harley-Davidson motorcycles.

The company recently generated an annual net income of $25 million on sales of slightly less than $800 million. Harley had met the challenge from Japan, learned from it, and came back from the depths of corporate despair. Slowly, but surely, more and more American companies are making the commitment to quality and increased productivity, and in the immortal words of one of America's greatest orators, Yogi Berra, "It ain't over till it's over."

Index

MIX
Papier aus verantwortungsvollen Quellen
Paper from responsible sources
FSC® C105338

If you have any concerns about our products,
you can contact us on
ProductSafety@springernature.com

In case Publisher is established outside the EU,
the EU authorized representative is:
Springer Nature Customer Service Center GmbH
Europaplatz 3, 69115 Heidelberg, Germany

Printed by Libri Plureos GmbH
in Hamburg, Germany